Articulating Citizenship

Civic Education and Student Politics in

Southeastern China, 1912–1940

Harvard East Asian Monographs 291

Articulating Citizenship

Civic Education and Student Politics in

Southeastern China, 1912–1940

Robert Culp

———

Published by the Harvard University Asia Center
Distributed by Harvard University Press
Cambridge (Massachusetts) and London 2007

Printed in the United States of America

The Harvard University Asia Center publishes a monograph series and, in coordination with the Fairbank Center for East Asian Research, the Korea Institute, the Reischauer Institute of Japanese Studies, and other faculties and institutes, administers research projects designed to further scholarly understanding of China, Japan, Vietnam, Korea, and other Asian countries. The Center also sponsors projects addressing multidisciplinary and regional issues in Asia.

Library of Congress Cataloging-in-Publication Data

Culp, Robert Joseph, 1966–
    Articulating citizenship : civic education and student politics in Southeastern China, 1912-1940 / Robert Culp.
        p. cm. -- (Harvard East Asian monographs)
    Includes bibliographical references and index.
    ISBN 978-0-674-02587-5 (cl : alk. paper)
    1. Civics, Chinese. 2. Citizenship--Study and teaching--China. I. Title.
    JQ1517.A2C85 2007
    323.6'50951209041--dc22

                                                        2007011926

Index by David Prout

    ∞ Printed on acid-free paper

Last figure below indicates year of this printing
16 15 14 13 12 11 10 09 08 07

To

Katie, Andrew, and

Elizabeth Mae

Acknowledgments

During a spring 1994 seminar discussion of a paper I was writing on the early Republican spread of social and cultural reform ideas, my adviser, Sherman Cochran, wondered aloud whether those concepts had made their way into textbooks and spread through schools. I started to wonder, too. Thus began the twelve-year sojourn that has become this book. I thank Sherm for starting me on that path and guiding me through each of its many turns. Along the way I have benefited from the guidance and generosity of many other teachers, colleagues, and friends.

In writing about the interactions between teachers and students, I have thought often about the teachers who have shaped my intellectual life. P. Steven Sangren and J. Victor Koschmann have both influenced my thinking and approach to scholarship, offering models for how to combine critical theoretical engagement with grounded primary research. In Shanghai, long discussions during the opening stages of this project with Professors Xiong Yuezhi and Li Huaxing of the History Institute of the Shanghai Academy of Social Sciences (SASS) challenged me to rethink the basic questions driving this project and helped me to find my way in Shanghai's rich collections of Republican-period sources. In Taipei, Lü Fangshang guided me in negotiating the Nationalist Party History Archive.

Although never formally my teacher, the late Stephen C. Averill was a mentor and friend who had a profound impact on me as a historian and student of China. Much of what I know about doing research, I learned working with Steve in the reading room of the Modern History Institute Library of Academia Sinica on Saturday mornings in 1990 and 1991. I regret that he will not see the completion of this project, to which he contributed so much.

As was true of the students discussed in this book, my ideas have been shaped as much by my peers as by my teachers. A remarkable cohort of fellow graduate students, several of whom were part of that seminar in the spring of 1994, shaped my experience at Cornell between 1992 and 1999. I especially thank Gardner Bovingdon, Sara Friedman, Jennifer Hubbert, Lu Yan, Tom McGrath, Adam Segal, and Li Zhang for their intellectual companionship. In the face of the isolation of sustained archival scholarship, they have made research and writing feel like a collective process.

The research on this book would have been impossible without the support of SASS. Li Yihai, Zhao Nianguo, and the staff at the SASS Foreign Affairs Office have proven indispensable at each stage of the research. I also thank the staff of the Lexicographical Publishing House Library, for their help, warmth, and good humor. I am grateful as well to the history department of Nanjing University and the Modern History Institute of Academia Sinica for hosting me as a research scholar in, respectively, 1995–96 and 1997. The collections and staffs at Cornell's Kroch Asia Library, the Harvard-Yenching Library, Columbia's C.V. Starr Library, and the University of California at Berkeley's East Asia Library and Center for Chinese Studies Library were also immensely helpful at various stages of this project.

I am also grateful to many alumni of the lower Yangzi region schools discussed here for sharing their memories and insights with me. I hope that I have done them justice and that I have not erred too often.

For better or worse, I am a dialogic thinker who learns best through engagement with others. As a result, critical commentary has been essential to this project at every stage. I especially thank Paul Cohen, Henrietta Harrison, Elizabeth Perry, Jeffrey Wasserstrom, and the anonymous readers for the Asia Center for their insightful critical readings of the entire manuscript in its final stages. Their commentary helped me resolve many thorny problems of argument and organization and rethink the goals of this project at a critical juncture. The remaining flaws and miscues are the result of my limitations, not their astute criticisms and suggestions.

Others who read and commented on specific parts of this manuscript or work related to it include Terry Bodenhorn, Shana Brown, Daniel Buck, Nara Dillon, Sara Friedman, Stevan Harrell, Jiang Jin, Rebecca Karl, Barry Keenan, Haiyan Lee, Mark Lincicome, Barbara

Mittler, Christopher Reed, Heidi Ross, Vivienne Shue, Patricia Stranahan, Eddy U, Frederic Wakeman, Jr., Jeffrey Wasserstrom, Wen-hsin Yeh, Peter Zarrow, and Li Zhang. I thank them all for their thoughtful questions and suggestions. Discussions at the Fairbank Center with Eileen Chow, Merle Goldman, and Ellen Widmer helped inspire my thinking on several key issues. I am also grateful to all my colleagues at Bard College who have posed stimulating questions and offered helpful comments in a faculty seminar and two History Program faculty colloquia. I extend special thanks to Myra Armstead, Sanjib Baruah, Richard Davis, Carolyn Dewald, Nara Dillon, Michael Donnelly, Tabetha Ewing, Laura Kunreuther, Mark Lytle, Gregory Moynahan, Gennady Shkliarevsky, Alice Stroup, and Li-hua Ying, whose thoughtful critical readings and engaged discussion about this work, as well as the craft of history and comparative social science, have made Bard a lively intellectual community. I also thank Kate Lawn for her careful editing of this work.

Two generous postdoctoral fellowships—the National Academy of Education/Spencer Postdoctoral Fellowship (2000–2001) and the An Wang Post-Doctoral Fellowship at the Fairbank Center for East Asian Research, Harvard University (2003–4)—made possible a comprehensive revision of this book. During each year I was fortunate to be part of a dynamic group of postdoctoral fellows who helped me grow intellectually. I thank Christian De Pee, Stephen Hershler, and William Schaefer for enlivening the 2000–2001 year at the University of California at Berkeley's Institute for East Asian Studies. I am grateful to the 2003–4 cohort of An Wang Fellows at the Fairbank Center—Cathryn Clayton, Shelley Drake Hawkes, Eileen Otis, Anne Reinhardt, Carlos Rojas, Shao Dan, and Shen Hsiu-hua—for their friendship, intellectual openness, and critical insight.

Research for this project in Taipei, Shanghai, Hangzhou, Nanjing, and other cities in the lower Yangzi region was supported by the following grant programs: the National Program for Advanced Study and Research in China, Committee for Scholarly Communication with China; a Fellowship for Dissertation Research Abroad from the American Council of Learned Societies/The Chiang Ching-kuo Foundation; a Foundation for Scholarly Exchange (Fulbright Foundation) travel grant; the National Endowment for the Humanities Summer Stipend program; and a Freeman Foundation Course

Development Research Grant. Over the years of research supported by these grants, I was fortunate to be able to share the joys and travails of library and archive work in China with Steve Averill, Julie Broadwin, Janet Chen, Megan Ferry, Tobie Meyer-Fong, Elisabeth Köll, Hongming Liang, Rebecca Nedostup, William Schaefer, and Helen Schneider.

Working on this project has given me added appreciation for the classroom as a context for intellectual exchange and helped me see how much of my scholarship is worked out in dialogue with students. I am grateful to all my students at Bard for their enthusiastic and creative engagement with the questions I bring to class. I extend special thanks to students in my seminars "The Politics of History" and "Political Ritual in the Modern World." If any of you ever read this book, you will see many of our discussions reflected in it.

This project has also helped me understand that schools are complex, socially embedded institutions, whose intellectual projects are sustained by the work of many who are neither faculty nor students. I thank the faculty secretaries at Bard College for their help with many aspects of this project. Thanks as well to the Inter-Library Loan staff at Bard's Stevenson Library for their patience in handling requests for Chinese-language materials. I am grateful to grandparents in Stuyvesant, New York, and Cleveland, Ohio, and to day-care providers in Newton, Massachusetts, and Annandale, Red Hook, and Rhinecliff, New York, for their dedicated care of Andrew and Lizzie.

Andrew and Lizzie have, quite literally, grown along with this book. Their love and laughter have been the most joyful gifts I could ever have imagined. The many times when they insisted that I read their books instead of my books, and that I walk with them in the woods instead of back to the office, have helped me keep Republican China in perspective. My love and greatest thanks go to my wife, Katie, who has believed in me and this project even when I have not. This book is dedicated to you, my family.

***

Some material from Chapter 3 appeared in a preliminary version in my article "Self-determination or Self-discipline? The Shifting Meanings of Student Self-government in 1920s Jiangnan Middle Schools," *Twentieth-Century China* 23, no. 2 (April 1998): 1–39. Portions of Chapters 2 and 6 previously appeared in "'China—The Land and Its

People': Fashioning Identity in Secondary School History Textbooks, 1911–1937," *Twentieth-Century China* 26, no. 2 (April 2001): 17–62. This material is included here with permission from *Twentieth-Century China*. Part of Chapter 4 was included in my article "Setting the Sheet of Loose Sand: Conceptions of Society and Citizenship in Nanjing-Decade Party Doctrine and Civics Textbooks," in *Defining Modernity*, ed. Terry Bodenhorn (Ann Arbor: University of Michigan, Center for Chinese Studies, 2002). It is reprinted by permission of the Center for Chinese Studies, University of Michigan. Parts of Chapter 5 were previously published in "Rethinking Governmentality: Training, Cultivation, and Cultural Citizenship in Nationalist China," *Journal of Asian Studies* 65, no. 3 (August 2006). They are reprinted with permission of the Association for Asian Studies.

My special thanks to Mike Hollingsworth for his careful work on the map of Jiangsu, Zhejiang, and the Lower Yangzi Macroregion. I am grateful to Feng Yiyin for permission to use Feng Zikai's drawing for the dust jacket illustration.

R.C.

Contents

Reference Matter

Tables, Map, and Figures

Abbreviations

MGRB   *Minguo ribao* (Republican daily; Shanghai edition)

SMA   Shanghai Municipal Archive

SMA2   Suzhou Municipal Archive

ZDLD   Zhongguo di'er lishi dang'anguan (China's Second Historical Archive)

ZKB 1929   Jiaoyubu zhong xiao xue kecheng biaozhun qicao weiyuanhui. *Zhong xiao xue kecheng zhanxing biaozhun* (Temporary curriculum standards for secondary and primary schools). 3 vols. Shanghai: Jiaoyubu, 1929.

ZKB 1933   Jiaoyubu zhong xiao xue kecheng biaozhun bianding weiyuanhui. *Gaoji chuji zhongxue kecheng biaozhun* (Curriculum standards for high schools and lower middle schools). Shanghai: Zhonghua shuju, 1933.

ZKB 1936   Zhongxue kecheng biaozhun bianding weiyuanhui [Jiaoyubu]. *Chu gao ji zhongxue kecheng biaozhun* (Curriculum standards for lower middle schools and high schools). 2 vols. Shanghai: Zhonghua shuju, 1936.

Articulating Citizenship

Civic Education and Student Politics in

Southeastern China, 1912–1940

Introduction

On July 10, 1912, Cai Yuanpei (1868–1940), first minister of education for the newly founded Republic of China, addressed scholars and educators gathered in Beijing for the Provisional National Education Conference (Quanguo linshi jiaoyu huiyi).[1] At this meeting called to establish a new national school system for China, Cai drew a stark contrast between education designed to produce subjects for a monarchy and education able to cultivate citizens of a republic. "The approach of republican education (*minguo jiaoyu*)," he said, "should consider what capabilities the person receiving the education himself will have to be able to fulfill certain responsibilities, and what kind of education will allow him to have those capabilities."[2] He then charged the conference with creating a form of education that would give China's people the capacities to fulfill their duties and enact their rights in society. In Cai's formulation, and in the minds of most Republican-era elites, creating modern citizens was now the central goal of Chinese education.

The process of using education to cultivate a modern citizenry had begun during the final decade of the Qing dynasty (1644–1911).[3] But the project took on new urgency as China became a republic after a rapid succession of revolutionary upheavals overthrew the imperial state during the last few months of 1911. Cai, in his brief lecture, mapped out his own vision of republican education, and

---

1. Cai Yuanpei, "Quanguo linshi jiaoyu huiyi kaihui ci."

2. Ibid., 15.

3. E.g., Ayers, *Chang Chih-tung and Educational Reform*; Borthwick, *Education and Social Change*; Bailey, *Reform the People*; Judge, *Print and Politics*, esp. chaps. 4 and 5.

the conference soon established a school system and set a standard curriculum.[4] But Cai's proposals and the new school system, rather than settling for good the issue of citizenship education, inaugurated two and a half decades of debate and experimentation. Intellectuals, educators, and political leaders of every stripe agreed with Cai that China's people could, and should, be taught the capacities necessary to be modern citizens.[5] However, over the following decades they came to argue fiercely over what those capacities should be and how they should be taught. Further, many of their carefully crafted policies and programs came to be transformed through educational practice and schoolhouse politics by textbook authors, teachers, school administrators, and students.

This book reconstructs the civic education and citizenship training in Republican-era lower Yangzi region secondary schools. It also analyzes how students used the tools of civic education encountered in their schools to make themselves into young citizens and explores the complex social and political effects of educated youths' civic action. In doing so, it helps to explain what citizenship meant in Republican China and why it took the forms it did.

Why Citizenship?

Why were late Qing and early Republican intellectuals, educators, and political leaders so concerned about citizenship? For these elites, the concept of citizenship seemed to offer answers to the crises of sociopolitical order and mounting imperialism that China con-

---

4. Cai called for a comprehensive approach to education composed of five elements: military-citizen education (*junguomin jiaoyu*); practical education (*shili zhuyi*); civic morality (*gongmin daode*); cosmopolitanism (*shijieguan*); and aesthetic education (*meiyu*). But he viewed civic morality as the foundation ("Quanguo lin-shi jiaoyu huiyi kaihui ci," 16). Cai had expressed a more cosmopolitan approach to education in the early months of 1912, but in the context of the education conference, nationalistic and civic educational ideals were foregrounded. See Bailey, *Reform the People*, 148–49, 152.

5. In the words of Paul Bailey, "the adoption of the Chinese school system so soon after the education conference of July–August 1912 had been a remarkably smooth affair in which Chinese educators had been unanimous in their desire to see an emphasis on general education within a fully coordinated system that would train a patriotic, hard-working and economically independent citizenry." See Bailey, *Reform the People*, 153–54.

fronted at the start of the twentieth century. In the eyes of critical intellectuals and reformist political leaders, many integral patterns of ethics and politics needed to be replaced or supplemented for the Qing empire to be reconstituted as a nation-state that could survive in the modern world. Late Qing and early Republican reformers and revolutionaries began the process of building a Chinese conception of citizenship by identifying these seemingly inadequate or limiting elements of the Qing sociopolitical order and describing contrasting qualities of civic action that they believed would transform China.

Fundamental to late imperial Chinese ethics and social organization was a form of personal morality (*si de*) that was organized around archetypal relations between individuals. The most important of these key social bonds were hierarchical relationships between fathers and sons, husbands and wives, and rulers and subjects.[6] To critical social theorists such as reformer Liang Qichao (1873–1929), this focus on personal, dyadic relations left relatively undeveloped the individual's sense of responsibility to society as a whole and thus needed to be supplemented with a new civic morality (*gong de*).[7] Civic morality was to relate the individual to society and be characterized by dynamic activity in public life that would lead to progressive social development.[8]

Late imperial society had an ideal of the morally refined, publicly engaged individual who worked for the social good, the members of the classically educated scholarly elite. But the scholarly elite was historically a small, highly privileged segment of society that took public action primarily through assuming an official position in the dynastic state or working locally in cooperation with the government to maintain order, implement state policies, and guide the people.[9] Late Qing and Republican reformers, by contrast, sought to generalize the responsibility for public service to all members of society and prescribe a wider range of social activities while also calling for popular

---

6. Brokaw, *The Ledgers of Merit*, chap. 4; King, "Individual and Group"; Munro, *Images of Human Nature*, 45–54; Rowe, *Saving the World*, 306–22.

7. Chang, *Liang Ch'i-ch'ao and Intellectual Transition*, 151–55.

8. Ibid., 177–89, 216; Schwartz, *In Search of Wealth and Power*; Wang, "Evolving Prescriptions for Social Life," 262–63.

9. E.g., Ch'u, *Local Government*, 175–85; Hsiao, *Rural China*, 145–257; Mair, "Sacred Edict"; Munro, *Images*, 144–47; Rowe, *Saving the World*.

sovereignty and political rights.[10] They further intended to reorient the late imperial moral elite's loyalty from its traditional focus on the emperor and the universal moral order of Neo-Confucianism to the bounded and sovereign nation-state, which became the "terminal community" for most late Qing and early Republican intellectuals.[11]

Thus, in a context of crisis, Chinese intellectuals, educators, and political leaders hoped to remake their society and polity by transforming the people into dynamic modern citizens.[12] Their basic understanding of citizenship is reflected in the terms most commonly used to represent it during the first decades of the twentieth century: *guomin* and *gongmin*.[13] *Guomin*, which was the term first used for "citizen" in China,[14] meant, literally, people (*min*) of the nation/state (*guo*). In the words of Shu Xincheng's popular civics textbook of the early 1920s, "The common people who live in a nation-state, as long as they are of that nationality (*guoji*), can all be called citizens (*guomin*)."[15] The term *guomin*, then, connoted the members of a horizontally interconnected and bounded national community and implied a primary identification with that community. *Gongmin* can be translated literally as "public people." At the most basic level the term meant those in the nation possessing civil rights (*gongquan*),[16] but it also described people engaged with community (*gong*) rather than private or personal (*si*) matters. In the words of another civics textbook from the 1920s, "Since a nation's citizens (*gongmin*) are produced from one nation's people, and they can enjoy all the various

---

10. See, for instance, Judge, *Print and Politics*, 83–88; and Zarrow, "Introduction," 5.

11. For the contrast between the nation as a bounded moral community and Confucian universalism in the thought of Liang Qichao, see Chang, *Liang Ch'i-ch'ao*, 157–64. Cf. Levenson, *Confucian China*, 1: 95–99.

12. Chang, *Liang Ch'i-ch'ao*, 150; Zarrow, "Introduction," 17.

13. For a discussion of the definitions of these two terms and their changing place in twentieth-century Chinese political discourse, see Goldman and Perry, "Introduction," 3–7. Although they see *gongmin* coming to displace *guomin*, educators and students continued to use *guomin* through the Republican period, but with the somewhat narrower meaning described here.

14. *Guomin* was a long-used character combination in imperial China that was reintroduced from Japan with modern associations of citizenship. See Liu, *Translingual Practice*, 308.

15. Shu Xincheng, *Chuji gongmin keben*, 1 (1923): 30.

16. Ibid.

rights within the nation, their feeling toward the nation should be uncommonly close, [such that] they are always responsible for developing the nation. . . . As [Theodore] Roosevelt said, 'A good citizen must pay attention to public matters (*gongzhong shiwu*).'"[17] As this passage, with its legitimizing reference to a recent U.S. president, suggests, the term *gongmin* described a person who was not solely immersed in dyadic, interpersonal relations after the manner of Confucian morality. Rather, the citizen felt morally responsible to the larger community of "the public" (*gongzhong*) and was willing and able to act for the public welfare. Thus, at a time when the Chinese state and any kind of collective social order seemed under dire threat, intellectuals and political leaders conceived of citizens as new kinds of social and political agents whose public action would rescue the national community. The project of remaking China's people as active citizens dedicated to the public welfare continued through the Republican period (1911–49) and into the People's Republic (1949–).

But the idea of citizenship, once introduced, raised as many questions as it answered. What were the boundaries of the national community in China, who were its members, and how was the nation constituted? What were the best institutions for political participation and the best modes of political action? How could the new "public" of national society be imagined? How, in other words, should people be interconnected if not through Confucian hierarchical relationships? How should people act in modern society and take responsibility for "public matters"? What was the proper culture and morality of the modern public (*gong de*)? Each dimension of citizenship as a new category of social and political action had to be negotiated and then taught during the first half of the twentieth century. Schools were one of the primary sites where this dialogue and instruction took place.

Citizenship as a Creative Process

Over the first half of the twentieth century, social and political theorists, as well as political activists, drew on competing strains of nationalism, liberalism, social Darwinism, anarchism, Marxism, and fascism, as well as the Confucian tradition, to formulate definitions

---

17. Gu and Pan, *Gongmin xuzhi*, 2: 52.

of citizenship. At the same time, pre-existing and emergent patterns of social and political action, ranging from gentry management, local self-government, and elite reform associations to mass demonstrations and grassroots political mobilization, offered diverse models of practical civic action.

Several scholars have begun to analyze how modern Chinese elites used the press and other media, new forms of civic ritual and public behavior, and various kinds of political organization to introduce competing conceptions of citizenship in modern China.[18] These approaches have added many pieces to an emerging mosaic of early Chinese citizenship, but focusing on civic education promises to sharpen our understanding of Chinese citizenship in a number of ways. First, because a diverse array of social elites, intellectuals, and political leaders sought to use civic education to present normative versions of citizenship to China's young people, school-based civic training captures the full spectrum of Republican citizenship discourse. Figures ranging from liberal intellectuals such as Jiang Menglin and Hu Shi to Chinese Communist Party activists such as Yun Daiying and Yang Xianjiang to conservative Nationalist Party ideologues Dai Jitao and Chen Lifu all participated in shaping civic education during the Republican period. Moreover, those who used schools as vehicles for their ideas of citizenship distilled those ideas and presented them in their most accessible form in textbooks, moral instruction, rules of conduct, and charters for student organizations. Consequently, analyzing civic education allows us to track with particular clarity changes in Chinese conceptions of citizenship and to explore why that citizenship took the particular forms that it did.

Further, Republican-era students were energetic social and political actors whose modes of civic action ranged far beyond the nationalist protests that have been the primary focus of scholars of modern China. Students ran literacy schools (Chapter 7) and formed self-government organizations to act out democracy (Chapter 3). They performed citizenship publicly in civic rituals by, variously, singing the national song, saluting the flag, passing out pamphlets, parading with lanterns and banners, marching in formation, and bowing to

---

18. E.g., Dunch, *Fuzhou Protestants*; Fitzgerald, *Awakening China*; Fogel and Zarrow, *Imagining the People*; Goldman and Perry, eds., *Changing Meanings of Citizenship*; Harrison, *The Making of the Republican Citizen*; Judge, *Print and Politics*.

images of Nationalist Party leader Sun Yat-sen (1866–1925) (Chapter 6). They called for and joined in military training during the 1930s (Chapter 5) and performed street-side dramas to raise national consciousness during the Sino-Japanese War (Chapter 7). Students' penchant for experimenting with the forms of citizenship that they encountered in schools provides an exceptional opportunity to explore how citizenship ideals took shape in social and political action during the Republican period.

Because schools were intended to give students comprehensive training in how to be citizens, they also became one of the only contexts in Republican China in which all the many dimensions of citizenship were introduced, juxtaposed, and related in anything resembling a systematic way. As with nation-building and state-building elites in other world contexts, modern China's intellectuals and political leaders grouped a vast array of ideas and actions under the rubric of citizenship. Everything from voting and military service to neat modern dress and teeth brushing was identified, at various times and by various proponents, as proper citizenship. But intellectuals and political leaders often concentrated their attention on one practice or another, seldom seeking to combine them in an integrated approach. As these multiple forms of citizenship intersected in secondary schools, educators, teachers, and students tried to relate them in ways that they found meaningful and practical. In this way, schools served as workshops in which educated Chinese elites experimented with how to fashion a coherent form of modern citizenship from a bricolage of borrowed foreign and retooled indigenous ideas and practices.

Reframing Citizenship

The very diversity of citizenship discourse in modern China, and indeed elsewhere, poses a challenge to the analyst. How can we conceptualize and relate the many different ideals and manifestations of citizenship in the modern world? Drawing on the rich comparative and theoretical literature on modern citizenship, I distinguish four dimensions, each of which can take a variety of forms. Two of the central concerns in that literature have been national identity, which has emerged as the dominant mode of collective identity over the past two centuries, and political participation and rights. A vast literature explores the multiple and competing ways that common civic association, and/or perceptions of shared race, ethnicity, belief, or culture, can inspire a sense of membership in a national community.[19] Contemporaneously, political theorists have revisited Euro-American conceptions and practices of political citizenship by exploring a central tension in Western thought between civic republicanism, which stresses community solidarity and direct participation, and liberal approaches, which emphasize individual freedom, civil rights, and mediated participation.[20] Their analyses delineate multiple approaches to political citizenship even in the Euro-American context, suggesting that there is no stable, normative model of political participation for modern citizens.

In contrast to this exclusively political focus, T. H. Marshall and Bryan Turner have challenged us to think of citizenship in social terms and to consider how socioeconomic inequality has limited some people's access to social membership and full participation in either the national community or the political process.[21] Complementing Marshall's and Turner's concern with social stratification and differential citizenship, anthropologists and cultural historians have explored how distinctions among cultural repertoires have contributed to varying forms of cultural citizenship.[22] Analysis of cultural citizenship assesses how ritual performance, differences in dress and taste, and embodied forms of demeanor and etiquette are essential to defining civic identities and relations of social and political power within and among modern nation-states.

Reviewing this theoretical and comparative literature, we can see modern citizenship as having four relatively distinct dimensions. My

---

19. E.g., Anderson, *Imagined Communities*; Balakrishnan, ed., *Mapping the Nation*; Balibar and Wallerstein, *Race, Nation, Class*; Brubaker, *Citizenship and Nationhood*; Chatterjee, *The Nation and Its Fragments*; idem, *Nationalist Thought and the Colonial World*; Gellner, *Nations and Nationalism*; Smith, *National Identity*.

20. Alejandro, *Hermeneutics, Citizenship, and the Public Sphere*; Beiner, ed., *Theorizing Citizenship*; Miller, *Citizenship and National Identity*; Mouffe, ed., *Dimensions of Radical Democracy*; Oldfield, *Citizenship and Community*.

21. Marshall, *Citizenship and Social Class*; Turner, "Contemporary Problems in the Theory of Citizenship."

22. Ong, "Cultural Citizenship as Subject-Making." Cf. Bourdieu, *Distinction*; Corrigan and Sayer, *The Great Arch*; Elias, *The Civilizing Process*; Kasson, *Rudeness and Civility*; Rosaldo, "Cultural Citizenship and Educational Democracy"; Williams, *Stains on My Name*.

shorthand characterizations of these four dimensions are national identity, political participation and rights, social membership, and cultural citizenship. Disaggregating citizenship discourse and practice in this way allows us to order conceptually how we approach citizenship education and youth civic action in Republican China. But it leaves us with the equally urgent problem of understanding how modern people in any given context relate these dimensions in meaningful and coherent ways and, in fact, act as citizens.

Given the complexity of the topic, scholars of modern China have understandably tended to focus on specific dimensions of citizenship. Henrietta Harrison has given us a groundbreaking and sophisticated account of early Republican etiquette, dress, and political ritual.[23] Several authors have traced new conceptions and dynamics of political participation during the Republican period.[24] A rich literature now details competing views of the parameters of China's national community.[25] Less clear is how these dimensions of citizenship related to one another at any given moment. Did they clash and contradict, confounding any coherent sense of what it meant to be a citizen in Republican China? Or, were there ways in which political, cultural, and social practices reinforced one another to constitute a unified pattern of Chinese citizenship?

This book argues that lessons about national identity, political participation, and the social order in lower Yangzi region secondary schools reinforced one another and promoted a coherent conception of republican citizenship, characterized by direct participation and practical action for the nation's welfare. History and geography textbooks described the Chinese nation as a unified, sovereign territory that had grown continuously over millennia but was now under threat from outside attack and was marked by ethno-cultural pluralism as well as economic unevenness and "backwardness" (Chapter 2). Textbooks and teachers argued that only committed action by China's citizens would guarantee the integrity of the nation's borders,

---

23. Harrison, *The Making of the Republican Citizen*. Cf. Dunch, *Fuzhou Protestants*.

24. E.g., Fung, *In Search of Chinese Democracy*; essays by Goodman, Perry, Strand, and Wasserstrom in Goldman and Perry, eds., *Changing Meanings*; Nathan, *Chinese Democracy*; Strand, *Rickshaw Beijing*.

25. E.g., Dikötter, *The Discourse of Race*; Duara, *Rescuing History*; Fitzgerald, *Awakening China*; Goodman, *Native Place*.

forge a unified national people, and allow it to develop economically. Paired with this developmental nationalism were civics textbook images of society as a functionally integrated organic body with citizens as the component cells (Chapter 4). The health of the whole social body and each of its parts was seen to be interdependent, providing a rationale for citizens to work for the greater social welfare. At the same time, when intellectuals, educators, and students experimented with forms of participatory democracy in student self-government organizations, they consistently adopted the model of civic republicanism, which stressed direct participation and concrete contributions by each member of the community (Chapter 3). The idea of citizenship as pragmatic action taken to build national strength and spark social transformation that emerged from these intersecting forms of instruction and training in lower Yangzi region schools became in turn one of the most persistent and distinctive features of twentieth-century Chinese political life.

The forcefulness of this message of active citizenship in schools of the Republican period has important implications for our understanding of twentieth-century Chinese society and politics. One, it suggests that a conception of the nation as defined by political boundaries and common civic action was at least as important as the racial, ethnic, or cultural formulations of national community that have been the focus of many scholars of modern Chinese nationalism. Further, it indicates that the publicly oriented civic activism characteristic of civic republicanism has been more fundamental to Chinese approaches to political citizenship than has the open debate and individual rights of liberal democracy, which many scholars suggest was never fully realized in Republican China.

The ideal of active citizenship in secondary education resulted from the ways that history and geography instruction, student self-government, and portrayals of society were articulated together.[26] I follow Stuart Hall in using "articulation" to connote the complex linkages or interrelations among diverse discourses, such as those of race, class, and gender, which can develop over time through political negotiation and social practice to give meaning and power to par-

---

26. Hall, "Signification, Representation, Ideology"; idem, "On Postmodernism and Articulation," 141–45. Cf. Slack, "The Theory and Method of Articulation in Cultural Studies."

ticular terms, ideas, and activities, such as citizenship. As people use concepts and enact patterns of practice, associations and connections are built up within and across discourses or arenas of social action so that terms in one discourse or actions in one field will evoke and reinforce those in other discourses or fields. They become, in other words, articulated, that is, both expressed and linked, together, creating a unified structure from discursive chains that operate independently. Competing social and political groups, in turn, seek to articulate, or rearticulate, concepts and practices in ways that define them as pivotal and powerful members of society and that identify themselves and their collective projects with widely accepted categories in influential systems of ideas. Thus, as we will see, many lower Yangzi region students sought to associate themselves with the ideal of the active citizen, which was produced through the articulation of discourses of political participation, social order, and national membership, as a way to claim social privilege and political agency. Moreover, from the time of the National Revolution (1926–28) through the start of the Sino-Japanese War (1937–45), the Nationalist and Chinese Communist parties competed to align their political programs with the vision of active republican citizenship, in part as a way to court student activists.

Teaching and Practicing Citizenship

Active republican citizenship formed the hard core of secondary-level civic education from 1912 through 1937, but it by no means exhausted it. Beyond lower Yangzi region schools' emphasis on direct civic action, civic education and citizenship training exposed students to varied conceptions of and approaches to citizenship. Civics (*gongmin*), moral cultivation (*xiushen*), or "party doctrine" (*dangyi*) classes presented contrasting and changing approaches to social reform and civic morality. History and geography textbooks contained tensions among ethnic, cultural, and territorial definitions of the Chinese nation. Classroom instruction, in turn, was combined with various kinds of formalized training, informal instruction, and independent exploration. Student self-government, Scouting activities, moral training, and civic ritual all carried diverse messages regarding norms of public behavior and trajectories of civic action. Routine practices of daily life related yet other messages. Lectures, reading groups, and independent reading also exposed students to new ideas about social reform and political action. These diverse avenues of instruction and training, which were the result of educators' and intellectuals' experiments with foreign and domestic ideas and pedagogy, meant that students consistently encountered a complex array of messages about citizenship, within which there were numerous tensions.

This pluralism, with its accompanying contradictions, meant that Republican Chinese schools could hardly act as "ideological state apparatuses," which Louis Althusser contends determine students' consciousness and social action by imposing the ideology and practices of the dominant society.[27] In fact, ethnographies of schooling that document student resistance warn us against ever assuming that the content of textbooks is directly transmitted to students or that schoolroom training will automatically shape student behavior in prescribed ways.[28] Accordingly, this study suggests that by presenting students with competing forms of civic education that sent mixed messages, lower Yangzi region schools constituted a loosely bounded universe of sanctioned approaches to citizenship from which students could draw strategically to fashion their own social identities and forms of civic action.[29] A vital part of this project, then, becomes reconstructing how students across the arc of the Republican period negotiated this diversity and chose to act as citizens.

The tensions in Republican-period civic education were particularly acute in instruction for female students. At base was the question, first debated by educators during the late Qing, of whether educated women would be, in Joan Judge's apt words, "citizens or mothers of citizens."[30] Across the Republican curriculum, female students at the lower Yangzi region's secondary schools received

---

27. Althusser, "Ideology and Ideological State Apparatuses." Cf. Bowles and Gintis, *Schooling in Capitalist America*.

28. E.g., Luykx, *The Citizen Factory*; Willis, *Learning to Labour*.

29. For one theory of how contradictions within schooling provide opportunities for autonomous student action, see Giroux, "Theories of Reproduction and Resistance." However, where authors like Giroux read student action primarily as a form of resistance, which it most certainly sometimes is, this study demonstrates ways in which student action can produce new ideas and patterns of social action.

30. Judge, "Citizens or Mothers of Citizens?" Cf. Bailey, "Active Citizen or Efficient Housewife?"; and McElroy, "Forging a New Role for Women."

complex signals, which were at times very gender specific and at others much more gender neutral, about what was appropriate behavior for them as educated modern women. But a prominent element in most lower Yangzi region women's schools were forms of civic training, such as Scouting and self-government, designed to prepare all students for full civic participation. Educators' and officials' persistent emphasis on these kinds of training created opportunities for female students to draw on their education to perform public civic action for the sake of the nation and thereby claim the "independent personhood" (*duli renge*) that Wang Zheng has identified as the basis of modern Chinese feminism.[31]

Localizing and Globalizing Citizenship

Using schools to instill citizenship was a national project in Republican China, but the process was most developed in the lower Yangzi region, an area dominated by the two provinces of Zhejiang and Jiangsu.[32] There, the cosmopolitan industrial and trading hub of Shanghai stood at the center of a dense urban network that spread commerce and culture to cities and towns throughout the region. Throughout the first half of the twentieth century, the region was a hothouse of innovation in modern education, as educators and social elites constructed modern schools and experimented with innovative curricula and pedagogies.[33] In terms of citizenship education, new modes of civic training, such as student self-government and Scouting, emerged early on and became most popular in this region. Likewise, because China's publishing industry was largely centered in Shanghai, modern textbooks and journals, and the ideas they carried, were

---

31. Wang, *Women in the Chinese Enlightenment*, 16–23, 172–86.

32. For accounts of the different ways in which education was used in projects of nation building, state formation, and economic development, see Bailey, *Reform the People*; Borthwick, *Education and Social Change*; Carter, *Creating a Chinese Harbin*, esp. chap. 2; Keenan, *The Dewey Experiment in China*; McElroy, "Transforming China Through Education"; Peake, *Nationalism and Education in Modern China*; Thøgerson, *A County of Culture*; and Tsang, *Nationalism in School Education in China*.

33. Bastid, *Educational Reform*; Chauncey, *Schoolhouse Politicians*; Culp, "Elite Association and Local Politics"; Keenan, *The Dewey Experiment in China*, 57–125; Yeh, *The Alienated Academy*; idem, *Provincial Passages*.

disseminated most easily, and penetrated most deeply, in this region.[34] Moreover, after the Chinese Nationalist Party made Nanjing its capital, party control of this region was most certain and direct, and the region's cities and towns became the natural objects of the party's periodic efforts at social, political, and, in turn, educational reform.[35] Because of these intersecting forces of change, the development and impact of various groups' approaches to citizenship education emerge most clearly when we focus our attention on the lower Yangzi region.

These self-conscious efforts to educate modern citizens merged with existing patterns of civic action and vibrant trends of social, cultural, and political change in the lower Yangzi region. The region's long history of scholarship, state service, and local gentry philanthropy generated a particularly strong ethos of scholar-elite activism that encouraged political participation and a commitment to social service by educated people.[36] The forms of civic association, political organization, and grassroots activism pioneered and enacted by the region's reformist elite also modeled for educators and students approaches to political participation that influenced the civic action and citizenship education in local schools.[37] In addition, treaty port foreigners' introduction of new norms of public decorum intersected in complex ways with the standards of indigenous elite culture. Together they generated a hybrid public culture that was perhaps more elaborated in this region than anywhere else in China because of the frequent interaction of indigenous elites and the foreign bourgeoisie, especially in the cosmopolitan metropolis of Shanghai.[38] This emergent public culture in turn shaped instruction in cultural citizenship in lower Yangzi region secondary schools.

---

34. Lee, *Shanghai Modern*; Reed, *Gutenberg in Shanghai*.

35. Cavendish, "The 'New China' of the Kuomintang," 181–82; Coble, *The Shanghai Capitalists*; Eastman, "Nationalist China During the Nanking Decade, 1927–1937," 147–48; Fewsmith, *Party, State, and Local Elites*; Geisert, *Radicalism and Its Demise*; Henriot, *Shanghai*; Tien, *Government and Politics in Kuomintang China, 1927–1937*; Wakeman, *Policing Shanghai*.

36. Brook, "Family Continuity and Cultural Hegemony"; Elman, *A Cultural History of Civil Examinations*, esp. 256–59; idem, *Classicism, Politics, and Kinship*.

37. Culp, "Elite Association and Local Politics"; Rankin, *Elite Activism and Political Transformation*; Schoppa, *Chinese Elites and Political Change*.

38. Goodman, "Improvisations on a Semicolonial Theme"; Honig, *Creating Chinese Ethnicity*; Yeh, *The Alienated Academy*, 49–88.

Because China's lower Yangzi region experienced very specific patterns of socioeconomic change and political incorporation during the Republican period, we must be wary of assuming educational trends there were automatically mirrored elsewhere. Rather, drawing on the growing corpus of studies of Republican education and youth politics that focus on diverse regions allows us to place specific changes in lower Yangzi region citizenship education in broader spatial context, tracking which were distinctive and which were more common nationally. At the same time, because the lower Yangzi region encompassed the national publishing center of Shanghai and the Nationalist capital of Nanjing, forms of citizenship education and youth civic action that developed within the region were sometimes disseminated nationally through print media and/or government policies. When this occurred, lower Yangzi region educators, schools, and students modeled practices that became nationally prominent.[39]

In this book, then, we consider how a complex universe of foreign ideas of citizenship and civic education was transmitted and circulated in China and how the social, cultural, and political legacies of one region of the country affected the reception and formulation of those ideas. In combining these two approaches, we place modern China in a comparative framework and demonstrate how in the discourse and practice of citizenship there was part of a global process of reshaping disparate communities of imperial subjects into national citizenries. Consequently, this study speaks to the ongoing debate in the Sinological literature regarding the extent to which foreign ideas and patterns of sociopolitical practice were imposed through global imperialism or variously resisted and/or transformed by Chinese thinkers and actors.[40] Although citizenship was certainly an imported category of social membership and political action, like hygiene (*weisheng*) or physical culture (*tiyu*), we will see that the

---

39. For instance, many Scouting manuals published for the national market were written by Scout leaders from this region, such as Cheng Jimei and Hu Liren, and reflected their activities. See Chapter 5. Moreover, the Nationalist Party developed certain forms of intensified military training in the lower Yangzi region and then extended them elsewhere. See Culp, "Rethinking Governmentality."

40. E.g., Duara, *Rescuing History*; Karl, *Staging the World*; Liu, *Translingual Practice*; Morris, *Marrow of the Nation*; Rogaski, *Hygienic Modernity*.

lower Yangzi region's modernizing elites were active agents in re-constituting it, in many different versions, for the context of Republican China. They crafted distinctive forms of citizenship both by re-framing foreign approaches to citizenship in light of pre-existing patterns of politics, society, and culture and by combining borrowed modes of citizenship in ways unparalleled in the Euro-American world. Reconstructing these complex patterns of parallel yet distinctive forms of civic action contributes to an ongoing comparative discussion of citizenship education and youth civic training in the modern world.[41]

* * *

The chapters of this book explore the creation of Chinese citizenship by reconstructing and relating specific dimensions of civic education and citizenship training in lower Yangzi region schools between the founding of the Republic in 1912 and the onset of the Sino-Japanese War in 1937. Because Japanese occupation of many of this region's core areas by the end of 1937 dramatically changed local conditions of schooling, that year marks a suitable end point for our discussion of Chinese elites' and political leaders' projects of republican civic education. Chapter 1 sets the institutional context for our analysis. It maps the school infrastructure of the lower Yangzi region, tracks educators' and Republican states' changing formulations of school curricula, and details the dynamics of textbook production by China's leading publishing companies.

Textbooks and school-based activities are the primary avenues for assessing, in Chapters 2 through 6, how educators, intellectuals, and political elites defined and taught citizenship. These chapters relate specific aspects of instruction, training, and school practice to the four dimensions of citizenship distinguished above: national identity, political participation, social membership, and cultural citizenship. Chapter 2, "Nation as Race, Culture, or Place?" maps the crosscutting tensions in secondary-level history and geography textbooks' representations of the national community. Students' experimentation with different modes of political participation in the model republican institutions of student self-government associations is the focus of Chapter 3. Civics and language textbooks portrayed for stu-

---

41. E.g., Knopp, *Hitler's Children*; Koon, *Believe, Obey, Fight*; Luykx, *The Citizen Factory*; Rosenthal, *The Character Factory*.

dents modern society, the social person, civic morality, and changing models of social reform. Chapter 4 unpacks these texts' complex and changing messages and relates them to the lectures, extracurricular reading, and study societies that provided alternative channels for learning about society. Students' cultural performance of citizenship is the focus of Chapters 5 and 6. Chapter 5 recounts how moral cultivation and training in hygiene, etiquette, and practical and military skills taught students varied and sometimes contradictory ways of acting as citizens in their daily lives. Chapter 6 illustrates how students played central roles in the dynamic and highly contested civic rituals of the Republican period, sometimes transforming them through their participation.

Chapters 2 through 6 all, to varying degrees, explore students' civic action, but Chapter 7 focuses explicitly on the changing ways students adopted, adapted, and combined elements from their schooling to act as republican citizens. It also gauges the Nationalist and Communist parties' relative success at harnessing students' civic action for their revolutionary projects. Although the onset of war with Japan seriously disrupted schooling in the region, many youths educated during the 1930s and before continued their efforts to enact citizenship beyond 1937. Chapter 7 accordingly tracks their activities through the early years of the war.

The Conclusion situates Republican citizenship in the longer temporal framework of twentieth-century Chinese history and traces vital continuities into the period of the People's Republic, after 1949. It suggests that the forms of active citizenship taught in lower Yangzi region schools resonated in the political dynamics of Maoist China.

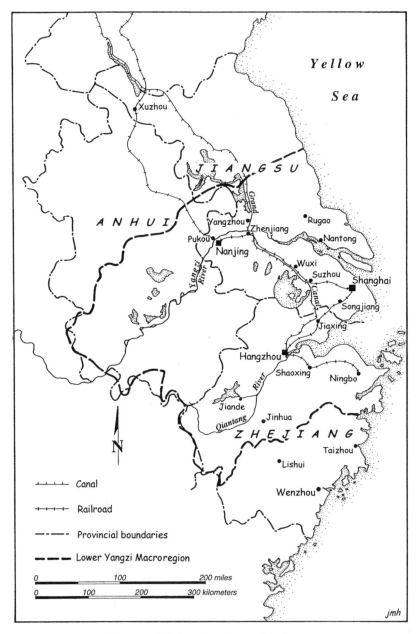

Yellow

Sea

Xuzhou

J I A N G S U

A N H U I

Rugao

Yangzhou

Zhenjiang

Pukou

Nantong

Nanjing

Wuxi

Suzhou

Shanghai

Songjiang

Jiaxing

Hangzhou

Shaoxing

Ningbo

Jiande

Jinhua

Qiantang

Z H E J I A N G

Taizhou

Lishui

Wenzhou

Yangzi River

Canal

Grand

River

Canal

Railroad

Provincial boundaries

Lower Yangzi Macroregion

0   100   200 miles

0   100   200   300 kilometers

N

jmh

Jiangsu and Zhejiang Provinces and the Lower
Yangzi macroregion, circa 1927

I

The Ideological
Infrastructure of Citizenship:
Schools and Publishing

Republican China's peculiar complex of citizenship ideals and practices spread and were contested within a specific institutional context constituted, in part, by schools and China's emerging publishing
industry. The school system expanded rapidly over these decades,
especially in the lower Yangzi region, through the combined efforts
of local elites, educational reformers, and various national and local
governments. Moreover, the organization and curricula of schools
changed frequently and sometimes dramatically as educators, intellectuals, and political parties sought to mobilize schools for various
social and political projects, including the creation of republican
citizens. The publishing industry, too, flourished during the decades
of the early Republic, often relying heavily on the large, growing, and
dependable textbook market. Publishers expanded capacity and
spread their marketing networks throughout the country, ensuring
that textbooks reached every region of the nation and every local
area within the core coastal provinces of Jiangsu and Zhejiang.

Within the institutional frameworks of schools and publishing
companies, diverse social and political groups engaged one another
over the meanings of citizenship. During the 1910s and 1920s, reformist intellectuals, educators, and leading academics drafted curriculum standards, wrote textbooks, established schools, and taught
in them in order to spread their ideas about citizenship and modern
society and politics. After 1927 the Nationalist government regulated

existing schools, produced centralized curricula, and enforced strict censorship regimes on private publishing companies, exercising a distinctive form of state power that operated through co-optation rather than either direct control and state monopoly or laissez-faire tolerance of social forces.

I. THE SCHOOL INFRASTRUCTURE

The primary focus of this book is the civic education and citizenship training in roughly two-dozen Chinese-run regular middle schools (*zhong xuexiao*) in the lower Yangzi region.[1] For the period of the 1910s and early 1920s, I also draw many examples from normal schools (*shifan xuexiao*) that trained primary school teachers. Between 1912 and 1923 normal schools were considered to be as academically rigorous as middle schools, and they constituted an equally important part of the overall complex of "secondary education" (*zhongdeng jiaoyu*).[2]

During the 1910s, under the first Republican-period school system, regular middle schools, normal schools, and higher-level technical schools each operated separately.[3] Regular middle and normal

---

1. The sample for this study encompasses both private and public secondary schools located in a range of cities and towns. Because many prominent schools clustered in the cities of the regional core, such as Shanghai, Suzhou, and Hangzhou, many examples are drawn from those areas, but schools from more peripheral areas, such as Jinhua, Xinchang, and Yangzhou, are also included. I exclude missionary schools from consideration because I am most concerned with analyzing how Chinese actors defined and promoted modern citizenship. Moreover, Chinese-run schools vastly outnumbered missionary schools. One missionary estimated in 1925 that there were roughly 200,000 students in missionary schools as compared to 6.6 million students in government-run schools of all levels (Graham, *Gender, Culture, and Christianity*, 166).

2. In terms of the kind and content of courses, normal and middle schools generally paralleled each other. See the discussion of the early Republican curriculum below. My analysis does not include technical or vocational schools (*shiye* or *zhiye xuexiao*). Technical schools' emphasis on practical training in contrast to academics—"practice" was to occupy two-fifths of the class time—complicates efforts to compare them to middle schools. See Bailey, *Reform the People*, 154–57.

3. Ibid., 154–55; *Zhejiang jiaoyu jianzhi*, 79, 121–22.

schools both held four years of classes, but normal schools also held a one-year preparatory course. In 1922, the National Federation of Education Associations, a professional organization of educators, and the Ministry of Education jointly developed the New School System (Xin xuezhi). Under this system secondary schooling was divided into lower middle school (*chuzhong*) and high school (*gaozhong*) sections, each of which had a three-year course of study.[4] From this period through much of the 1930s, high schools often became comprehensive, offering courses in specialized academic areas, teacher training, and various technical fields, with consolidation of normal schools and some higher-level vocational schools into existing middle schools occurring at varying rates across the region.[5] The practical result was a dramatic decrease in the number of students choosing the teacher-training track. Most students instead focused on regular academic training, which could help them advance to tertiary schooling.[6]

After the founding of the Nationalist government in 1927, a central objective of Nationalist Party educational policy was to revitalize teacher-training education and to emphasize vocational training and practical education over academic training and college preparatory courses of study.[7] The University Council[8] and the Ministry of Education reversed the New School System's consolidation of normal and regular secondary schools by promulgating separate laws for each type of school and promoted their separate operation. The ministry set guidelines for establishing normal schools at the lower

---

4. Keenan, *The Dewey Experiment*, 65–66; Li Huaxing et al., *Minguo jiaoyushi*, 131–52; Zhu Youhuan et al., eds., *Zhongguo jindai xuezhi shiliao*, part 3, 2: 804–7.

5. See *Zhejiang jiaoyu jianzhi*, 102–3, 122.

6. Ibid., 122; "Shinianlai zhi zhongdeng jiaoyu gaishu," 12–13.

7. "Shinianlai zhi zhongdeng jiaoyu gaishu," 12–14; Linden, "Politics and Higher Education in China," 111–13, 175–92, 196–217.

8. The University Council system was an attempt by Cai Yuanpei and other liberal educators within the Nationalist Party to set up an autonomous system for academic and educational administration under the aegis of the Nationalist government. The centralized University Council soon succumbed to intraparty factionalism and was replaced in August 1928 by the Ministry of Education, which directed national educational policy for the rest of the Nanjing decade. The provincial-level university districts were replaced with departments of education the following year. See Linden, "Politics and Education in China"; idem, "Politics and Higher Education," 47–76; and *Zhejiang jiaoyu jianzhi*, 21–26.

middle and high school levels. It also encouraged local areas to set up rural normal schools (*xiangcun shifan*) and abbreviated teacher-training programs. The Nationalist government backed these proposals with guidelines to increase support for or to shift funding toward normal and vocational schools.[9] Despite these efforts, normal school education never really regained parity with middle schools. Regular middle schools remained dominant in prestige, numbers, and numbers of students through 1937.[10] Consequently, discussion here of civic education during the 1930s concentrates primarily on regular middle schools.

Secondary schools were categorized as provincial, county, and private schools. Many of Jiangsu and Zhejiang's earliest middle and normal schools were established during the last decade of the Qing in prefectures—the level of administration between the province and the local county—by reform-minded officials and local elites. They often converted academies, state-sponsored institutions traditionally used to support candidates competing in the imperial examinations to select government officials, into modern-style schools or arranged taxes, contributions, or land grants to support new schools.[11] After provincial departments of education were formed in 1917, provincial secondary schools came under the jurisdiction of those departments.[12] The Bureau of Education (Jiaoyuju) in the special municipality (*tebieshi*) of Shanghai operated at an administrative level parallel to provincial departments of education, directly administering all municipal (*shili*)

---

9. "Shinianlai zhi zhongdeng jiaoyu gaishu," 1, 3, 13–16.

10. Regular middle schools constituted roughly 60 percent of the total number of secondary schools in the nation from 1928 to 1936 and continued to educate nearly 75 percent of the secondary students in the country for most of the decade. Normal schools generally constituted 25 to 30 percent of China's secondary schools during this period but on average taught approximately 17 percent of the students (Shen Guanqun, "Woguo zhongdeng jiaoyu zhi shi," 43). Regionally, regular middle schools predominated in Jiangsu, Zhejiang, and Shanghai, where they constituted, respectively, 60 percent, 73 percent, and 83 percent of the secondary schools in 1935 (Jiaoyubu tongjishi, *Quanguo zhongdeng xuexiao yilan biao*, tables 2, 3, and 31).

11. Li Huaxing et al., 618–19. For examples within the region, see Mao Liyuan, "Wo suo zhidao de Jiangsu shengli Suzhou Zhongxue"; *Zhejiang jiaoyu jianzhi*, 78–79, 120–22; and Zhejiang sheng zhengxie wenshi ziliao weiyuanhui, *Zhejiang jindai zhuming xuexiao he jiaoyujia*.

12. See Yeh, *Provincial Passages*, 137; and *Zhejiang jiaoyu jianzhi*, 20–24.

schools and overseeing all private schools.[13] During the course of the Republican period, an increasing number of counties also established lower middle schools, which were administered locally.

Private citizens and groups also established secondary schools in great numbers.[14] In some places, particularly Shanghai, private schools constituted the vast majority of local schools.[15] Shanghai's concentration of commercial wealth and reformist intellectuals, combined with inspiration from foreign models of schooling and opposition to foreign imperialism, encouraged social elites to establish their own modern schools.[16] For instance, Shanghai's prominent Minli Middle School was established in 1903 by four brothers of the Su family to honor the wishes of their wealthy merchant father, who believed, like many others, that education was necessary to save the nation.

Private schools such as Minli flourished across the region during the first three decades of the twentieth century, in part because of the relative freedom afforded by unstable and limited governments. But after 1927 the Nationalist Party's national, provincial, and municipal governments sought to influence them through co-opting strategies such as the distribution of supplementary funds and requiring registration with local authorities.[17] In Shanghai, the Nationalist Party also sought to influence private schools by placing party members or sympathizers on private schools' boards of managers (*dongshihui*).[18]

Intellectuals, educators, and reformist elites played a prominent role in establishing and running secondary schools, whether the schools were administered privately or by the province, municipality, or county. The chapters that follow are replete with examples of

---

13. Henriot, *Shanghai*, 185–202. Initially the bureau had jurisdiction only over private schools in the Chinese-controlled portion of the city, but after 1931 it also oversaw all Chinese-run schools in the International Settlement as well.

14. E.g., Yang Ronan, "Ji Zhenhua nüzi zhongxue"; Zhejiang sheng zhengxie wenshi ziliao weiyuanhui, *Zhejiang jindai zhuming xuexiao*.

15. In Shanghai, 93 percent of secondary schools were privately run in 1935. Jiaoyubu tongjishi, *Quanguo zhongdeng xuexiao yilan biao*, table 31.

16. Hang Wei, "Shanghai zhongdeng xuexiao de fazhan." For Minli Middle School, see Wang and Wu, "Minli zhongxue jianshi."

17. For the distribution of public funds to private schools, see Jiangsu sheng jiaoyuting, *Jiangsu jiaoyu gailan*, 1: 157. For rules regarding middle school registration, see Jiaoyubu, ed., *Di yi ci Zhongguo jiaoyu nianjian*, 2: 34–35.

18. E.g., Pudong zhongxue xiaoshi bianxie zu, "Pudong zhongxue jianshi," 215–16.

elite participation in schooling, but the career of Jing Ziyuan, the educator, revolutionary, and cultural reformer from Zhejiang, demonstrates with particular clarity these groups' involvement in both public and private secondary schools. Jing served as principal for Zhejiang Provincial First Normal School for over a decade and hired as teachers the young cultural reformers Xia Mianzun, Liu Dabai, and Chen Wangdao.[19] After being forced by conservative provincial officials to leave Zhejiang First Normal in 1920, Jing subsequently cooperated with a local elite in his home district of Shangyu to start in 1922 the experimental Chunhui Middle School, which included on its faculty prominent young intellectuals such as Xia Mianzun, Feng Zikai, and Zhu Ziqing. In 1923, Jing simultaneously became principal of Ningbo's Zhejiang Provincial Fourth Middle School, undertaking educational reform there.[20] This kind of active involvement in local education by prominent intellectuals and educators of national or regional reputation was common throughout the lower Yangzi region.[21]

During the Republican period, through the combined efforts of social elites, local educators, and national, provincial, and municipal governments, the numbers of schools and students in the lower Yangzi region expanded steadily and significantly, especially during the Nanjing decade (1927–37). (See Appendix A.) By the late 1910s, lower Yangzi region middle and normal schools constituted between 12 and 15 percent of the total nationwide, and the region's middle and normal school students constituted anywhere from 15 to 20 percent of the national whole. Striking, though, is the relatively small aggregate numbers of secondary students both within the region and nationwide, even into the 1930s. Given a conservative estimated national population of roughly 400 million during the Republican period, the small numbers of middle and normal school students—reaching roughly 100,000 within the region and about 500,000 nationwide before the start of the Sino-Japanese War in 1937—suggest that secondary education was still rather exclusive and limited. Further, a large proportion of regional schools were concentrated in core areas, or those with the greatest

---

19. Yeh, *Provincial Passages*.

20. Zhejiang sheng zhengxie wenshi ziliao weiyuanhui, 148–49, 206–9.

21. E.g., Southeast University Affiliated Middle School's faculty included the prominent educational reformers Guo Bingwen, Liao Shicheng, and Shu Xincheng (Liao Shicheng et al., eds., *Shixing xinxuezhi*, 409–13).

population density and most developed commerce and infrastructure, which in Jiangsu meant to the south of or adjoining the Yangzi River.[22]

Students came mostly from within the region, with schools in the regional core drawing disproportionately from areas south of the Yangzi.[23] Schools in relatively peripheral areas of the region, such as Jiangsu Provincial Yangzhou Middle School, Zhejiang Provincial Jinhua Middle School, and schools in Nanjing drew more students locally or served primarily as points of collection for students scattered over the regional periphery.[24] Students generally ranged from twelve to twenty years of age throughout the period between 1912 and 1937.[25] But a relatively high proportion of students were in their late teens and early twenties, especially during the early Republic, when secondary education served as a way for young adults trained in the classics to convert their learning into a marketable credential.[26] By the Nanjing decade the average age of secondary students gradually drifted toward the middle teens, but a significant proportion of

---

22. Jiang Weiqiao, *Jiangsu jiaoyu*, 27–31.

23. Jiangsu shengli Shanghai zhongxue chuban weiyuanhui (1936), 307; (1933), 287; *Jiangsu shengli Wuxi zhongxue gailan*, "Xuesheng jiguan fenbu tu" (Chart of the distribution of student native place); *Jiangsu shengli Wuxi zhongxue shifanke gaikuang*, "Xuesheng jiguan bijiaobiao" (Comparative chart of student native place); *Shanghai Minli zhongxue sanshizhou jiniankan*, 50.

24. In Nanjing, students from Anhui made up 22 percent of the student body in private secondary schools (*Nanjing shi shiba niandu jiaoyu gaikuang tongji*). For Yangzhou and Jinhua middle schools, see, respectively, "Jiangsu shengli Yangzhou zhongxue niaokan" (A bird's-eye view of Jiangsu Provincial Yangzhou Middle School), 40; and Zhejiang shengli Jinhua zhongxue chuban weiyuanhui, *Zhejiang shengli Jinhua zhongxue yilan*, "Ershisi niandu di'er xueqi gao chu zhong xuesheng jiguan tongji biao" (Statistical chart of the native place of upper and lower middle school students for the second semester of the 1935 school year).

25. E.g., *Nanjing shi shiba niandu*; *Suzhou zhongxue gaikuang*, 24.

26. In 1918, 229 of 274 (83.5 percent) students at Suzhou's Jiangsu Provincial First Normal School were 18 or older; the average age of students there was 19; and there were 25 students aged 23 or older (*Jiangsu shengli diyi shifan xuexiao yaolan*, "Zai xiao shengtu nianling tongji biao" [Statistical table of the ages of the students at the school] [May 1918]). At Zhejiang Provincial First Middle School in 1923, 84 percent of the students were age 18 or older (*Zhejiang shengli diyi zhongxuexiao ershiwu nian jinian ce*, "Quanxiao xuesheng nianling baifen bijiao biao" [Table comparing the percentages of the ages of the whole school's students]). Cf. Liao Shicheng et al., *Shixing xinxuezhi*, 366.

students were still in their late teens and early twenties.[27] Well into the Nanjing decade, a fairly large proportion of the students at middle and normal schools were young adults rather than children or young adolescents. Because of this age structure, Republican period secondary students often acted as adults in the political sphere, and they were treated as adults by party and government authorities.

Because of the expense of secondary schooling, most students came from families belonging to the emerging urban professional classes or the well-to-do rural elite. Relative cost is always difficult to gauge, but the price of secondary education in Republican China appears to have been rather high. At Shanghai's private Pudong Middle School, tuition and board from 1924 to 1927 was 67 yuan (Chinese silver dollars)[28] per semester.[29] At Zhejiang Provincial Ninth Middle School in rural Jiande in 1926, average annual fees for normal-course students were 40 to 50 yuan, whereas middle school students' average tuition and fees ranged from 60 to roughly 100 yuan.[30] During the Nanjing decade the combined cost of tuition, fees, and room and board at middle schools ranged widely between 52 yuan and upward of 130 yuan a year.[31] Students in normal schools or programs of study were exempt from tuition, and sometimes any fees at all, on the

---

27. At Jiangsu Provincial Shanghai Middle School, students eighteen or older constituted roughly 63 percent of the students in 1928 but only 31 percent and 29 percent, respectively, in 1933 and 1936 (Zhongyang daxuequli Shanghai zhongxuexiao mishuchu chuban weiyuanhui, comp., *Zhongyang daxuequli Shanghai zhongxuexiao yilan*, Gezhong tongji, 10; Jiangsu shengli Shanghai zhongxue chuban weiyuanhui [1933], 289; [1936], 308). Only approximate figures can be read from the charts.

28. The term "yuan" became the official unit designation for Chinese paper currency starting in 1935, but it was colloquially used to connote the silver dollar throughout the early Republican period. I follow Wen-hsin Yeh in using yuan to refer to Chinese silver dollars for the pre-1935 period as well (*Alienated Academy*, 195). According to Sidney Gamble, USD 1 in gold was worth 2.24 Chinese silver dollars in 1927 and 4.54 Chinese silver dollars in 1931 (*How Chinese Families Live*, 3).

29. Pudong zhongxue xiaoshi bianxie zu, 213.

30. "Shengli dijiu zhongxuexiao shicha baogao shu," 11.

31. E.g., *Jiangsu jiaoyu gailan*, 212, 224, 256, 261, 270, 289, 299, 307, 315; *Shanghai Minli zhongxue sanshizhou jiniankan*, 94. Some public and private schools also had modest scholarship programs. See *Shanghai Minli zhongxue sanshizhou jiniankan*, 94; *Zhejiang jiaoyu jianzhi*, 115.

expectation that they would serve as teachers in local schools after graduation.[32] But ultimately the ability to pay even the minimal fees of 40 to 50 yuan of subsidized normal schools or teacher-training courses of study would still have limited the student body in most secondary schools to children of families whose income was in roughly the top 10 to 15 percent of the urban national average.[33]

Given that most middle school students came from relatively elite families, there were distinct intraregional variations in their family backgrounds. In commercial cities in the affluent regional core, such as Shanghai and Wuxi, a majority or plurality of students came from merchant or professional backgrounds.[34] By contrast, in the more rural areas of the regional periphery, such as Jinhua in Zhejiang, a majority of students were from agricultural (*nong*) backgrounds.[35] Given the significant cost of secondary education as described above, the high proportion of students with backgrounds designated "agricultural" in schools of the regional periphery were most likely from landlord families or at least from the "rich peasant" stratum of the agrarian elite.

From whichever of these social groups students came, Republican-period secondary education seems to have fed graduates primarily into the emerging professional classes. Statistics charting post-graduate career trajectories for Jiangsu during the 1910s and 1920s, for instance, show three main paths for secondary school graduates: further study, teaching careers, and careers in business or govern-

---

32. Cong, "Localizing the Global," 304; Zhu Youhuan et al., part 3, 2: 448–49.

33. Gamble found that in Beijing only families earning more than 100 silver dollars a month were likely to pay 40 to 50 silver dollars a year on educational expenses and that these families were only 13 percent of his sample (3–4, 165–69, 335). This method of gauging the relative cost of education is borrowed from Yeh, *Alienated Academy*, 195–99.

34. At Shanghai Middle, for instance, in 1933 and 1936, 53 percent of the student body characterized their family background as "commercial" (*shang*) (Jiangsu shengli Shanghai zhongxue chuban weiyuanhui [1933], 288; [1936], 305). For Wuxi, see *Jiangsu shengli Wuxi zhongxue gailan*, "Xuesheng jiating zhiye bijiaotu." Cf. *Shanghai Minli zhongxue sanshizhou jiniankan*, 50.

35. At Zhejiang Provincial Jinhua Middle School, whose students came mostly from the rural periphery, most students were from agricultural backgrounds (52 percent) (Zhejiang shengli Jinhua zhongxue chuban weiyuanhui, "Ershi niandu di'er xueqi geji xuesheng jiazhang zhiye zhi baifenbi").

ment administration.[36] Many secondary graduates (over 80 percent from normal schools and 13 to 20 percent from regular middle schools) became teachers or educational administrators, filling in the lowest tier of China's growing nonofficial professional class. Much smaller proportions of secondary graduates took jobs in the new urban professions, joining banks, Chinese companies, foreign firms, and transportation companies, or pursued careers with the railroad, telegraph, post, port affairs, and customs offices, and in the fields of publishing and the news media. The roughly 50 percent of regional middle school graduates who went on in school prepared to join China's new scholarly elite, moving toward careers in technical fields, elite schools, and higher-level government service. Even when students came from diverse backgrounds, their secondary school qualifications seem to have allowed them to enter key sectors of the modern professions, with a heavy concentration becoming teachers in local modern schools.

II. CURRENTS OF CURRICULUM

Three major curricular transitions occurred in Chinese secondary schools between the 1911 Revolution and the start of the Sino-Japanese War in 1937: first, in 1912, after the founding of the Republic; second, in 1922 and 1923, with the promulgation of the New School System; and third, in 1928, after the founding of the Nationalist government. At each moment of transition, intellectuals, educators, and government officials interacted in different ways to craft the curriculum. Each new curriculum triggered complex changes in the content of courses in history, geography, civics, and national language (*guowen*) that constituted the core of civic education within the formal curriculum.

The Late Qing Curriculum

Chinese-run, modern, Western-style education in China began not with the founding of the Republic in 1912, but in the final decade of the Qing dynasty. The weight of multiple foreign and domestic pres-

---

36. "Jiangsu sheng ge zhongdeng xuexiao sannian jian biyesheng chulu zhuangkuang tongji biao," 5–9, 14–15; "Jiangsu ge zhongdeng xuexiao qinian jian biyesheng chulu zhuangkuang bijiao biao," 7–10, 16–17.

sures pushed reformist elites and high-ranking Qing officials to call for adoption of modern-style schools and an end to the examinations in Confucian classics that had been used for more than a millennium to recruit government officials.[37] Their goals for establishing a new school system were to create a patriotic, loyal, and morally grounded citizenry that would have the technical skills and modern knowledge needed to compete in the global marketplace. The result of their efforts was the 1904 School System, which was patterned closely on the Meiji-era (1868–1912) Japanese school system.[38]

Review of the late Qing middle school curriculum supports Hiroshi Abe's and Li Huaxing's claims that the 1904 Curriculum followed Zhang Zhidong's dictum of "Chinese learning as the essence, and Western learning for utility" (*zhong xue wei ti, xi xue wei yong*).[39] By introducing a full array of courses in the Western-style natural and social sciences, the 1904 middle school curriculum departed significantly from the curricula of the late imperial academies, even from that of the late Qing reformed academies.[40] Students in the new middle schools studied geography, law, finance (*licai*), biology, physics, chemistry, and mathematics.[41] Students also studied history, as had their predecessors in late imperial academies, but history courses in the new middle schools adopted a more Euro-American approach, focusing on the recent past, exploring issues of immediate political importance, and presenting students with national narratives of countries in Asia, Europe, and the Americas.[42] In addition, foreign language study was a significant part of the curriculum, claiming 6 to 8 classroom hours out of a 36-hour week.

Just as striking, though, are some of the continuities with late imperial education. Together courses in the classics (*dujing jiangjing*) and Chinese literature (*zhongguo wenxue*) constituted over a third of the curriculum (12 to 14 out of 36 hours of instruction per week),

---

37. Ayers, *Chang Chih-tung*; Bastid, *Educational Reform*; Bailey, *Reform the People*; Li Huaxing et al., part 1, chap. 4; Peake, *Nationalism and Education*.

38. Abe, "Borrowing from Japan"; Li Huaxing et al., part 1, chap. 4.

39. Abe, 65–66; Li Huaxing et al., 83–84.

40. On late Qing academies, see Keenan, "Lung-men Academy."

41. Hayhoe, "Cultural Tradition and Educational Modernization," 51; Li Huaxing et al., 618–19; Zhu Youhuan et al., part 2, 1: 382–93.

42. Zhu Youhuan et al., part 2, 1: 385–86.

with classics instruction focusing on the *Zhouli* (Rites of Zhou) and the *Chunqiu zuozhuan* (The Zuo commentary on the *Spring and Autumn Annals*). In addition, the moral cultivation course that was to teach secondary students social ethics used as its primary text Chen Hongmou's *Wuzhong yigui* (Five sourcebooks), which had been used as a primer in Confucian ethics since the late eighteenth century.[43] Both courses in moral cultivation and reading the classics reinforced late imperial patterns of social relations and political loyalties rather than encouraging the new ethics associated with Euro-American citizenship that reformers like Liang Qichao had been promoting since the mid-1890s.

The 1904 Curriculum made no provision for girls' or young women's education, on the assumption that girls would continue to be educated in their fathers' households to prepare for a life in the home. But the proliferation of private girls' schools, the perceived need to train female teachers for kindergartens and girls' primary schools, and the nationalist goal of educating women to prepare children for citizenship all pushed the Qing court to issue regulations for girls' primary and normal schools in 1907.[44] Girls' normal schools, equivalent to secondary-level boys' lower normal schools, became the highest level of education available to women between 1907 and 1912. The curriculum at these schools included more moral cultivation and less science instruction than the curriculum for males, no readings in the classics, and training in household skills. This curriculum was consistent with a form of education designed to prepare women to manage their homes "scientifically" in accord with modern standards of efficiency as "good wives" (*liangqi*) and to raise their sons to be modern, morally upright citizens as "wise mothers" (*xianmu*).[45] Female normal school students were prepared to be teachers, but they were also expected to serve as exemplars of traditional female virtues.[46]

---

43. Ibid., 383–84.
44. Cong, "Localizing the Global," 90–99, 117–20. Cf. McElroy, "Forging a New Role," 350, 355–56.
45. Bailey, "Active Citizen," 319 and *passim*; Cong, 121–24; Judge, "Citizens or Mothers of Citizens?"
46. McElroy, "Forging a New Role," 353–55.

The Early Republican Curriculum

In educational circles, the post-1911 republican fervor fed directly into efforts to use education to train a new generation of Chinese citizens. The tone for this project was set in 1912, by the Conference of Education, which was dominated by educational reformers from Zhejiang and Jiangsu and overseen by the European-trained educator and Revolutionary Alliance member Cai Yuanpei, whose opening address started this book.[47] These educators formulated a curriculum that in its general outline would stand from 1912 until the formulation of the New School System in 1922, despite efforts by Yuan Shikai to reintroduce classical texts and to promote nationalism and military preparedness.

At the secondary level, educators designed and the Ministry of Education promulgated a four-year course of study for both middle schools and normal schools, with the latter having a one-year preparatory course (*yuke*).[48] In the words of Paul Bailey, the goal of the middle school curriculum was "to complete general knowledge and create an all-around citizen" who could be an active participant in a modern republican polity rather than an imperial subject.[49] Normal school students were to be taught "to understand the fundamental source of nation formation [*ming jianguo zhi benyuan*] and to fulfill the duties of citizenship [*jian guomin zhi zhifen*]," and their instruction emphasized physical education and moral cultivation in preparation for becoming teachers.[50] The primary focus of citizenship education in the early Republican curriculum lay in classes in moral cultivation, national language, history, and geography.

Normal and middle school students, male and female both, studied one hour per week of moral cultivation.[51] In their final year, middle and normal school students also studied legal and economic systems for two hours each week. The content of law and economics classes was mostly a factual recitation of the constitution, the main institu-

---

47. Bailey, *Reform the People*, 139–67; Li Huaxing et al., part 1, chap. 5; part 3, chap. 2; Peake, *Nationalism and Education*, 72–107.

48. Zhu Youhuan et al., part 3, 1: 352–61; 2: 438–60.

49. Bailey, *Reform the People*, 155.

50. Zhu Youhuan et al., part 3, 2: 438.

51. Students in normal schools studied two additional hours of moral cultivation per week during the one-year preparatory course.

tions of the state, and an introduction to basic concepts of legal theory and market-based economics. The core of citizenship training for these students came in moral cultivation classes. These classes taught students, in turn, personal conduct, how to treat others, and their duties to the nation, society, their families, themselves, humankind, and all living creatures while also giving them a general introduction to the study of ethics and "the special characteristics of [China's] ethics."[52]

National language constituted one of the core classes in middle and normal schools throughout the Republican period. The tone for this focus on language training was set in the 1912–13 curriculum, which dedicated a great deal of class time to Chinese language classes. Male middle school students studied national language seven hours per week during their first two years, and five hours per week for the last two.[53] Language classes were intended to teach students how to read a vast stretch of Chinese literature, interpret sophisticated texts, and express their own ideas.[54] By teaching classical Chinese literature for both practical and cultural literacy rather than as a guide to timeless truths of the cosmic and moral order, Republican-period Chinese-run schools contributed to a process of decanonization that missionary schools began during the 1890s and the Qing government extended further with the abolition of the imperial exams in 1905.[55]

Further, after the first years of the Republic, an increasingly divisive issue among secondary educators was whether or how much to temper reading and writing in classical language with exposure to the new, written vernacular. The issue was usually negotiated independently at each school, for language teachers often selected readings for their classes themselves, rather than relying on a single, mass-produced textbook.[56] Students could be exposed to broad variations

---

52. Zhu Youhuan et al., part 3, 1: 352, 359.

53. Female students studied national language one hour less per week during the second year. Normal school students studied one weekly classroom hour more national language than middle school students of the respective gender, but in a different configuration, with ten weekly classroom hours of language study compressed into the preparatory course.

54. Zhu Youhuan et al., part 3, 1: 352.

55. Dunch, "Science, Religion, and the Classics," 18, 23, 31; Elman, *Cultural History*, chap. 11.

56. E.g., *Shanghai qiuxue zhinan*, 2: 8–11; Shen and Xiao, eds., *Jiangsu sheng Yangzhou zhongxue*, 19–28; Zhu Youhuan et al., part 3, 1: 405.

in readings. At Shanghai's exclusive private Nanyang Middle School, for example, readings from classical texts such as the *Zuozhuan* (Zuo commentary), *Shiji* (Records of the Grand Historian), *Hanshu* (History of the Han dynasty), and classical philosophy predominated, exposing students to a syllabus that paralleled the curricula of the late Qing middle schools and academies.[57] At other schools, such as Zhejiang First Normal, adoption of vernacular language readings sometimes became the focus of a crusade by progressive educators. There, principal Jing Ziyuan hired New Culture Movement advocates Xia Mianzun, Liu Dabai, and Chen Wangdao to teach language, leading to the rapid introduction of vernacular literature.[58] Similarly wide variations in reading materials could happen even within the same school.[59] Because social and cultural reformers often adopted the written vernacular to express their ideas, assigned essays and fiction in this style could expose students to new ideas related to socialism, gender equality, and customs reform.[60]

Classes in history and geography presented to students some of the diverse ideas about national identity that were circulating in early twentieth-century China. In both boys' and girls' middle schools, each year students received two hours per week of history and geography instruction for each discipline.[61] Significantly, Chinese history (*benguo lishi*) and foreign history (*waiguo lishi*) were given equal time but were divided into separate courses. This division carefully segregated the narrative of Chinese political, cultural, and social development from the normative account of Euro-American modernization and world domination that was presented in world history classes.[62] Geography, too, was divided into separate Chinese and foreign geography courses, which were supplemented by classes

---

57. *Nanyang zhongxue*, 17.

58. Shen Ziqiang et al., eds., *Zhejiang yishi fengchao*, 121–22, 351–52; Yeh, *Provincial Passages*, 151–59.

59. E.g., Wang Chichang, "Shifan xuexiao disan, si xuenian guowen jiaoxue de guanjian," 39–54; Zhang Shengyu, "Wunian lai duiyu guowen jiaoxue shang zhi yixie jingyan," 3–4.

60. Schwarcz, *The Chinese Enlightenment*.

61. In boys' and girls' normal schools, students received in total six or seven weekly hours each of history and geography that were configured in slightly different ways during the first three years of the regular course of study.

62. Zhu Youhuan et al., part 3, 1: 353, 359.

dedicated to natural and human geography. Together history and geography classes presented students with normative visions of the nation's community and territory and taught them where they fit within the national order.

Although there were relatively few distinctions by gender in the core courses of study in the early Republican curriculum, several differences in the curricula of boys' and girls' secondary schools suggest the different capacities educators and officials promoted for male and female student-citizens. Female middle school students were to study less foreign language, mathematics, and physical training (*ticao*)— omitting the one hour per week of military training—than their male counterparts. Instead, girls' middle schools were to teach their students home economics and gardening for two hours per week for the final three years of school, and sewing for two hours per week throughout.[63] Early Republican educators and educational officials also continued to emphasize the importance of teaching traditional feminine virtues in girls' schools.[64] These distinctions between the curricula for boys' and girls' secondary schools suggest that for many educators secondary education was to prepare girls for household-centered activities and boys for more public roles. Yet there had also been a significant convergence between boys' and girls' secondary curricula since the first modern curricula of the late Qing. After 1912 female students learned many skills that prepared them for careers outside the home, and they studied courses in history, geography, language, and moral cultivation that exposed them to many of the same messages about citizenship as their male counterparts.

The New School System

In the decade between 1912 and 1922, Chinese educational circles experienced a series of tectonic shifts. The wide-ranging critiques of indigenous social institutions and cultural patterns in the New Culture Movement (1915–25) and the mass nationalism of the May Fourth Movement of 1919 transformed the social and political

---

63. Ibid., 354–55. For home economics courses at many girls' missionary schools during the 1910s, see Graham, *Gender, Culture, and Christianity*, 52–57.

64. E.g., Bailey, "Active Citizen," 328–29; McElroy, "Forging a New Role," 361.

climate. These changes drove a reassessment of society, politics, and the forms of citizenship that students should be taught. In the political disorder of the early Republican period, associations of professional educators at the provincial and national levels played an increasingly important role in defining the objectives and methods of education.[65] Moreover, growing numbers of Chinese educators were going to the United States, especially to Teachers College at Columbia University, to study modern education. Many returned to China inspired by the movement for child-centered education focused on problem solving and social improvement led by John Dewey and Paul Monroe, both of whom made extended visits to China between 1919 and 1921.[66] In this transformed environment, professional educators held a series of national meetings, which led to a reorganization of the educational system in 1922 and a complete revision of primary and secondary curriculum standards in 1923.[67] The aims of the New School System reveal the profound influence of U.S.-style progressive education and the new social and political currents that had enveloped educators during the previous decade. The new system was to "adapt to the needs of social evolution," "develop a spirit of popular education" (*pingmin jiaoyu*), "plan for personality development" (*gexing zhi fazhan*), "be attentive to citizens' economic capacity" (*jingjili*), "be attentive to life education" (*shenghuo jiaoyu*), "make education spread easily," and "allow each locality space for flexibility."[68]

In line with the goals of gender equality that were central to the New Culture Movement, the New School System made few distinctions between the curriculum standards for boys' and girls' secondary schools. The clearest gender distinction in the curricula came for lower middle school handicrafts (*shougong*) classes, especially in the area of skills training (*jineng*).[69] The boys' curriculum concentrated on woodwork, metalwork, and design skills that were increasingly important in China's growing industrial economy. The girls'

65. Bailey, *Reform the People*; McElroy, "Transforming China"; Keenan, *The Dewey Experiment*, 60–61; Peake, *Nationalism and Education*.

66. Keenan, *The Dewey Experiment*, 55–125.

67. Li Huaxing et al., part 1, chap. 6, esp. 131–41.

68. Zhu Youhuan et al., part 3, 2: 804–5.

69. Quanguo jiaoyuhui lianhehui xinxuezhi kecheng biaozhun qicao weiyuanhui, *Xinxuezhi kecheng biaozhun gangyao*, Chuji zhongxue: 28–32.

handicraft curriculum stressed feminine-gendered household skills such as being able to cook for a family and prepare a banquet, to sew, embroider, and make lace, and to tend a garden. These distinctions in handicraft training, which continued under the Nationalist government during the Nanjing decade, prepared male and female students for different places in the industrial economy. But the overall uniformity of the new curricula after 1922 suggests that intellectuals, educators, and government leaders intended for boys and girls to learn to be modern citizens in similar ways.

At the lower middle school level, the new curriculum also combined previously separate classes in moral cultivation, law, and economics into a class on civics (*gongminke*) of six weekly credit hours spread over the three years of lower middle school.[70] Civics class was still to give students a foundation in common moral knowledge and relate a basic understanding of political and economic life. However, it was also to give students a detailed exposition of the organization of social life, which was seen to include both interest groups and local self-government organizations, in addition to the family, school, and nation. The course also included a segment on "social issues" (*shehui wenti*), which discussed concerns related to industrial labor, poverty, and gender inequality.[71] High schools had a required course in social issues that met for three hours per week for the first year.[72] This emphasis on education about society reflected Chinese intellectuals' growing awareness during the May Fourth period of society as a sphere distinct from the state and the family, and as a space for struggle and contention.[73]

The change in the language curriculum was equally profound, with increasing emphasis on vernacular literature and works dealing with social and cultural reform. Pioneering vernacular-language writer Ye Shengtao was responsible for drafting the lower middle school standards, and leading literary reformer Hu Shi drafted the high school standards.[74] The lower middle school curriculum reading standards

---

70. Quanguo jiaoyuhui lianhehui xinxuezhi kecheng biaozhun qicao weiyuan-hui, 7–8.

71. Ibid., Chuji zhongxue: 1–5.

72. Ibid., Gaoji zhongxue: 16–8.

73. E.g., Dirlik, *Anarchism*.

74. Quanguo jiaoyuhui lianhehui xinxuezhi kecheng biaozhun qicao weiyuan-hui, 4.

mandated that a minimum of three-quarters of the first year, half of the second year, and one-quarter of the third year be focused on vernacular language study.[75] Though the curriculum did not dictate what texts were to be used for the close readings parts of the course, Hu Shi compiled a list of general readings suggestions. These included much vernacular literature and many texts with a cultural reformist agenda, such as fiction by Lu Xun, Zhou Zuoren's translated fiction, translations of Ibsen's plays, and collections of essays by Hu Shi, Liang Qichao, and Zhang Shizhao. The content of the readings proposed by Hu and Ye suggests that they, and the other educational reformers involved in setting the new curriculum standards, intended to give lower middle school language instruction a firm push toward the promotion of the vernacular language movement and social and cultural reform.[76]

History and geography continued to be core classes in lower middle schools under the 1923 curriculum. The new standards recommended that lower middle school students study eight credit hours each of history and geography.[77] However, a major difference from the earlier curriculum was that the lower middle school curriculum specifically encouraged the integration of instruction in Western and Chinese history and geography, with equal stress on each. Incorporating Chinese history and geography as one part of what were really courses in global history and geography placed China in both spatial and temporal hierarchies of cultural, economic, and political development. These implicit hierarchies revealed China's relative "backwardness" according to the standards set by Europe and the United States and eroded a sense of national distinctiveness.[78]

Significantly, despite the stress on integrated history and geography instruction in the New School System curriculum standards, publishers continued to produce mostly separate world and Chinese history and geography textbooks throughout the 1920s. Classroom

---

75. Ibid., Chuji zhongxue: 12–15.

76. Ibid., Chuji zhongxue: 15–16.

77. Ibid., Xinxuezhi kecheng gangyao zong shuoming: 7. The high school curriculum mandated nine credit hours of cultural history (*wenhuashi*) but did not make geography a required class for high school students (Gaoji zhongxue kecheng zonggang: 3, 19).

78. Ibid., Chuji zhongxue: 5–11. For a more extensive discussion, see Culp, "'China—The Land and Its People.'"

instruction in Chinese and foreign history and geography seems to have remained largely segregated as well. Although some schools, such as Southeast University Affiliated Middle School in Nanjing, taught "integrated" (*hunhe*) geography and history classes—making them required courses for first- and second-year lower middle school students, respectively—even teachers there taught world and Chinese geography and history separately in other courses.[79] In addition, many other schools continued to separate Chinese history and geography from world or Western history and geography during this period.[80] This separation enabled textbooks and teachers to highlight the distinctive character of China's historical development and national territory.

The Nanjing Decade

The Nationalist Party, because of its distinctive ideological foundation and its ambivalence toward the social and cultural reform movements of the May Fourth period, moved to reorient Chinese education soon after forming a semicentralized state in Nanjing in 1927. Sun Yat-sen's theories were adopted as the ideological core of the educational system at the Third Party Congress in 1929.[81] Party and government leaders, with cooperation from leading educators, revised secondary curriculum standards three times during the Nanjing decade, with new curriculum standards issued in 1929, between 1932 and 1934, and in 1936.[82] The government furthered its efforts to unify educational content according to a common standard in 1932 and 1933 when it instituted nationwide comprehensive examinations for graduation from primary and secondary schools.[83]

The new Nationalist government made its boldest effort to politicize secondary education by instituting a regime of party doctrine

79. Liao Shicheng et al., 151–60.

80. E.g., *Jiangsu shengli diyi shifan xuexiao xuesheng ruxue xuzhi*, 20, 22; *Shanghai Minli zhongxuexiao yichou nian zhangcheng*, 44–53; Shen and Xiao, 19–28; Zhongyang daxuequli Yangzhou zhongxue chuban weiyuanhui, *Yinianlai zhi Yangzhong*, 42–48.

81. *Zhongguo Guomindang disanci quanguo daibiao dahui huiyi jilu*, 139–41.

82. Li Huaxing et al., 455–62; ZKB 1936, 1: 1–9.

83. Jiaoyubu, "Zhong xiao xue xuesheng biye huikao zhanxing guicheng," 45; Li Huaxing et al., 629. For a local example of comprehensive exam questions, see Zhejiang sheng jiaoyuting, *Zhejiang sheng ershier niandu di'er xueqi*.

or Three Principles of the People (*sanmin zhuyi*) classes based on party leader Sun Yat-sen's eclectic 1924 lectures on nationalism, political organization, and management of modern socioeconomic systems. The party had started experimenting with instituting education in Sun's thought for primary and secondary students during the period of its reorganization in Guangdong before the Northern Expedition.[84] But factional tensions complicated the process of systematizing education based on the Three Principles of the People.[85] Standards passed in the summer of 1928 and January 1929 established basic curricula for party doctrine classes, which then became a required class, replacing civics.[86] The textbooks published for use in party doctrine and Three Principles of the People classes closely followed Sun's lectures, his other writings, and—for high school texts—Nationalist Party policy statements and orthodox party histories.[87]

However, many party leaders and supporters were dissatisfied with the reception of Three Principles of the People and party doctrine classes in local schools. They also expressed concerns regarding the propriety of segregating Party leader Sun's teachings into a distinct class rather than making it the organizing principle for all education about society and politics.[88] Ultimately, separate classes in the Three Principles of the People were abandoned in favor of a return to the more flexible and comprehensive civics classes, with new

---

84. Chen Jinjin, "Kangzhan qian Guomindang de jiaoyu zhengce," 180; Lü Fangshang, *Cong xuesheng yundong*, 321–26; Zhongyang zhixing weiyuanhui qingnianbu, "Zhonghua minguo Guominzhengfu gonghan, Di bashijiu hao" (November 18, 1925) Dangshihui, Wubu dang'an, no. 10349.

85. Linden, "Politics and Education," 771–72; Zhongguo Guomindang zhongyang zhixing weiyuanhui xunlianbu, *Jiaoyu zongzhi biaozhun ji shishi fang'an (cao-an)*, 21–22.

86. Zheng, "'Partification,'" 46; Zhongguo Guomindang Zhejiang sheng zhixing weiyuanhui xunlianbu, ed., *Dangyi jiaoyu*, 2: 19–22, 67–71; Zhongguo di'er lishi dang'anguan, ed., *Zhonghua minguoshi dang'an ziliao huibian*, 2: 1073–75; ZKB 1929, 2: 2; 3: 4.

87. E.g., Guo and Wei, *Gaozhong dangyi*; Hu Yuzhi, *Minzu zhuyi*; Lou Tongsun, *Minquan zhuyi*; Su Yiri, *Minsheng zhuyi*; Wei and Xu, *Chuzhong dangyi*; Zou Zhuoli, *Sanmin zhuyi jiaoben*.

88. Zhongguo di'er lishi dang'anguan, 2: 1082–84, 1090–96. Teaching materials appropriate for the new civics classes were not in fact introduced until at least the 1933–34 academic year. Formal curriculum standards for civics classes were promulgated on August 10, 1934. See Guoli bianyiguan, *Guomin zhengfu chengli yilai*, 1–2.

curriculum standards promulgated in 1934 and then again in 1936.[89] This new civics class followed the general pattern of the civics curriculum introduced in the 1920s. Ranging between one and two hours of class a week, it introduced lower middle school students to basic social institutions, modes of government and political participation, basic economic systems, and the civic mores and responsibilities of citizens.[90] High school civics, which also varied between one and two hours of class per week, focused on contemporary social issues, systems of government, economic mechanisms, legal structures, and ethical systems.[91] Although the new civics curriculum still made liberal references to Sun's thought and Nationalist Party policy, it was no longer organized explicitly according to the Three Principles of the People.

National language curricula of the Nanjing decade largely paralleled the standards of the New School System curriculum of 1923.[92] National language continued to be a major part of the required secondary curriculum, with lower middle school class time ranging from five to six hours a week and high school class time ranging from four to five hours a week. Teachers could still choose the materials for both intensive and general readings, and they often selected readings from a publishing catalog, or developed their own textbooks or systems for coordinating teaching materials.[93] Consequently, even though the

---

89. ZKB 1936, 1: 1–9.

90. "Ling fa gao chuji zhongxue gongmin kecheng biaozhun," 173–74; ZKB 1936, Chuji zhongxue kecheng biaozhun: 1–7. The earlier standards called for two hours of class per week during students' first and second years in lower middle school and one hour per week during the third year; the 1936 standards made civics uniformly one hour per week for all three years of lower middle school.

91. "Ling fa gao chuji zhongxue gongmin kecheng biaozhun," 171–73; ZKB 1936, Gaoji zhongxue kecheng biaozhun: 1–11. The 1934 standards mandated two hours of class per week for all three years of high school; the 1936 standards called for two hours of class per week during the first year of high school and then one hour per week during the second year and the first semester of the third year.

92. ZKB 1929, 2: 1–4; 3: 9–13; 1933, "Chuji zhongxue guowen kecheng biaozhun"; "Gaoji zhongxue guowen kecheng biaozhun"; 1936, Chuji zhongxue kecheng biaozhun: 23–31; Gaoji zhongxue kecheng biaozhun: 35–41.

93. Comprehensive catalogs from Kaiming Bookstore and Beixin Book Company gave teachers the option of compiling their own textbooks of selected readings, which the local branch stores of the publisher would collate and bind

party's curriculum standards encouraged teachers and publishers to introduce party-centered materials into language class readings, teachers and publishers could choose to present students with varied readings that strayed far from party-sanctioned topics. Language textbooks of the 1930s were correspondingly eclectic, including readings from classical and vernacular literature, early Republican-period social and cultural reform writings, essays promoting science and technology, as well as materials promoting the Nationalist Party's ideology and rule.[94]

After 1928, secondary-level history classes ran between twelve and fourteen weekly classroom hours distributed over three years for both lower middle school and high school.[95] Significantly, though, all Nanjing-decade curricula called for Chinese and world history to once again be taught as separate courses, with more weight sometimes given to Chinese history, in an effort to present China's history as an autochthonous process.[96] Chinese history curricula during this decade had two clear foci: explaining the origins of imperialism and how it had affected China's modern history; and tracing China's ethnic and cultural development throughout history, showing how those trends of development formed the basis of present-day China. Both topics were meant to build students' "national spirit" (*minzu jingshen*) and describe a path for China's future development, to which students were expected to contribute. Similarly, separate Chinese and foreign geography classes, which were taught two hours per week for all three years of

---

for the students (*Kaiming huoye wenxuan zongmu*; Jiang and Zhao, eds., *Beixin wenxuan*). For an example of a language textbook composed of readings selected by a school's language teachers, see Ji Tongyao, comp., *Guowen*.

94. E.g., Zhu Jianmang, *Chuzhong guowen*; Zhu and Song, *Chuzhong guowen duben*; Fu Donghua, *Guowen*.

95. ZKB 1929, 2: 25–41; 3: 41–63; 1933, "Chuji zhongxue lishi kecheng biaozhun"; "Gaoji zhongxue lishi kecheng biaozhun"; 1936, Chuji zhongxue kecheng biaozhun: 95–111; Gaoji zhongxue kecheng biaozhun: 121–44.

96. For lower middle schools under all the Nanjing decade curriculum standards, class time was divided into four semesters of Chinese history and two semesters of world history. For high schools, class time in Chinese and world history was evenly divided under the 1929 and 1936 curricula, whereas under the 1933 curriculum, students studied eight hours of Chinese history and six hours of world history.

school,[97] had two points of emphasis: to promote nationwide economic and cultural development according to Sun Yat-sen's vision, and to characterize international relations and situate China in the world.[98]

Following the precedent of the New School System, the bulk of the secondary curricula during the Nanjing decade applied equally to male and female students. But gender distinctions did infiltrate the curricula in minor ways. For example, home economics courses that focused on teaching female students scientific methods of family management remained part of the manual labor training (*laozuo*) curriculum for lower middle schools.[99] Moreover, female high school students often received nursing training rather than full military training, and physical education curricula varied in minor ways according to gender.[100] Yet the overall uniformity in curriculum standards for male and female secondary students meant that they continued to receive many of the same messages about citizenship during the Nanjing decade.

* * *

Educators, intellectuals, and the Nationalist Party sought to use secondary school courses in civics (or moral cultivation or party doctrine), language, geography, and history to relate to students fundamental ideas about the meaning and practice of citizenship, and they crafted curriculum standards to carry those ideas. History and geography classes were to introduce students to conceptions of national community and identity. Moral cultivation, civics, party doctrine, and language classes were to relate contemporary ideas about mod-

---

97. ZKB 1929, 2: 43–49; 3: 64–71; 1933, "Chuji zhongxue dili kecheng biao-zhun"; "Gaoji zhongxue dili kecheng biaozhun"; 1936, Chuji zhongxue kecheng biaozhun: 113–24; Gaoji zhongxue kecheng biaozhun: 145–56. Under the 1929 high school geography curriculum standards, high school students were to study three hours per week of Chinese geography during the first semester of their first year and three hours per week of foreign geography during the second semester of their first year.

98. E.g., ZKB 1936, Chuji zhongxue kecheng biaozhun: 113.

99. E.g., ZKB 1933, "Chuji zhongxue laozuo (jiashi) kecheng biaozhun."

100. ZKB 1933, "Gaoji zhongxue junshi kanhu kecheng biaozhun." For more on female students' nursing training, see chap. 5. For differences in physical education curricula, see ZKB 1929, 2: 122, 125; 3: 101–3, 107–9, 111, 116; 1933, "Chuji zhongxue tiyu kecheng biaozhun," 2–5, 7–8.

ern society, social membership, and civic culture and morality. In addition, civics and party doctrine courses were to explain the institutions of political participation and patterns of self-government. But curriculum standards only provide a broad outline of the content of instruction. To understand more precisely what ideas of citizenship spread to students in local schools, we must consider the textbooks students read and understand how they were produced and disseminated.

III. MEDIATING MODERNITY: TEXTBOOK PUBLISHING AND THE SPREAD OF IDEAS

Textbooks were the main vehicles for transmitting the content of China's Republican-period curricula to students in its modern schools. Because textbooks were powerful cultural media, valuable political resources, and profitable commodities, various groups competed to control their content, production, and distribution. In Republican China, private publishing companies' efforts to expand markets led to the rapid and far-flung expansion of distribution networks. Market expansion allowed textbooks, publishers' most consistently profitable products, to reach students in local communities throughout China. At the same time, new systems for compiling and reviewing textbooks shaped their contents. During the 1910s and 1920s, large editing departments and the contract system enabled leading intellectuals to use textbooks as vehicles to spread their ideas with little state intervention, even as publishers turned a handy profit. By contrast, after 1927 the Nationalist government used strict review and censorship to increasingly standardize textbook content.

Textbooks and the Publishing Industry

Missionaries started publishing textbooks in China during the mid-nineteenth century,[101] but Chinese publication of modern-style textbooks began during the last decade of the Qing with the transition from the examination system to new forms of modern schooling. During the first years of the twentieth century, Shanghai's Nanyang

---

101. Dunch, "Science, Religion, and the Classics," 13–15.

Academy (Nanyang gongxue) and Wenming Book Company made the first moves toward publishing books that approximated textbooks. However, both publishing concerns were soon displaced in the textbook market by the Commercial Press (Shangwu yinshuguan), which was formed in 1897 by a group connected through kinship ties, study at the same missionary school, and work at the same printing house. The Commercial Press's 1904 "Newest Textbook" series (*zuixin jiaokeshu*) was the first set of Chinese books designed specifically to fit appropriate textbook content to the different levels of schooling—lower primary, upper primary, and secondary.[102] The Commercial Press dominated textbook publishing, the core of its business, through the end of the Qing.[103]

The Zhonghua Book Company (Zhonghua shuju) emerged to challenge Commercial Press dominance in textbook publishing starting in the first year of the Republic. Lufei Kui and other former Commercial Press stalwarts anticipated the immanent fall of the Qing and, immediately after the 1911 Revolution, issued a set of Republic-suitable textbooks, stealing dominance in the textbook market from the Commercial Press for a short time. Zhonghua pressed its temporary advantage by setting up branches throughout the country just as the Commercial Press had always done, expanded its staff and capital base, and established a publication list focused largely on education.[104] Though the Commercial Press quickly responded, publishing the well received "Republican Textbook" series (*gongheguo jiaokeshu*), it now had to face a well-positioned competitor.[105] During the companies' intense two-way competition between 1912 and 1922, Commercial Press's "Republican Textbook" series sold an estimated 70 to 80 million volumes.[106] If Zhonghua sold a comparable number of textbooks during that period, together the two publishers were producing somewhere close to 15 million volumes of textbooks per year.

---

102. Ji Shaofu, *Zhongguo chuban jianshi*, 298–301, 319–20.

103. Dai Ren, *Shanghai Shangwu yinshuguan*, 14.

104. Ji Shaofu, 301, 324–25; *Minguo sannian chun Zhonghua shuju gaikuang*, 2–8, 11–15; Reed, 225–31.

105. Ji Shaofu, 302; Chen Yuan et al., eds., *Shangwu yinshuguan jiushinian*, 63; Dai Ren, 15.

106. Dai Ren, 15.

In 1917, Zhonghua cofounder Shen Zhifang broke away to form World Book Company (Shijie shuju), which initially grew on the basis of its publications of popular fiction. World emerged as a major competitor to the Commercial Press and Zhonghua Book Company during the early 1920s, when it began to concentrate on publishing textbooks.[107] As Christopher Reed has demonstrated, cutthroat competition in the textbook market among Commercial Press, Zhonghua, and World characterized Shanghai's publishing sector during the 1910s and 1920s.[108]

During the 1920s and 1930s a number of other, smaller publishers, including Dadong Book Company, Beixin Bookstore, and Kaiming Bookstore, emerged and attempted to acquire some share of the textbook publishing market. Kaiming Bookstore, which was started in 1926 by the brothers Zhang Xichen and Zhang Xishan, proved to be the most resilient of these new publishers. Xichen had been the editor of Commercial Press's *Eastern Miscellany* (*Dongfang zazhi*), and he drew support for his new company from prominent young figures in leftist literary circles. Counting among its staff a number of former educators and New Culture intellectuals, including Xia Mianzun, Feng Zikai, and Ye Shengtao, Kaiming Bookstore made its greatest mark in publishing for young people, producing several popular youth journals and collections from the late 1920s to 1949.[109] Textbooks, too, came to be a key element of Kaiming's youth publications. By the mid-1930s its numbers of approved secondary textbooks were approaching those of the three leading publishers, although they remained focused mostly in the areas of language, math, and science.[110] By the end of the 1940s, textbooks supposedly accounted for 62 percent of Kaiming's sales.[111] In general, the Commercial Press, Zhonghua Book Company, World Book Company, and Kaiming Bookstore were the leading forces in textbook publishing during the Nanjing decade. Together they

---

107. Ji Shaofu, 328–30; Reed, 245–47; Zhu Lianbao, "Guanyu Shijie shuju de huiyi," 56, 57.

108. Reed, chap. 5.

109. Wang Zhiyi, *Kaiming shudian jishi*, 1, 6.

110. Guoli bianyiguan, *Guomin zhengfu chengli yilai*, 17–24; "Ling fa zhengshi kecheng biaozhun gongbu hou shending zhi zhong xiao xue jiaokeshu biao."

111. Wang Zhiyi, 100–101.

published more than 80 percent of the middle school textbooks approved by the government between 1929 and 1936.[112]

During the Republican-period competition among the top publishing companies, each publisher sought to claim market share, in part, by expanding distribution networks to make its products available to more consumers.[113] As a result, during the two decades after the founding of the Republic, private publishing houses, whose main offices were in Shanghai, developed nationwide distribution systems, with branch offices in major cities and sales agreements with local bookstores throughout the country.[114] Within each region, distribution networks extended from branch offices in local cities to brokers and local bookstores in smaller towns.[115] Brokers and local bookstores mediated between local schools and publishers' branch offices or large bookstores in local cities. As textbooks traveled through these marketing networks, they spread new ideas about citizenship from cosmopolitan Shanghai to teachers and students in local cities and towns.

Within these commercial networks, books were marketed and sold like any other commodity, using common techniques of advertising, networks of relationships, price cuts, and kickbacks. Newspapers, magazines, and books themselves were often filled with adver-

---

112. The precise figures are 81 percent between 1929 and 1933 (Guoli bianyiguan, 17–24) and 88.7 percent between 1933 and 1936 ("Ling fa zhengshi kecheng biaozhun," 269–72).

113. For more extensive discussion of publishers' distribution networks, see Culp, "Mediating Modernity."

114. Wang Zhiyi, 9; Liu Dajun, "Zhongguo jindai tushu faxing tixi de jubian," 75–76. By the early 1930s, Commercial Press had 36 branch offices and World Book Company had 20 (Dai Ren, 49, 53–54). As of 1914, Zhonghua already had 23 branch offices (*Minguo sannian chun Zhonghua shuju gaikuang*, 9). The publishers' use of regional marketing centers and marketing techniques paralleled techniques used in other industries. See Cochran, "Commercial Penetration and Economic Imperialism."

115. This description of local distribution networks for publishers in the lower Yangzi region is based on interviews with three men who all had lifelong careers in publishing that began before or during the Sino-Japanese War. In the lower Yangzi region, major publishers had branch offices in Lanxi (Commercial Press and World Book Company only), Nanjing, Hangzhou, and Wenzhou (Zhonghua and World only).

tisements for individual textbooks or series.[116] Following a practice started by missionaries during the late Qing, publishing companies also sent catalogs directly to local bookstores and schools.[117] Publishers routinely gave local retailers special prices for exclusively marketing their books, and discounts were a common way to grab market share.[118] At the local level, retailers and brokers relied on methods such as "relationships" (*guanxi*), treating, and kickbacks to facilitate sales of books.[119] In all these ways, the newest textbooks reached students in local schools through the mechanism of the market.

Making a Medium: Compiling, Licensing, and Censorship

Using very different resources and strategies, intellectuals and government officials worked within this highly commercialized system to shape textbooks' content. Between 1912 and 1937 processes of textbook compilation and licensing changed dramatically. Broadly speaking, during the 1910s and 1920s, elite intellectuals hired on contract by the publishing houses used textbooks to spread their ideas with little state interference. By contrast, after 1927 the Nationalist Party instituted an effective censorship regime that increasingly standardized textbook content according to state-approved curricula. This shift from a production process dominated by reformist intellectuals to one indirectly supervised by the Nationalist Party significantly affected the content of textbooks.

Throughout the Republican period, textbook writing, compiling, and editing was most often undertaken directly by staff editors in a publishing house's editing department or an arrangement was made by contract for someone outside the publisher to draft a manuscript that would then be used as a textbook. The major publishers all had editing departments of at least several dozen writers, editors, and proofreaders. In the early 1930s, for instance, Zhonghua Book Company's editing department had over one hundred employees who

116. Liu Dajun, 79; and interview. For examples of textbook advertisements, see *Shenbao* September 14, 1911; October 20, 1911; December 26, 1915; January 1, 1917; July 6, 1917; July 27, 1919; May 14, 1923; May 16–17, 1923.

117. Liu Dajun, 78–79.

118. Ibid., 75, 77; Wang Zhiyi, 104; and interview.

119. Wang Zhiyi, 104; Yeh, *Schoolmaster Ni Huan-chih*, 203.

were divided into offices for editing textbooks, regular books, and dictionaries.[120] In many cases, textbook compiling was one of the primary tasks of the editing department.[121]

The editors working in these departments were full-time employees, and in the case of textbook editing, they were often former secondary school teachers who were recruited because of their teaching experience.[122] Publishers also recruited new staff for training on the basis of open examinations.[123] Several well-known textbook authors, such as Zhonghua's Jin Zhaozi, tested into publishing companies or were recruited for training in their youths and spent most of their working lives in the publishing industry.[124] These compilers and editors were part of an emerging class of lower-level urban professionals, similar to journalists and modern school teachers, who found in "publishing circles" (*chubanjie*) a professional identity and a site for literate activity in the rapidly changing social and cultural context of the Republican period. Because many of these new professionals had studied in modern schools, where they were influenced by the ideas introduced by prominent reformist intellectuals and modern-style academics, the textbooks they wrote often contained digested or translated versions of ideas circulating in elite intellectual discourse.

At the same time, publishers often hired more prominent intellectuals on a short-term basis or arranged to have them draft a textbook on contract.[125] The well-known historian and folklorist Gu Jiegang, for instance, worked briefly for Commercial Press during the mid-1920s, cowriting Chinese history and national language middle school textbooks.[126] Similarly, Kaiming Bookstore contracted with U.S.-trained writer Lin Yutang to write an English reader that reportedly earned him 300,000 yuan in royalties over twenty years.[127] This kind of arrangement was mutually beneficial. As the estimate of

---

120. Ji Shaofu, 326. For organization of editing departments, see *Minguo sannian chun Zhonghua shuju gaikuang*, 5; Dai Ren, 51.

121. Zhu Lianbao, 56.

122. Chen and Chen, eds., *Shangwu yinshuguan jiushiwunian*, 195–201; Ji Shaofu, 330; Wang Zhiyi, 3; Zhu Lianbao, 53–54.

123. Zhu Lianbao, 53.

124. Jin Zhaozi, "Wo zai Zhonghua shuju sanshi nian," 16–18.

125. Shu Xincheng, *Wo he jiaoyu*, 321–22; Zhu Lianbao, 62–63.

126. Chen Yuan et al., 295–97.

127. Wang Zhiyi, 100–101.

Lin's royalties from his English textbook suggests, working for a publisher could garner intellectuals a significant income during a time of economic turmoil in China. At the same time, publishers could sell textbooks bearing the name of a leading intellectual figure. The resulting textbook would accrete cultural capital from its author's fame and maybe some cachet if the author became embroiled in a high profile "war of pens."

Besides earning income, by writing secondary textbooks prominent reformist intellectuals such as Cai Yuanpei, Hu Shi, and Shu Xincheng could spread their ideas directly to young people. Moreover, a new generation of specialized, often Western-trained academics used secondary textbooks to promote their fields' approaches and insights.[128] Tao Menghe, who was trying to establish sociology as a modern academic discipline in China; Gu Jiegang, He Bingsong, and Chen Hengzhe, who sought to reform Chinese historical methodology and discourse; and Ge Suicheng and Zhang Qiyun, who introduced the modern science of geography to China, all wrote important secondary textbooks, which presented students with conceptions of citizenship drawn directly from the latest academic debates.

During much of the 1910s and 1920s, textbooks could effectively relate the ideas of reformist intellectuals directly to secondary students in part because the textbook-licensing process was lax and easily manipulated. In 1912, 1914, and 1916, the Ministry of Education promulgated rules requiring that textbooks accord with the regulations for different levels of education, stipulating that all textbooks should be submitted to the ministry for review. In theory, the ministry could require publishers to make changes to their textbooks, which would then have to be reapproved.[129] However, anecdotal evidence related to the textbook review process during the 1910s and 1920s also suggests that it was flexible, informal, and highly susceptible to manipulation through personal connections (*guanxi*). For instance, during the late 1910s, Zhonghua Book Company and Commercial Press used connections to the frequently changing ministers of education to gain advantage for their textbooks, most likely by

---

128. E.g., Chen Hengzhe, *Xiyangshi*, "Xu": 1–4.
129. Liu Zhemin, *Jin-xiandai chuban xinwen fagui huibian*, 68–71, 75–77.

facilitating the approval process and courting recommendations for their books.[130] Likewise, during the 1920s World Book Company had its textbooks reviewed by leading figures in Beijing's educational circles such as Hu Renyuan, who had previously served as president of Beijing University, so that their names could be invoked to facilitate the review process.[131]

After 1927 the textbook review process tightened considerably. As with earlier regimes, the Nationalist government quickly promulgated regulations requiring submission of textbooks for approval.[132] Moreover, new curriculum standards often contained full lists of topics for the courses in question, providing a strict framework against which the textbooks produced by particular publishers could be assessed. The Ministry of Education publicized standards for how it would review textbooks, as did the Central Training Department of the Central Executive Committee, which was responsible for reviewing party doctrine textbooks. Both sets of guidelines mixed criteria of orthodoxy regarding party ideology with criteria regarding the suitability of a given book to the level of student for which it was intended.[133] The Nationalist government also established a state-sponsored trade association for Shanghai publishers that served to reinforce state regulation.[134] Detailed curriculum standards coupled with regular review of textbooks and increasing institutional oversight led to a progressive standardization of textbooks over the course of the Nanjing decade.

In order to review textbooks, the University Council organized a Book Review Committee (Daxueyuan tushu shencha weiyuanhui; 1927–29). Then the Ministry of Education formed, in turn, the Office for Editing and Review (Bianshenchu; 1929–32) and the National Institute for Compilation and Translation (Guoli bianyiguan; 1932–49). The latter two organizations were both intended to review textbooks submitted by private publishers and to edit standardized government textbooks that would then be published by farming out

---

130. Ji Shaofu, 331.

131. Zhu Lianbao, 57.

132. Liu Zhemin, 356–58, 362–64, 384–85. The regulations were revised in 1929 and 1935.

133. Liu Zhemin, 363–66.

134. Reed, 224.

manuscripts to commercial publishers.[135] However, in practice it seems that during the Nanjing decade review of existing and newly published textbooks became both offices' main priority. For instance, during the five years before the Sino-Japanese War, the Institute for Compilation and Translation edited only 70 titles, most of which were college textbooks, reference works, and academic books; it began to edit primary and secondary school textbooks in earnest only in 1942.[136] Between 1932 and the start of the war in 1937, it was sent more than three thousand textbooks and related educational materials for review.

How effective was Nanjing government oversight of textbook publishing? During its tenure from 1929 to 1932, the Office for Editing and Review closely supervised textbook publishing, but it also faced pressure to review and approve materials quickly to avoid a dearth of textbooks. In May 1931, for instance, the office emphasized the need to establish efficient procedures to allow books that were largely trouble free to move through the review, revision, and approval processes easily so that they could be published quickly.[137] At the same time, though, the office was concerned with checking all kinds of textbooks in detail for accuracy, appropriateness, and political orthodoxy. For example, in one report reviewers pondered whether it was appropriate to use the term "bandit" (*fei*) when referring to the Nian Rebellion.[138] The detailed issues the office discussed suggest that the Nationalist government kept close tabs on textbooks.

---

135. Yang Changchun, "Guoli Bianyiguan shulüe," 198–200. Jiaoyubu bianshenchu, *Bianshen huiyi jilu*, contains many instances of the office setting an agenda of editing textbooks itself (see 2: January 11, March 25, July 11, and September 12, 1929). It is unclear how many of these projects, if any, were carried out. Party leader Chen Lifu established the Zhengzhong Book Company (Zhengzhong shuju) in 1930, and in 1931 the Nationalist Party's Central Executive Committee assumed management of it. Zhengzhong, which began publishing middle school textbooks in 1935, might have provided an alternative to government agencies' publishing of textbooks. See Ch'en, *The Storm Clouds Clear*, 101.

136. Yang Changchun, 200–202. Cf. Zhu Lianbao, 58–59.

137. Jiaoyubu bianshenchu, 2: May 8, 1931. See also 1: January 31, 1929, which includes a proposal that old high school textbooks be adopted for continued use to counteract a shortage of books.

138. Ibid., 1: February 12, 1930. Cf. 1: May 10, 1929; 2: May 8, 1931; September 24, 1931.

The impression of close government oversight is reinforced when we look at specific cases of textbooks that were banned or revised during this decade. In one example, in 1935 the Central Executive Committee's Propaganda Committee and the Ministry of Education imposed substantive revisions on Lü Simian's popular *Vernacular Chinese History (Baihua benguoshi)*, which was often used as a supplemental history text in middle schools. The Propaganda Committee took exception to Lü's judicious praise of the Song minister Qin Hui and his criticism of the General Yue Fei, who at the time was widely celebrated as an uncompromising patriot in Nationalist Party propaganda material. The Commercial Press was ordered to revise the offending passages before the book could be approved for sale, and it complied immediately.[139] This and other cases of books that were banned or had revisions imposed on them indicate the power of the Nationalist state's censorship apparatus when applied to textbooks.[140]

In many cases, publishers responded quickly to the changes imposed by government censors, suggesting that the Nationalist government was able to require a high level of compliance with its curriculum standards and political expectations. Because textbooks were a dependable source of income, publishing companies had a powerful incentive to comply with uniform curriculum standards in order to ensure smooth production and sale of these books.[141] Perhaps the best evidence that such measures were effective was the fact that by the early 1930s, the textbooks produced by the major publishing houses were becoming increasingly uniform in their organization and content. Because government review was pervasive and effective, there were hard constraints on the ideas that textbooks could present during the Nanjing decade.

---

139. Wu xian jiaoyuju miling no. 3 to Zhenhua Girls' Middle School (May 17, 1935) SMA2 Jia5.422.

140. Gu Jiegang and Wang Zhongqi's popular Commercial Press Chinese history textbook was banned when it raised the ire of Dai Jitao (Chen Yuan et al., 297–98; Guoli bianyiguan, 10; Hon, "Ethnic and Cultural Pluralism"). Cf. Zhongguo di'er lishi dang'anguan, 2: 1116–18.

141. See Jin Zhaozi, "Wo zai Zhonghua shuju sanshinian," 17–18. Christopher Reed also suggests that Shanghai's main publishers accommodated themselves to the Nationalist regime for the sake of profits and security (205–7).

IV. CONCLUSION

The combined efforts of intellectuals, local elites, private publishing companies, and, after 1927, the Nationalist government produced schools and textbooks, the infrastructure of citizenship education, in the lower Yangzi region. A hybrid group of New Culture reformers, modern-style academics, and professional editors and educators, ran schools, worked through professional associations to draft curriculum standards, and wrote textbooks. In these activities, they interacted closely with private business interests, local elites, and the Nationalist government. Through their cooperative efforts, intellectuals and educators, at many different status levels, used the power of the modern press to spread their ideas to a wider public and worked in schools under state supervision to create a new generation of modern Chinese citizens. However, such cooperation also entailed accommodation to the requirements of the publishers and the party. This accommodation often affected the practice of education, for publishers expected profits and the Nationalist Party imposed political expectations.

The Nationalist Party's efforts to manage education through co-optation of existing schools and private publishing companies contrasted clearly with late imperial educational practices, even though the Nationalist Party shared with imperial dynasts the ultimate goals of ideological uniformity and centralized political control. Late imperial states dictated educational, ideological, and political orthodoxy by establishing canonical versions of classical texts, printing them, distributing them to all government schools, and making them the basis for government examinations.[142] Yet once these standardized, government-approved teaching materials were produced and distributed, dynastic rulers allowed much of the educational process, outside of state-run schools such as provincial academies, to proceed through private channels such as tutors, informal local schools (*si-shu*), lineage schools, and private academies.[143] The Ming and Qing

---

142. Brokaw, "On the History of the Book in China," 17–18; Elman, *Cultural History*, chaps. 1, 2.

143. Elman, *Cultural History*, chap. 5; Elman and Woodside, eds., *Education and Society*, 11, 525–28; Leung, "Elementary Education"; Rawski, *Education and Popular Literacy*, chap. 2.

governments tended either to manage educational institutions directly and control publication or to allow great latitude for social forces.

The Nationalist government paralleled late imperial states in introducing examinations as a way to impose uniform educational standards, but it also intervened in the production of knowledge and the educational process in new ways. It provided general curriculum standards and then reviewed textbooks generated by private publishing companies rather than compiling and publishing orthodox teaching materials itself. Moreover, in addition to sustaining state-run secondary schools, the Nationalist government also actively regulated and supervised private schooling through registration, review, and funding assistance. As we shall see in the following chapters, it further mandated from the center patterns of organization, administration, and training for both public and private schools. Through this approach the Nationalist government sought to mobilize educators, intellectuals, and private publishers to pursue its goals. But co-optation also made the Nationalist Party dependent on others' actions and built on the precedents of earlier modes of education, publication, and youth training. These forces had the potential to transform state policy in practice.

2

Nation as Race, Culture, or Place?

The National Community in History

and Geography Textbooks

One alumnus's most vivid memory of his days at Jiangsu Provincial Yangzhou Middle School between 1933 and 1937 was of his history teacher recounting China's gradual loss of Vietnam, Taiwan, and the Northeast during the nineteenth and twentieth centuries. An older alumnus, who had attended Yangzhou Middle between 1927 and 1933, similarly recalled how his geography teacher had portrayed the loss of territory to imperialism from its grand extent during the Qing by using the poignant metaphor of a mulberry leaf being nibbled away at the edges. In reciting these 60-year-old messages of territorial loss and collective danger, these former students evoked the nationalism that—in many guises and registers, both in the classroom and in mass-produced textbooks—was one of the central themes of Republican China's secondary-level history and geography instruction.

Throughout the Republican period, leading historians and geographers, as well as an emergent group of professional editors in the editorial departments of publishing houses, used the methods of modern historical and geographical discourse to write secondary-level history and geography textbooks with nationalist messages. These texts encouraged students to conceive of China as a nation of horizontally connected and sovereign citizens (*guomin*) rather than as a collection of disparate peoples in random far-flung territories joined together under the Qing empire. They cultivated this sense of "imagined community" through narratives of a unified Chinese people that had evolved over ages and through descriptions of an iconic, cohesive national territory whose fixed boundaries corresponded to an idealized version of the Qing "family property" that the Republic had inherited.

But as these authors wove diverse strands of nationalist geography and historiography into their textbooks, they presented students with varied images of the nation. History textbook authors used multiple strategies—some based on race and others on culture, some homogenizing and others pluralistic—to construct narratives of a historically rooted and continuously evolving, cohesive national community. Parallel to these culturally or racially based accounts of national unity, history and geography textbooks also portrayed territorial cohesion and boundedness as the basis for imagining a unified Chinese nation. As these competing portrayals of the nation—racial, cultural, and territorial—found expression in textbooks and classrooms they provided lower Yangzi region students with diverse resources for imagining and rhetorically constituting the nation.

At the same time, history and geography textbooks also suggested that China was more of a potential than a fully realized nation-state. The putative ethno-cultural[1] unity history textbooks described was challenged by geography textbooks' detailed accounts of China's ongoing cultural, religious, and linguistic diversity. History and geography textbooks also portrayed China's economic unevenness and imperialist threats to its borders so as to suggest that territorial unity and sovereignty depended on ongoing projects of defense and development (*fazhan*). These courses, then, provocatively described a gap

---

1. I use "ethno-cultural" and "ethnicity" to connote the cluster of marking qualities encompassed by the Chinese term *minzu*, which can be translated variously as "race," "nation," or "ethnic group." *Minzu* cannot be equated simply with race, since custom, culture, and language have been seen as essential parts. See Sun, *San Min Chu I*, 8–11. Yet *minzu*, by incorporating the character *zu* (clan), indicates that genealogical continuity was seen as fundamental to these groups' cohesion over time. Moreover, many Republican-period authors used phenotypic difference as a way to distinguish between *minzu*. When textbook authors use specific terms for race (*zhongzu*) or culture (*wenhua*), or isolate purely phenotypic or cultural markers to distinguish between social groups, I use "race" or "culture" to characterize those distinctions. When they use the broader, omnibus term *minzu* and/or fail to specify the qualities distinguishing social groups, I use the terms "ethno-cultural," "ethnic group," and "people" (as in, "the Tibetan people" [*zangzu*]).

between China's uneven and "backward" (*luohou*) economy, its cultural diversity, and its embattled state, on one hand, and the objectives of economic modernity, full national integration, and geopolitical security, on the other. This gap provided a rhetorical justification for projects of socioeconomic development, political and cultural unification, and defense of the borders. The nation's incompleteness, as textbooks and teachers described it, served as a call to student-citizens to take part in civic action that would make the nation, even as civic participation itself would define what it meant to be a citizen.

I. THE NATION AS ETHNO-CULTURAL COMMUNITY: ROOTEDNESS, CONTINUITY, UNITY

By projecting a unified community into the deep historical past and tracing its progressive development, modern forms of narrative history can help to create a sense of the nation as rooted in time immemorial and as evolving but continuous.[2] The sense of continuity and rootedness encouraged by historical narrative can contribute to making the current national community seem natural, cohesive, and taken for granted. Chinese history textbooks of the Republican period were written as national histories that sought to establish the unity and continuity of the Chinese people. Significantly, though, the textbooks published between 1912 and 1937 varied considerably in the narrative strategies they used to create a sense of national cohesion and historical depth. One of the central forces driving changes in the dominant narratives of national unity and continuity were changes in China's historical profession itself. As Chinese historians adopted and adapted Western historical methods of evaluating sources and constructing arguments and narratives, their own accounts of Chinese history shifted accordingly, moving away from mythic accounts of the ancient past to rely on seemingly more dependable written sources and archaeology.

---

2. See, for instance, Chatterjee, *The Nation and Its Fragments*, chaps. 4, 5; Tanaka, *Japan's Orient*; and Zerubavel, *Recovered Roots*. Prasenjit Duara, by contrast, argues persuasively that national history is always caught between establishing the ancientness and continuity of the nation, on one hand, and describing its emergence as a new, modern nation-state, on the other. See Duara, *Rescuing History*, chap. 1.

All the history textbooks of this period accommodated the Republican-period orthodoxy that China's national people was a "union of five peoples in the Chinese Republic" (*wuzu gonghe*). Yet they differed over whether they portrayed those people as united as a common racial group or as the product of complex processes of ethno-cultural assimilation. Even among the textbooks that stressed a long process of assimilation, there were two competing accounts of that process. One version stressed that ethnic integration had been a one-way process of cultural homogenization or "Sinicization" (*han-hua*). The other version described assimilation as a complex inter-mingling of equally constitutive races and cultures. These two narratives stood in tension across the decades of the Republican period. Which of them took a more prominent position in textbook narratives had important ramifications for what secondary students were taught about Chinese national identity. At stake in these competing visions of the national past was whether Chinese society would be seen as plural and continually evolving or as fixed, continuous, and dominated by the majority Han people.

The most direct way to establish the cohesion and deep historical roots of the current national community was to project a picture of unity back into the ancient past, as Zhong Yulong did in his New System textbook, published in 1914 and 1915. At the start of his textbook, Zhong stated his aim to reverse a history of Han disdain for the other ethnic groups that composed the Chinese nation—here, Mongol, Manchu, Muslim (Hui), and Tibetan—by stressing a spirit of unification and by presenting evidence of how and when other peoples had contributed to progress (*jinhua*) in China's past.[3] The primary way Zhong's narrative worked to establish this sense of unity was by positing a common racial origin of these different peoples. All of these groups, Zhong asserted, were of the "yellow race" and came from a Western place of origin beyond the Pamir Mountains in Central Asia. Differentiation among the peoples came only as a result of having lived in different environments and evolved under different cultural conditions.[4] This alleged common racial origin allowed Zhong to assert a deep historical foundation for the national unity that subsumed what had become five apparently distinct eth-

---

3. Zhong Yulong, *Benguoshi*, Bianji dayi: 1.
4. Ibid., 1: 3–4.

nic groups under the single designation of the "Chinese nation" (*zhonghua minzu*) during the Republican period. Later, at the beginning of the final volume of his textbook, Zhong asserted that the basis for interaction among and reunification of the five ethnic groups was established during the modern period, when the Han and other ethnic groups had all been conquered by the Manchu rulers of the Qing empire.[5]

Textbooks of the 1910s also sought to establish the deep historical roots of the Chinese people by recounting myths of the sage kings, the legendary heroes credited with founding Chinese culture, which portrayed their activities in the Yellow River valley in deep antiquity. For instance, Zhao Yusen asserted that, although earlier legends could not be proven, accounts from the Yellow Emperor's (Huangdi) time forward were reliable, leading him to recount the Yellow Emperor's territorial unification and the activities of his successors, Yao, Shun, and Yu, as history.[6] Zhong Yulong likewise used stories of the mythical kings to recount the creation of Chinese culture and the settlement of the Central Plains by the predecessors of the Chinese people, also without clearly identifying them as legends or myths.[7] These accounts worked to project the origins of a proto-national Chinese community far back, before the beginning of the historical record, making it seem almost primordial.

Changes in the historical profession during the following decades shifted the methodological baseline for history writing in China, as Chinese historians adopted critical methods of source evaluation drawn from both Western historiography and China's own philological tradition.[8] By raising doubts about the reliability of ancient myths, the critical methods of the new history problematized accounts that sought to establish with certainty the ancient origins and unity of the Chinese people. The first educational expression of these methodological changes in Chinese historiography came with the publication of a lower middle school history textbook by Gu

---

5. Ibid., 3 (1914): 1.

6. Zhao Yusen, [*Gongheguo jiaokeshu*] *Benguoshi*, 1: 4–6.

7. Zhong Yulong, *Benguoshi*, 1 (1915): 4–11.

8. Duara, *Rescuing History*, chap. 1; Schneider, *Ku Chieh-kang*; Wang, *Inventing China*.

Jiegang[9] and Wang Zhongqi (see Appendix B) in 1923 and 1924. Gu was a leader in the movement for a new history in China, and he used this textbook in part to advance one of his main agendas: questioning the reliability of ancient textual sources.

In their textbook, Wang and Gu offered a trenchant critique of the mythological accounts on which arguments about the origins of the Chinese people were based. They began the book by categorizing foundational stories, such as the Judeo-Christian biblical story of the flood and China's Pangu legend about the origins of the world, as myths that could not be taken literally. More importantly, they also criticized different theories about the origins of the Hua, or Chinese, people by saying that they rested on unreliable sources.[10] They thus relegated any search based on questionable ancient texts for the origins of the Chinese people to the realm of mythmaking, not history. By extension they made problematic any efforts to argue for a simple and certain racial unity of all "Chinese" people in the prehistoric past.

In addition, Wang and Gu's discussion of the myths of the ancient sage kings, the Shang dynasty, and even the early Zhou dynasty also reflected a markedly critical relationship to historical sources that was characteristic of new forms of historical writing. In each case, Wang and Gu argued that foundational myths could not be taken literally. Rather, they asserted that accounts of the mythical ancient emperors Huangdi, Fuxi, and Shennong reflected collective experiences and social practice in ancient society and/or that particular legends had been invented much later and perpetuated to support particular social institutions and political claims.[11] By revealing how ancient myths had been constructed to serve particular political interests and to explain long-term transitions in the prehistoric past, Gu and Wang eroded those myths' power for establishing with certainty the prehistoric roots and cultural antecedents of China's people. Significantly, later textbooks of the Nanjing decade also generally assumed a critical perspective toward myths about the ancient sage kings. Some, such as a 1933 textbook by longtime

---

9. The definitive account of Gu Jiegang's life and work is Schneider, *Ku Chieh-kang.*

10. Gu and Wang, *Benguoshi,* 1 (1927): 9–11, 18–19.

11. Ibid., 23–39.

Commercial Press editor Fu Weiping (see Appendix B), revealed that the ancient legends made contradictory statements about the origins of the Han people.[12]

Rather than grounding the modern unity of the Chinese people in a static and mythic ancient origin, Gu and Wang portrayed it as the product of continual evolution, through which the Hua people changed as they assimilated other peoples. But two contrasting narratives of assimilation stood in tension in Gu and Wang's textbook. At numerous points in the text they adopted the rhetoric of Sinicization. For instance, they defined the Hua people (*huazu*) as the "main elements" (*zhuyao de fenzi*) of the Chinese people. Moreover, they characterized the Hua as an original and continuous ethnic group by saying, "They had a deep source of culture, so although they were conquered politically by other peoples several times, the result was always that the conquerors were assimilated into the conquered."[13] Their later description of the Central Asian invaders', the Wuhu's, abandoning their languages and customs for those of the Hua during the Northern and Southern dynasties (420–589 CE) reinforced this narrative of Sinicization.

In tension with an account of Sinicization in Gu and Wang's book was a second narrative of the Chinese people's being transformed by outside influences. They summarized their conception of the historical changes of the Chinese people by saying, "However, we have to acknowledge that the Chinese people (*zhongguo de minzu*) certainly went through many changes and experienced many outside influences."[14] They reinforced this generalization with examples and statements later in their textbook. For instance, even though they described the Wuhu's adopting "Chinese" culture during the Northern and Southern dynasties, they also characterized that time as a period when a moribund and stagnant Chinese culture had been

---

12. Fu Weiping, *Benguoshi*, 1: 9, 12–13. Cf. Fu Weiping, 1: 1–40; Yao Shaohua, *Chuzhong benguo lishi*, 1: 1–13; Zheng Chang, *Xin zhonghua benguoshi*, 1 (1930): 6–15. Several authors framed their discussions of these myths by discussing the problematic and unreliable nature of the ancient texts, and they peppered their accounts with qualifying phrases such as "legend has it" (*xiangchuan*) and "it is said" (*jushuo*).

13. Gu and Wang, *Benguoshi*, 1 (1927): 10, 12.

14. Ibid., 11.

revived by absorbing "new ethnic and spiritual blood."[15] "Having passed through two or three centuries of stimulation and acceptance or rejection [of different cultures during the Northern and Southern dynasties], each side's intrinsic strengths were blended together (*huxiang tiaohe*) and gave birth to a new culture. That the Sui and Tang could flourish as great unified empires, and that they both arose in the North, was due to this assimilation of different cultures."[16] In these instances, Gu and Wang portrayed China's culture and people as having been significantly transformed by outside peoples and cultures.[17] Such accounts of ethno-cultural transformation from outside forces reflect the emphasis on ethnic pluralism within the Chinese nation that was characteristic of Gu's monographs and other historical writings.[18]

Tensions between competing narratives of assimilation persisted in textbooks of the 1930s.[19] The 1934–35 version of the high school textbook written by Lü Simian (see Appendix B), who was one of Republican China's leading historians, suggests how many textbook authors during the Nanjing decade juxtaposed varying accounts of ethno-cultural interaction. On one hand, Lü described China as having benefited greatly from the influx of the new cultural elements from India that had entered China along with Buddhism. In addition, he noted how the Western Regions (*xiyu*) had contributed many new forms of music to China, and how "Hu," or Central Asian, customs had changed Chinese habits of dress and daily life.[20] On the other hand, these examples of powerful outside cultural influences were juxtaposed with an overarching account of non-Han peoples being "Sinicized" and subsumed within Han culture, which had maintained continuity with its "origin" over millennia.

Our China in the past absorbed (*xihe*) many different peoples. Because we often were in contact with other peoples, we were able mutually to temper

---

15. Ibid., 20.

16. Ibid., 164. Gu and Wang also described Buddhism as rescuing Chinese thought and culture from stagnation under Confucian orthodoxy (ibid., 90).

17. Cf. Zhao Yusen, *[Xinzhu] Benguoshi*, 1 (1924): part 2: 73–74, 175–76.

18. Duara, *Rescuing History*, 42–43; Hon, "Ethnic and Cultural Pluralism"; Schneider, *Ku Chieh-kang*, 260–66.

19. Fu Weiping, 2: 153; 4: 142–43; Zhu Yixin, *Zhushi chuzhong benguoshi*, 1: 64; 2: 43, 90.

20. Lü Simian, *Benguoshi*, 1 (1934): 194–97.

(*cuili*) one another, adopting others' strengths in order to supplement our weaknesses; although we were civilized very early, the circumstances were often new. Further, because our traditional culture was superior, its power to assimilate was great. Although it changed frequently, it did not lose its origin (*benlai*). . . . In sum, the formation of China (*zhonghua de liguo*) was focused around the Han people (*yi hanzu wei zhongxin*). Either they used political power to control other peoples, or they used the power of culture to convert (*ganhua*) other peoples. That is, at times the Han people's political power was not competitive, and it was temporarily conquered by other ethnic groups, but because its cultural level was high, other peoples had to accord with its methods of rule (*zhifa*). Then, having passed through some time, they would still become assimilated into the Han people.[21]

By taking the Han ethnic group to be the "core" (*zhongxin*) of China, Lü attempted to account for the encompassment of other peoples and to acknowledge the contributions they might have made to Chinese culture while crafting an image of a unified ethno-cultural subject that remained historically constant and was rooted in the ancient past. The Han people might develop because of encounters with other cultures and peoples, and they might come to include other groups that were remade in their image, but they would not be changed fundamentally by those outside influences. This narrative of Sinicization stood in obvious tension with the portrayal, discussed above, that stressed the renewal and transformation of the Chinese people by outside influences.

Other textbooks of the Nanjing decade focused almost exclusively on a narrative of Sinicization and gave little sense of how outside peoples and cultures might have transformed the Chinese people. Textbooks by Zhonghua Book Company staff editors Zheng Chang (see Appendix B) and Yao Shaohua,[22] for instance, described the Wuhu during the Northern and Southern Dynasties and the middle period Central Asian dynasties of the Liao (915–1125), Xia (1032–1227), and Jin (1115–1234) as having fully and smoothly adopted Chinese culture, completely Sinicizing. Both textbooks also described the Manchu Qing rulers becoming Sinicized during the seventeenth and eighteenth centuries as they adopted Han culture and governing

---

21. Ibid., 5–7. Cf. Ibid., 2 (1934): 252–53.

22. In addition to compiling textbooks, Yao Shaohua served as editor for the journals *New China* (*Xin zhonghua*) and *Chinese Education Circles* (*Zhonghua jiaoyujie*). See Yu and Liu, eds., *Lufei Kui yu Zhonghua shuju*, 8.

techniques in order to rule China.[23] These examples together sketched out a vision of assimilation as a one-way adoption of a stable and unchanging Han culture by non-Han peoples, who were assumed to be less culturally sophisticated.

Further, many Nanjing-decade textbooks turned to the new scientific methodology of archaeology to establish the prehistoric origins of the Han people.[24] For instance, Yao, in his 1933 textbook, asserted that "during the past ten years or so, most archaeologists have discovered prehistoric remains in this area, and the human bones excavated there are very similar to those of today's Northern Chinese. Obviously, that the Han people during ancient times were settled in the Yellow River valley is a certainty."[25] Archaeology was used in these textbooks to establish a genealogical connection between a biologically continuous Han people and the region of the Central Plains along the Yellow River, an area long believed to be the cradle of ancient Chinese civilization. It seemed to provide a much more "scientific" and thus certain version of a foundational narrative of Han rootedness and continuity than did discredited ancient legends. Moreover, archaeology extended the origins of the Han people's presence in the Yellow River valley much farther back into antiquity than had the questionable myths on which early Republican authors such as Zhao Yusen and Zhong Yulong had relied.

Though the narrative of the Han people's ancient origins, cultural superiority, and one-way assimilation of others became dominant in educational circles under the Nationalist Party during the 1930s, this version of Chinese history was never monolithic or unchallenged. A vision of the evolving Chinese people as plural and composite was always available to Chinese teachers and students, even if it was the minority view. At one level, mass-publication textbooks, such as the one by Fu Weiping, who suggested at points that the Chinese people had been generated through a complex process of racial and cultural integration, were still available and used during

---

23. Yao Shaohua, 1 (1933): 72–73, 83; 2 (1933): 63–65; 3 (1934): 14; Zheng Chang, *Xin zhonghua benguoshi*, 1 (1930): 67, 101–13; 2 (1931): 4. Cf. Yao Shaohua, 1 (1933): 77; Zheng Chang, *Xin zhonghua benguoshi*, 1 (1930): 81–82.

24. For Fu Sinian's introduction of archaeology to the newly formed Academia Sinica, see Wang, *Inventing China*, 121–30.

25. Yao Shaohua, 1 (1933): 4. Cf. Zheng Chang, *Xin zhonghua benguoshi*, 1 (1930): 14.

the 1930s.[26] In addition, teachers in local schools, through the great flexibility they had in choosing teaching materials, could stress a narrative of hybridity over one of homogeneity. For instance, the mimeographed lecture notes teachers distributed to lower middle school students at Jiangsu's prominent Jiangsu Provincial Suzhou Middle School in 1934 were none other than a verbatim reproduction of Gu and Wang's lower middle school textbook of 1923, which the Nationalist Party had banned in 1929.[27]

II. THE CHALLENGE OF DIFFERENCE

In contrast to Chinese history textbooks' narrative of assimilation and inevitable national unity, Republican-period Chinese geography textbooks raised serious doubts about ethno-cultural cohesion. Most specifically, in the human geography sections of the textbooks, authors' discussions of the composition of the Chinese national community revealed it to be incredibly pluralistic and resistant to reductionist formulas of unity. The textbooks recorded prominent differences in custom, language, religion, and phenotype that confounded any possibility for imagining an ethno-cultural community that cohered in the present. In fact, geography textbook authors suggested the tremendous obstacles in the way of Chinese ethno-cultural unity by revealing that the supposedly homogeneous Han majority was nearly as divided internally as it was distinct from other ethnic groups.

Most geography textbooks of the Republican period included in their discussions of the Chinese national community some effort to establish its intrinsic unity, but the basis for national cohesion varied widely among textbooks. Moreover, claims for unity were often brief and token, dwarfed by the long expositions of cultural, linguistic, and religious differences that were recited within these same textbooks. One suggested basis for the unity of the national people was their racial similarity. For instance, Commercial Press editor Xie Guan (see Appendix B), in his "Republican Textbook" series geography textbook of 1913, paralleled history textbooks of the 1910s and argued

---

26. Jiangsu Provincial Suzhou Middle School, for instance, used Fu's book in its Chinese history classes during the mid-1930s. See Gao and Zhang, "Shishi ting ban benguo lishi."

27. Wang Gaiyai, *Benguoshi*.

that there existed a racially unified Chinese community: "Our country's people generally belong to the Mongolian race (*Mengguliya zhong*). Since historical times, because of various relationships and unions, [we have] formed Asia's greatest and strongest people."[28] This formulation posited a deep unity of race under manifest differences of culture and language. However, race, as Xie formulated it here, was a connection of dubious value for *national* unity, since Japanese people, too, were considered by convention to be part of the Mongoloid, or "yellow," race.[29] During the Nanjing decade, some textbooks shifted the focus of this racial approach to unity by repeating Sun Yat-sen's assertion that China was a "genealogical nation" (*guozu*) because the assumedly homogeneous Han people constituted the vast majority of the population.[30]

Yet even these textbooks, which sought a physical, and thus seemingly certain, basis for unification of the Chinese people, described multiple levels of cultural, linguistic, religious, and often even phenotypic difference. This overwhelming evidence of ethnic and cultural diversity suggested the many practical ways in which the Chinese community was internally divided, despite supposedly being unified by sharing a common racial background.

The extent of that diversity was demonstrated most clearly in the textbooks' discussions of language. Chinese geography textbooks showed great differences in language type among Han and various non-Han languages. They also discussed in detail the dialect variations within spoken Chinese language.[31] Liu Huru's 1927 textbook included a dialect map that graphically represented how diversity in language fragmented the Chinese national community by marking each language or dialect area with a different pattern. The result was a map that looked like a patchwork quilt.[32] (See Fig. 2.1.) Many of the authors echoed Zhonghua Book Company editor Li Tinghan

---

28. Xie Guan, *Benguo dili*, 2: 138. Cf. Ge Suicheng, *Chuzhong benguo dili*, 4 (1934): 7; Li Tinghan, *Benguo dili*, 1 (1914): 58–59.

29. See, for instance, Xie Guan, *Waiguo dili*, 1: 70.

30. Liu Huru, *Xin shidai benguo dili*, 1 (1927): 17–19; Wang and Zhou, *Benguo dili*, 4: 109.

31. Ge Suicheng, 4 (1934): 22–25; Li Tinghan, 1 (1914): 61–63; Liu Huru, 1 (1927): 109–14; Wang Zhongqi, *Benguo dili*, 1 (1924): 142–53; Xie Guan, *Benguo dili*, 2: 145–46.

32. Liu Huru, 1 (1927): 111.

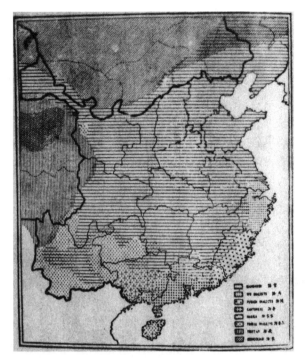

2.1 "Map of China's Dialects." The checkered pattern, especially along the coast from Jiangsu south, highlights China's linguistic diversity. Liu Huru, *Xin shidai benguo dili*, 1 (1927): 111.

in openly lamenting the adverse effects of linguistic diversity on national unity: "Places separated by 100 li already have distinctive local dialects, whereas at a distance of 1,000 li people meet and talk only to stare uncomprehendingly."[33]

Authors also described in detail China's religious pluralism, often claiming it as a badge of the Republic's enlightened policy of religious freedom (*zongjiao ziyou*). Yet many authors also noted that religious beliefs and practices created differences in custom between the Hui, Mongol, and Tibetans and the other groups making up the Chinese nation. New Culture Movement intellectual Wang Zhongqi noted, for instance, that the Hui differed from the Han because of their belief in Islam, which caused the Hui to have distinctive

---

33. Li Tinghan, 1 (1914): 61. One li was equivalent to one-third mile.

customs.[34] Many authors also identified belief in Lamaism, or Tibetan Buddhism, as one of the features that established the distinctiveness of the Tibetan and Mongolian peoples. Yet several textbooks further marked that religion as a form of baseless superstition (*mixin*) that kept Tibetans and Mongolians ignorant and benighted.[35]

Parallel to distinctions of language and religion were distinctions of culture, character, and phenotype. Several authors drew attention to the nomadic pastoralism of the Mongols, which, it was asserted, shaped their special martial character, and to their perceived cultural idiosyncrasies, such as the milk- and meat-centered diet and tent housing.[36] (See Figs. 2.2 and 2.3.) Leading geographer Ge Suicheng (see Appendix B) also pointed out how Tibetans' gender division of labor differed from that of the Han and how they often lived in stone houses. Other sections described the Tibetan diet and their loose and heavy clothing.[37] Some textbooks further noted physical differences between Han, Mongol, Tibetan, and Turkic (*tujue*) peoples by describing how they varied in terms of stature, skin color, and facial features.[38]

As with their discussions of China's diverse languages, however, geography textbooks described cultural and phenotypic differences not just between Han and non-Han groups but also within the supposedly unified Han people. Most textbooks listed a set of stereotypic differences in character, body type, and custom between Northern and Southern Han Chinese.[39] More significantly, several textbooks from the Nanjing decade described in some detail the distinctive ethnic subgroups, such as Hakka, Li, She, and Dan, which were included within the supposedly homogeneous Han Chinese people.[40]

---

34. Wang Zhongqi, 1 (1924): 150–53, 158. Also see Ge Suicheng, 4 (1934): 10–11; Liu Huru, 1 (1927): 120–28; and Xie Guan, *Benguo dili*, 2: 144.

35. Liu Huru, 1 (1927): 120–23; Wang Zhongqi, 1 (1924): 153, 156; Xie Guan, *Benguo dili*, 2: 144–45, 147–48.

36. Ge Suicheng, 4 (1934): 10; Wang Zhongqi, 1 (1924): 149–50.

37. Ge Suicheng, 1 (1933): 89; 4 (1934): 11. Ge further described the Miao, Yao, and other Southwestern peoples as being even more culturally distinct from the Han. See 2 (1933): 43–44; 4 (1934): 11–12.

38. Liu Huru, 1 (1927): 22–30.

39. E.g., Liu Huru, 1 (1927): 23; Wang Zhongqi, 1 (1924): 144–45; Xie Guan, *Benguo dili*, 2: 143–44.

40. Ge Suicheng, 2 (1933): 41–43; 4 (1934): 9; Wang and Zhou, 1: 73–74, 4: 110.

2.2 "Mongol Nomads." The livestock and tent housing in this illustration identified Mongol society as nomadic. This pictorial summary of Mongol culture differentiated it from Han culture and indexed its relative level of civilization within a universal evolutionary time scheme. Xie Guan, *Benguo dili,* 1: 79.

2.3 "Mongolian People." As has often been the case with cross-cultural representations in the modern world, images of women serve to mark cultural distinctiveness, in this case the difference between the Mongol and Han cultures. The prominent headdress of the woman on the left, the clearly unbound bare feet of the woman on the right, along with all the women's behavior—sitting on the bare ground and smoking—illustrated multiple differences between Mongol and Han culture. Liu Huru, *Xin shidai benguo dili,* 1 (1927): 25.

Difference, then, along many dimensions—linguistic, religious, cultural, and phenotypic—proliferated within the geography textbooks, overwhelming their often feeble claims for the unity of the Chinese people as a national community. Further, as evidenced by their careful detailing of the distinct ethnic subgroups within the Han majority, there seems to have been more exposition of difference within the later textbooks of the Nanjing decade, a period when calls for the unity of the national community were urgent and constant. What accounts for this exposition of internal pluralism and differentiation in these textbooks that were ostensibly dedicated to instilling ideas of national cohesion?

The expression of internal distinction provided Chinese governments and elite intellectuals with justifications for the normalizing and unifying agendas they directed toward particular subgroups in their efforts to create a cohesive and homogeneous Chinese national community. Each of the axes of distinction described above corresponded to a cultural program, orchestrated by Han elites or a Han-dominated state, that was geared toward erasing or downplaying that distinguishing characteristic and imposing a Han-centered cultural or linguistic norm. However, geography textbooks' exhaustive accounts of pluralism simultaneously served to defer any sense of certainty about the unity of the national community into the distant future.

The symbiosis between a form of cultural difference emphasized in geography textbooks and a normalizing nation-building project was clearest in regard to language. For instance, in Xie Guan's textbook, the identification of linguistic diversity fed directly into a rallying cry for the project of linguistic unification. "Because of the breadth of our territory and obstructions of rivers and mountains, our country's languages are varied in type, even to the extent that within 10 li there are small differences and within 100 li great differences. Various perceptions of division come from this. Although there is a common standard pronunciation, it has not been able to spread in a short time. If we do not quickly plan a method for unification [of our language], it will be difficult to experience the positive results of [national] unity (*tuanjie*)."[41] Textbooks written after the national language (*guoyu*) movement took shape in the 1920s were even

---

41. Xie Guan, *Benguo dili*, 2: 145. Cf. Li Tinghan, 1 (1914): 61.

more explicit in their advocacy of aggressive implementation of the program.[42] Portrayals of linguistic diversity provided a backdrop and justification for programs of language unification, with the dominant dialect of Han Chinese language providing the "national" standard.

Cultural and religious diversity similarly worked to trigger calls for assimilation to a norm defined by reference to the dominant Han culture. Prominent geographer Zhang Qiyun's (see Appendix B) 1926–28 textbook offered the clearest combination of a description of cultural differentiation and a rallying cry for Han-based cultural assimilation. In his discussion of China's peoples, Zhang included a table that outlined the attributes of each ethnic group and their roles in Chinese history.[43] At the end of each description Zhang offered an assessment of the relative level of integration of different groups, making clear the goal was assimilation to a Han Chinese norm. The Han were considered the "main body" (*zhuti*) that assimilated others, and the Manchus were described as having been almost completely integrated, but the remaining groups were portrayed as still awaiting Han assimilation. Zhang's discussion of the relationship between Han and Miao at the end of this section of his textbook made the assimilative project explicit:

Besides this, the uncivilized Miao-Man people in the Southwest probably number 1.5 million people. Their tribes are numerous, and their customs confused. They maintain the customs of high antiquity and are totally incompatible with the Han people. Eliminating their barbarism and changing their customs and habits is the responsibility of the Han people. In sum, the work of assimilation by our [Han] forebears was already 80 to 90 percent completed; the remaining 10 to 20 percent is left to the efforts of future generations (*houqizhe*).[44]

Zhang, here, explicitly transposed cultural difference between the Han and Miao into a temporal difference between backwardness and "civilization." By associating any difference with backwardness, Han intellectuals and states could justify a project of assimilation according to their own cultural standards in terms of introducing other

---

42. E.g., Ge Suicheng, 4 (1934): 22–25; Liu Huru 1 (1927): 109–20.

43. Zhang Qiyun, *Benguo dili*, 1: 46–48. The same table was reproduced in Liu Huru, 1 (1927): 20–22.

44. Zhang Qiyun, 1: 49.

peoples to the seemingly universal value of "civilization." This equation of Sinicizing with civilizing was repeated throughout other geography textbooks.[45] For example, Ge Suicheng declared that as members of the Chinese nation (*zhonghua minzu*), the peoples of the Southwest should be helped to become more "civilized" (*kaihua*).[46] These textbooks promoted the universalizing project of "civilizing" and standardizing diverse peoples' culture, language, and behavior according to Han linguistic and cultural norms.

III. THE NATION AS GEO-BODY

Geography and history textbooks, in addition to their efforts to define the Chinese nation as an ethno-cultural or racial community, worked together to produce an image of the Chinese nation as a geobody. That is to say, they formulated an image of China as a cohesive territorial unit with fixed boundaries and unified sovereignty.[47] Moreover, in ways characteristic of nationalist geography elsewhere, they sought to naturalize China's territorial unity by projecting it backward into the deep historical past and, in some cases, by associating arbitrary political borders with seemingly natural boundaries.[48] Metaphorical and cartographic representations of the national territory also gave it an iconic quality that enhanced its sense of unity and tangibility. The rhetorical and symbolic construction of China as a bounded and sovereign territory helped to underwrite two geopolitical arguments that have been fundamental to twentieth-century Chinese nationalism. One is the perceived need to reclaim territory deemed to have been "lost" to imperialist powers. The other is the drive to integrate bordering areas of questionable sovereignty into the centralized polity.

The territorial unity these textbooks constructed was in all cases a paradigmatic one. It marked out, in other words, an idealized national territory that did not correspond to the "real" borders of the contemporary Chinese state, which throughout this period was internally fragmented and had little, if any, functional control over out-

---

45. E.g., Li Tinghan, 1 (1914): 60.
46. Ge Suicheng, 4 (1934): 11–12; 2 (1933): 41–44.
47. Thongchai, *Siam Mapped.*
48. Hooson, ed., *Geography and National Identity.*

lying provinces, let alone distant border regions such as Mongolia and Tibet.[49] In fact, the idea of the territorially bounded state itself was relatively new to most of the Chinese public, and textbooks' maps and other images of the Chinese territory were one medium that introduced to students the model of the fixed national space. Moreover, insofar as textbooks sought to establish the historical continuity of China's bounded national space, they necessarily projected backward into the past a recent vision of the territorial state, for the late imperial conception of the state had seldom been premised on ideas of fixed borders and uniform sovereignty.

Given recent work on Qing-era empire building, these last two statements require explanation and qualification. Laura Hostetler has demonstrated that Jesuit missionaries used new cartographic techniques of surveying and to-scale mapping to produce comprehensive maps of the Qing domain during the Kangxi era. She also argues convincingly that, in a context of competitive global empire building, these maps delineated clearly a bounded imperial territory, which foreshadowed the Chinese nation-state.[50] However, as John Fairbank and others have argued, the dominant late imperial geopolitical vision into the nineteenth century was one of layers of graduated political and ritual relations between the emperor and various rulers and peoples radiating outward, with diminishing power, from a centralized imperial capital.[51] Because of ambiguous conceptions of territorial sovereignty, in China as in other parts of the world, efforts to establish a nation-state required, for the first time, clearly delineating national boundaries to establish the extent of the nation's territorial sovereignty.[52] The 1910 Pianma Incident, in which Qing

---

49. See Sheridan, *China in Disintegration*, for an exposition of the fragmentation of the Chinese state throughout the Republican period.

50. Hostetler, *Qing Colonial Enterprise*, Introduction, chap. 2. Cf. Perdue, "Boundaries."

51. Fairbank, ed., *The Chinese World-Order*. Cf. Hevia, *Cherishing Men from Afar*, chap. 2. James Hevia describes Qing patterns of sovereignty as negotiated hierarchical relations between Qing rulers and a "multitude of lords" rather than uniform territorial sovereignty. Significantly, Hostetler notes that one version of the Kangxi-era domain maps distinguishes between China proper (*neidi*) and the farthest extent of the Qing imperium, leaving ambiguous what would constitute "China" (*Qing Colonial Enterprise*, 75–76).

52. For one account of this process in Europe, see Sahlins, *Boundaries*.

officials and the British Burmese government negotiated from
scratch the Yunnan-Burma boundary, exemplifies the fluidity of
Qing/Chinese conceptions of territorial boundaries as late as the
early twentieth century.[53]

Moreover, pictorial maps with plentiful text, rather than to-scale,
mathematical mapping, continued to predominate in Qing China
well into the nineteenth century.[54] Such maps focused on character-
izing and establishing relations among political and physical places,
situating China at the cultural center, rather than marking out clear
boundaries that would constitute fixed territorial units.[55] Just as importantly, maps of the bounded imperial state such as the Kangxi at-
las were only for specialized use by Qing officials and to send clear
messages to foreign rulers, not for mass consumption. Thus, even if a
limited number of elite scholar-officials had conceived of the Qing
empire as a bounded territorial state, only circulation of definitive
and serialized national maps in popular media such as textbooks during the early twentieth century made the image of the geo-body rec-
ognizable to a Chinese mass public, and thus fundamental to Chinese national consciousness.[56]

In ways characteristic of nation-building projects elsewhere,
then, Republican-period Chinese popular geography first constituted China as a delimited and coherent national territory in
ideal terms through geographic and cartographic representations
rather than by mapping the chaotic existing geopolitical reality.[57]
Each Chinese geography textbook of this period began by presenting
descriptions of the Chinese national territory as an expansive, cohesive, bounded whole. Most often they took as the basis for the
contemporary national territory the most optimistic boundaries of
the Qing dynasty during its final decades and claimed all areas associated with the former empire—"the remaining vestiges of [the
Qing] family property," in Wang Zhongqi's words—as part of the
legitimate territory of the modern Chinese nation-state, with slight

---

53. McGrath, "A Warlord Frontier," 4–8.
54. Hostetler, *Qing Colonial Enterprise*, 16–17; Howland, *Borders of Chinese Civilization*, 188–92; Smith, "Mapping China's World," 71–73.
55. See, for instance, Smith, "Mapping China's World."
56. Peter Perdue makes a similar observation. See Perdue, "Boundaries," 286.
57. Thongchai, *Siam Mapped*, 130.

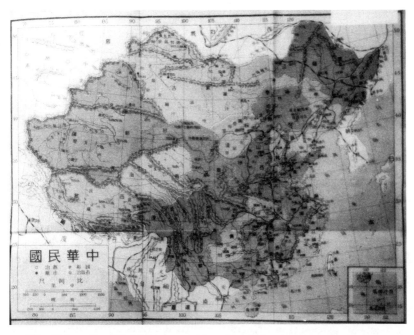

2.4 "The Chinese Republic." This map, with its bounded and cohesive territory, offers a canonical image of the Chinese national geo-body as it was represented during the Republican period. Wang Zhongqi, *Benguo dili*, 1 (1924): 52–53.

variations.[58] Textbooks staked out these expansive boundaries for Republican-period national states even though many areas they included had either been claimed by outside powers or had declared independence, such as Mongolia and Tibet, leaving a national territory that in fact was much smaller. (See Fig. 2.4.) The descriptions and maps of China's territory presented in Republican-period textbooks contributed to a normative conception of the national territory that underlies many Chinese people's mental maps of the Chinese nation-state to this day.

Both history and geography textbooks sought to give historical depth and continuity to these expansive national borders by projecting them back into the deep historical past. Chinese history textbooks identified the Qin (221–206 BCE) and Han (202 BCE–220 CE) dynasties as having set the first fixed national boundaries in the

---

58. Ge Suicheng, 1 (1933): 15–16; Liu Huru, 1 (1927): 6–8; Wang Zhongqi, 1 (1924): 50–53; Wang and Zhou, 1: 25–28; Xie Guan, *Benguo dili*, 1: 46.

ancient past, over two thousand years before. They then presented a teleological narrative of expansion of the legitimate boundaries of the ideal imperial state by describing the far-flung Central Asian empires of the Tang (618–906) and Yuan (1260–1368) dynasties, which had led to the full realization of the paradigmatic national territory under the Qing.[59] Lü Simian captured this teleological narrative in compressed form with the following statement: "In sum, when the Han and Tang were flourishing, they were able to encompass what today are Mongolia and Xinjiang. As for Tibet being part of China, that is a matter of the Yuan and Qing periods."[60] These historical accounts asserted that China had been a territorial state with fixed borders deep in the ancient past, and that those borders had been extended progressively by successive dynasties, leading to the continental-scale Qing territory. In this way, the paradigmatic Qing boundaries that served as imagined parameters of the current national territory were shown to be the organic outgrowth of a continuous and ancient history.

Some textbooks sought to naturalize the national boundaries still further by associating arbitrary political boundaries with defining natural features.[61] For instance, Ge Suicheng near the beginning of his 1933 Chinese geography textbook described the national territory as bounded by the sea, rivers, and mountain ranges that "naturally" separated it from other nations or colonies. He portrayed the Ussuri and Heilong Rivers, Altai Mountains, Congling Mountains, Himalayas, and the Pacific Ocean as shaping China's "natural" periphery. Ge's description was reinforced by an accompanying map, which showed primarily these features and a corresponding national boundary.[62] (See Fig. 2.5.) Similarly, Lü Simian's high school history textbook from 1934 drew geographical parameters for the national

---

59. Fu Weiping, 1: 87–90, 2: 308–9, 3: 37–39; Gu and Wang, 1 (1927): 62–63, 99, 163–68; 3 (1926): 4–5; Jin Zhaozi, *Chuji benguoshi*, 1: 46; 2: 57; Zheng Chang, *Xin zhonghua benguoshi*, 1 (1930): 4–5. For geography textbook narratives of territorial expansion, see Liu Huru, 1 (1927): 6–8; Wang Zhongqi, 1 (1923): 50–53.

60. Lü Simian, 1 (1934): 10.

61. Drawing correlations between arbitrary political boundaries and supposedly limiting natural features is a common strategy of national geography. See Lowenthal, "European and English Landscapes," 22.

62. Ge Suicheng, 1 (1933): 15. Also see Wang Zhongqi, 1 (1924): 58–59.

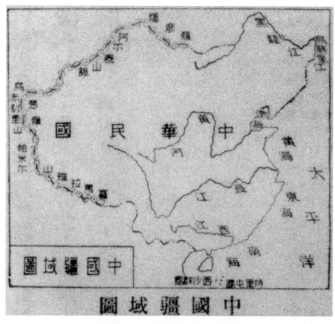

2.5 "Map of China's Territory." River systems, mountain ranges, and the sea serve as the bases for tracing the "natural" geopolitical borders of the Chinese nation-state. Yet borders facing French and British colonial Southeast Asia and the Mongolian steppe are sketched in despite the absence of iconic physical features. Ge Suicheng, *Chuzhong benguo dili*, 1 (1933): 15.

territory by naturalizing both the core from which it began and the extent to which it reached throughout history. The Ridge of Asia, he noted, split off East Asia from the rest of the continent, forming a "natural" unit. At its core was the Yellow River valley, where Chinese culture had supposedly started.[63] By drawing an original core and a limiting periphery, Lü set the stage for the teleological narrative of expansion from core to periphery that was recounted in his and other Republican-period history textbooks, and he established that periphery as the "natural" national boundary.

Parallel to the fixing of broad idealized boundaries for the nation was the assumption that the territory contained within those boundaries was unified and characterized by uniform sovereignty throughout. Many geography textbooks presented the political

---

63. Lü Simian, 1 (1934): 8–9.

nation in paradigmatic terms, with a functioning central government and a full complement of provinces and special administrative regions.[64] These descriptions of Republican political systems with a full array of administrative entities were corroborated by the geography textbooks' many maps that always presented China as a cohesive national whole. (See Figs. 2.4 and 2.6.) Yet all these textbooks were written and published during periods when there was no functional central government or when the government's control over many parts of the paradigmatic national territory was compromised or contested. They described, in other words, a political national space that did not exist in fact.

Textbooks reinforced the perception of a nation that was both bounded and coherent by using images and shapes that separated the national territory from surrounding areas to which it was connected and made it seem like a thing in itself. In many textbooks, the national map was presented in isolation, distinct from surrounding countries and territories. This effect was produced either by coloring the nation to distinguish it from surrounding areas or by presenting it as a cutout figure.[65] (See, for instance, Figs. 2.4, 2.5, 2.6, 2.8, 2.9, and 2.10.) This strategy of presentation identified the national body with an iconic shape that could be circulated as a sign for the nation.[66] Further, in some textbooks the national territory was described metaphorically as being like a "begonia leaf" or a "mulberry leaf."[67] In others, it was described abstractly as a triangle.[68] These metaphorical concretizations further helped to define the national territory as a fixed shape and an icon in ways parallel to other mod-

---

64. See Ge Suicheng, 4 (1934): 37–40; Li Tinghan, 1 (1914): 22–23, 71–72; Xie Guan, *Benguo dili*, 1: 48–50, 2: 148–51. Significantly, Wang Zhongqi's textbook, which was written during the period of China's greatest fragmentation in the 1920s, noted that Republican sovereignty over Tibet and Outer Mongolia was very uncertain and that provincial militarists had undermined fiscal centralization (1 [1924]: 189–95).

65. E.g., Xie Guan, *Benguo dili*, 1: 42–43; Zhang Qiyun, 1: 78–79; Zhong Yulong, *Benguoshi*, 3 (1914): 140.

66. Anderson, *Imagined Communities*, 175.

67. Liu Huru, 1 (1927): 1; Wang Zhongqi, 1 (1924): 58–59. The mulberry leaf image obviously would have resonated poetically with the vision of China as a silk-producing place.

68. Liu Huru, 1 (1927): 8–9; Xie Guan, *Benguo dili*, 1: 45.

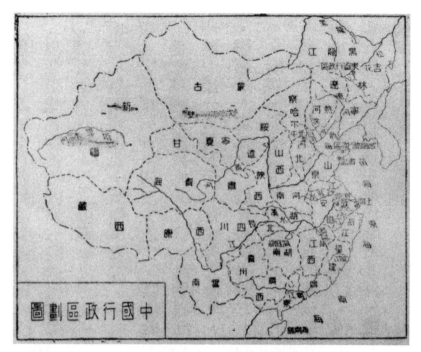

2.6 "Map of China's Administrative Districts." This political map presented a full array of provinces as part of a single geopolitical entity. Absent was any indication that border areas such as Tibet, Xinjiang, Mongolia, and Manchuria were not firmly under the control of the Chinese government or that China proper was experiencing some degree of internal political fragmentation. Ge Suicheng, *Chuzhong benguo dili*, 1 (1933): 17.

ern nations.[69] The circulation of national map as icon encouraged the sense that the national territorial body was something knowable, whole, indivisible, and tangibly real.

Geography and history textbooks' formulations of the national territory lent to the Chinese nation a level of coherence, unity, and continuity that was absent from those same textbooks' descriptions of the ethnically and culturally diverse national community. The fixed nature of the national boundaries provided an alternative, potentially pluralistic definition of the national community by identifying it with everyone who resided within those geopolitical borders. Textbooks' descriptions of unified sovereignty over all the provinces, administrative regions, and municipalities within those borders

---

69. The pre-eminent example is the French hexagon. See Smith, "The Idea of the French Hexagon."

created a sense of the nation's political cohesion, despite the ethnic and cultural complexity of its peoples. Moreover, the projected deep historical roots of those borders and a "Chinese" state's supposedly constant sovereignty over the territory they enclosed made the nation seem continuous and stable.

At the same time, textbooks' constructions of the nation as continuous, expansive, and unified provided the justification for projects of anticolonialism and consolidation of border regions. Because the textbooks identified the unified Chinese territorial body with the extensive imagined boundaries of the late Qing empire, places associated with that empire and later occupied by foreign powers or declared to be independent could only be imagined as "lost" territory that had to be "recovered" in order to reconstitute that idealized wholeness. Further, continued threats to border areas imagined as integral parts of the national whole evoked calls for the consolidation of Han Chinese state control over those areas to ensure the cohesion and security of the nation as a whole.

History and geography textbooks during the 1920s and 1930s helped to make visible the danger of the depletion of territory envisioned as part of an idealized national whole by circulating the so-called Map of China's National Humiliation (*Zhongguo guochi ditu*). The map outlined the paradigmatic Qing-era boundaries while shading in the surrounding territories that had been "lost" to the imperialist powers and dating the time of their loss.[70] (See Fig. 2.7.) The shading created a stark before-and-after effect, dramatizing the sense of the national territory's "shrinking." The dates on the "lost" territories established the recent and sudden nature of that shrinkage. By portraying the paradigmatic national territory as surrounded by shaded areas "lost" to outside powers, the map fostered anxiety about possible further territorial degradation and inspired concern for shoring up the current boundaries against it.

In general, Chinese history textbooks throughout the Republican period made the Qing government's surrender of territory, leaseholds, and sovereign rights one of the central narratives of their discussions of modern history. Likewise, Wang Zhongqi, in his 1924–25

---

70. See Jin Zhaozi, *Chuji benguoshi*, vol. 2; Zheng Chang, *Xin zhonghua benguoshi*, 2 (1931): 36–37; Zhu Yixin, 3: 63. The map was sometimes also included in geography textbooks as well. See, for example, Ge Suicheng, 3 (1934).

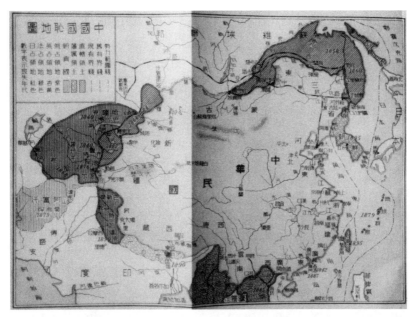

2.7 "Map of China's National Humiliation." Shaded areas surrounding China's current (imagined) bounded, sovereign territory fostered a sense of progressive deterioration and impinging threat, as dates documented the time of each territory's cession. The shaded areas also established a historical claim to surrounding areas that had supposedly once been included in a bounded national geo-body but had since been wrested away by imperialist powers. Left ambiguous was whether areas marked as previous "directly administered territory" (*zhixia lingtu*) or those labeled "tribute states" (*chaogong guo*) and "vassal state territory" (*fanshu lingtu*) were to be considered part of that historical claim. Ge Suicheng, *Chuzhong benguo dili*, vol. 3 supplement.

geography textbook, presented a powerful recounting of the disasters accompanying the Qing's encounters with foreign imperialism: "Since the loss in the Opium War, there was a precedent for ceding territory to other people, surrendering Hong Kong to England as thanks for making peace. From this point on the national situation was ruined and the territory contracted day by day."[71] Wang went on to detail the territory lost and concession areas surrendered with each defeat and treaty. The narrative of decline from an imagined past ideal was emphasized even more in Nanjing-decade geography textbooks, which after 1932 dedicated a chapter to recounting

---

71. Wang Zhongqi, 1 (1924): 51–52. The same paragraph was repeated in Liu Huru, 1 (1927): 7–8. Cf. Li Tinghan, 1 (1914): 20; Xie Guan, *Benguo dili*, 1: 46–47.

the "loss" of Korea, the Ryukyus, Taiwan, Annam, Thailand, and Burma.[72]

The rhetoric of loss fed directly into calls for mobilization to recover the nation's "original territory." For instance, in his early Republican textbook, Xie Guan translated China's geopolitical crisis into a call for civic action: "Thus, the four strong nations of Russia, Japan, England, and France are adjacent to our country's terrain. Since the opening of trade, the land we have lost has reached 1/5. . . . Recovering the old territory lies with the [nation's] citizens."[73] The positing of "old territory" from which pieces had been carved away created a rhetorical imperative for recovery. Such calls for recovery of a paradigmatic original territory escalated after the Mukden Incident in 1931, when Japanese forces occupied the three northeastern provinces and established the puppet state of Manchukuo. Nanjing-decade texts recounted Japan's occupation of the rich agricultural land and growing industrial sector of Manchuria in conjunction with statements about the need for the state and the people to struggle for its recovery.[74] Rhetoric and images that established ongoing threats to and transgressions of a pre-existing geopolitical totality provided a ready justification for the intense anti-imperialism of the Republican period.

Left ambiguous by these textbooks was whether the territory to be "recovered" and/or protected was to be the paradigmatic geo-body of the last decade of the Qing empire that was inherited by the Republic or the much larger territory associated with the Qing dynasty at its height. By including territories that the Qing had ceded in nineteenth-century treaties to Russia, and by representing former feudatories and tribute states such as Korea, Annam, Nepal, Burma, and Afghanistan as "lost" territories, some textbook narratives and the Map of China's National Humiliation opened the possibility of those areas being claimed by a Chinese state as having been part of the Qing "family property." At the very least, areas once considered "directly governed territory" (*zhixia lingtu*), such as Taiwan, parts

---

72. Ge Suicheng, 3 (1934): 85–113; Wang and Zhou, 4: 101–7.

73. Xie Guan, *Benguo dili*, 1: 47. Cf. Li Tinghan, 1 (1914): 20. Wang Zhongqi's textbook of the 1920s concentrated on recovery of treaty rights and leasehold areas (1 [1924]: 198–99; 2 [1925]: 29).

74. Ge Suicheng, 3 (1934): 1–33; Wang and Zhou, 3: 91–97.

of northern Manchuria, and portions of Turkic Central Asia, were implicitly included in the national geo-body, providing a justification for reclaiming these areas. Yet in plotting the paradigmatic national territory, most Republican-period geography textbooks, as we have seen (e.g., Figs. 2.4, 2.5, and 2.6), took the boundaries of the last decade of the Qing as standard, implicitly releasing national claim to these other areas. The two constructions of national territory encouraged two distinct forms of geo-nationalism: a defensive version that sought to consolidate the borders ultimately inherited from the Qing, and a more aggressive, expansionist vision that sought to reclaim China's place as a continental empire by absorbing territories under Qing influence at the dynasty's height.

Even when focused on the more modest version of the national geo-body, a narrative of anti-imperialist protection of an already assumed national territory provides an implicit justification for the state's consolidation of control over territories that are really of ambiguous sovereignty, as Thongchai Winichakul has astutely suggested.[75] Calls for vigilance regarding border defense in peripheries such as Xinjiang, Mongolia, and Tibet legitimated efforts to consolidate Han Chinese authority in those areas. Thus, Xie Guan first described British and Russian political and economic incursions into Tibet, Mongolia, and Xinjiang and then called for Han Chinese to build better transport into these areas and improve the administration in order to consolidate state control.[76] Similarly, Ge Suicheng used reference to British and Russian imperialist threats of interference in so-called border regions to justify calls for greater Han Chinese control.

The races (*zhongzu*) are complex, the border defenses are vacant, and the foreign powers wait on the side and watch for an opportunity, continually thinking to come and seize [China's western areas]. If one day there is an alarm, the danger is unimaginable. Consequently, today we should urgently promote the acculturation of the Mongols, Hui, and Tibetans so that they are not lured by the imperialists, move [Han] inhabitants to the border areas for colonization, do our utmost to develop barren land along the borders, and build more railroads for the transmission of news. Then the wilderness of the western areas will become a paradise of farming and herding, and the

---

75. Thongchai, *Siam Mapped*, 147–48.
76. Xie Guan, *Benguo dili*, 2: 74–75, 121–22, 135.

Hui and Tibetans will all become good citizens of the Republic. This will have the two benefits of promoting what is profitable and providing for the borders.[77]

Here, the evocation of the imperialist threat to China's supposedly bounded and cohesive national territory provided the justification for control over areas that some might have argued were not really parts of the Chinese state. Anti-imperialism to preserve an assumed national whole legitimized the state's control over all of its ostensible parts.

Anecdotal evidence suggests that many lower Yangzi region history and geography teachers effectively and powerfully translated into the classroom parallel images of the national territory as under threat and in need of defense. In interviews, alumni from schools across the region independently recalled their teachers' emphasis on the deterioration of China's borders and the corresponding need for national mobilization. According to a woman who studied at Shanghai's Wuben Girls' Middle School during the early 1930s, her elderly geography teacher frequently recounted the loss of China's coastal treaty ports, using maps similar in design to the Map of China's National Humiliation, to highlight what national territory had already been ceded. In addition to the two examples from Yangzhou Middle cited at the start of this chapter, another Yangzhou Middle alumnus from the years 1926–29 described history classes that emphasized Japanese aggressiveness toward China. A Hangzhou High School alumnus recalled that during the early 1930s geography teacher Wang Mengshou discussed territorial losses to Western imperialists during the Qing era in order to contextualize the current Japanese threat. A fellow student at Hangzhou High described history lectures during the mid-1930s that highlighted imperialist incursions and the need to resist Japan and apotheosized historical figures such as Yue Fei, who had resisted foreign invaders. History and geography teachers throughout the region echoed and emphasized dominant themes in mass-publication textbooks when they vividly portrayed the fragility of the nation's territorial integrity and the need for resistance to foreign invasion.[78]

---

77. Ge Suicheng, 3 (1934): 61; also see 56–57, 83.
78. In separate interviews, alumni from the late 1920s through the mid-1930s of Shanghai's private Minli and Nanyang middle schools and Suzhou Middle also

IV. THE PROJECT OF DEVELOPMENT

Geography and history textbooks together portrayed the national territory as bounded, unified, and historically continuous, thereby encouraging a sense of national cohesion and stability among secondary students. Yet geography textbooks also described how the national territory was internally differentiated on the basis of variations in topography and climate, which together formed natural regions with diverse modes of economic and social life. The textbooks structured the relations among the multiple regions of the Chinese nation in two distinct but overlapping ways. They described a functional complementarity between the nation's different parts, and they also ranked the regions along a developmental hierarchy. Euro-American nations set the basic standards for the scale of development in terms of industrialization, mechanized transport, and trade. Together these two modes of relating China's regions and connecting China to the world reinforced the perception of national territorial unity while at the same time charting a clear developmental trajectory that was to lead to ever greater integration, progress, and power.

Geography textbooks throughout the Republican period divided Chinese territory into numerous natural regions. Most textbooks organized these divisions around seemingly clear topographical markers such as river basins and plateaus.[79] Geographer Zhang Qiyun pushed this technique even further by offering a theoretical justification for dividing China into 23 natural regions.[80] (See Fig. 2.8.) In order to mitigate the possibility that regional differentiation could generate a sense of fragmentation, authors tried to establish the functional interdependence of the different regions. Wang Zhongqi at the start of his textbook made the case normatively that different natural environments produced distinctive resources and benefits: mountains provided timber and minerals; well-hydrated plains were ideal for agriculture; plateaus had grass for pastoralism; and oceans

---

emphasized how history and geography teachers frequently described the territorial incursions of imperialism and the threat from Japan.

79. Ge Suicheng; Xie Guan, *Benguo dili*; Wang Zhongqi.

80. Zhang Qiyun, 1: 76–79. Zhang explicitly followed Weng Wenhao's pioneering efforts in constructing a region-sensitive geography for China. See Zhang Qiyun, 1: Zixu, 3.

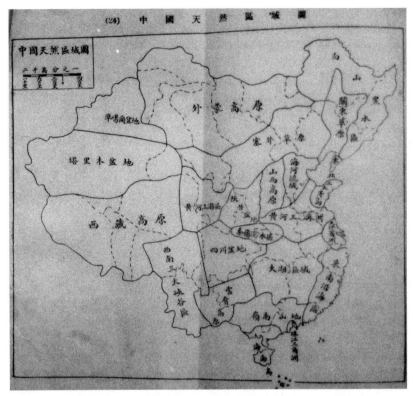

2.8 "Map of China's Natural Regions." Leading young geographer Zhang Qiyun's map distinguishing 23 natural regions threatened to exacerbate a sense of territorial fragmentation. But his and other geography textbook accounts emphasizing the economic interdependence of the various regions served to reestablish a sense of unity and cohesion. Zhang Qiyun, *Benguo dili*, 1 (1928): fig. 24.

sustained fish and facilitated transportation and trade.[81] The diversity of benefits offered by distinctive environments was then represented in his more focused discussions of China's regions. When compared side by side, their complementarity became clear, as in Zhang Qiyun's chart describing China's natural regions.[82] The Yangzi Delta was an exceptionally rich area for farming, but it lacked minerals. By contrast, the Shanxi plateau was rich in coal deposits, but it suffered from dry

---

81. Wang Zhongqi, 1 (1924): 6–12.

82. Zhang Qiyun, 1: 79–90. For similar, if less elaborated, interregional comparisons, see Ge Suicheng, 1 (1933): 18; Li Tinghan, 1 (1914): 23–25; Liu Huru, 1: 133–47; Xie Guan, *Benguo dili*, 1: 77–88.

conditions for farming. Northern Manchuria had extensive forests, and the Mongolian plateau had broad grasslands for herding, but both areas were short of suitable farmland. The textbooks demonstrated that China's interregional diversity generated a complex functional economic unity; each area was an essential part of the nation because of the products it could provide.[83]

However, all regions were not considered to be of equal value. Rather, geography textbooks organized regions hierarchically along the lines of "civilization" (*wenming*) and "development" (*fazhan*). Different modes of economic organization and production were ranked as more or less advanced on the basis of an evolutionary view of world history, which became hegemonic during the late Qing and Republican periods.[84] This historical meta-narrative, which was incorporated into Republican-period secondary-level world history textbooks, portrayed nomadic pastoral, agricultural, and modern industrial economies as successive stages of development.[85] This account of linear progress from one mode of production to another translated easily into spatial hierarchies both within and beyond China's national space. In a recent essay, Thongchai Winichakul demonstrates how Thai elites at the end of the nineteenth century imagined a hierarchy of civilization (*siwilai* in Thai) that spanned from the rural and mountainous interior of Thailand itself to the industrialized societies of the modern West.[86] Republican-period Chinese geography textbooks constructed a similar hierarchy.

Geography textbook discussions of China's regions clearly differentiated between more developed coastal and lowland areas and less developed inland regions.[87] Areas such as the middle and lower

---

83. This stress on unity through socioeconomic complementarity closely paralleled that of other national geographies. See Claval, "From Michelet to Braudel," 48–52.

84. Dirlik, *Revolution and History*; Duara, *Rescuing History*, chap. 1; Pusey, *China and Charles Darwin*; Schwartz, *In Search of Wealth and Power*.

85. E.g., Chen Hengzhe, *Xiyangshi*, 1: 10–29; 2: 286–96; Fu Yunsen, *Lishi jiaokeshu*, 1 (1924): 39–72; Jin Zhaozi, *Chuji shijieshi*, 2: 107–9; idem, *Xin zhonghua waiguo shi*; Li Jigu, *Li shi chuzhong waiguoshi*.

86. Thongchai, "The Quest for 'Siwilai.'"

87. See *passim* in the gazetteer (*difangzhi*) sections of the following textbooks: Ge Suicheng; Li Tinghan; Liu Huru; Wang Zhongqi; Wang and Zhou; Xie Guan, *Benguo dili*; Zhang Qiyun.

Yangzi regions along with the Pearl River Delta and Manchuria were sometimes called the "essence (*jinghua*) of the nation." Many textbooks described them as fertile alluvial plains that were interlaced with waterways that facilitated transportation and identified them as China's most developed commercial and industrial areas. By contrast, mountainous inland areas and the high, dry plateaus of North and Northwestern China were portrayed as lacking the proper conditions for agriculture because of insufficient water and/or infertile land. They also had limited trade and industry because of impenetrable mountains (west and southwest) or inadequate infrastructure (north and northwest).

By identifying nomadic pastoralism as the main occupations in many of these inland areas, or stating that people there "half farmed and half herded" for their livelihoods, the textbooks clearly marked the evolutionary distance between the rural interior and the coastal or river plains. Some textbooks explicitly identified the lifestyles of people in these areas as "backward."[88] But often "backwardness" was signaled in more subtle ways, such as with the observation that Mongols relied on their animals for subsistence and lived a mobile existence, pitching their tents wherever there was grass and water. Mode of production served to index a people's stage of development, as with Ge Suicheng's comment that "in Xikang, they have still not broken away from the lifestyle of the nomadic age (*youmu shidai*) of prehistoric times (*shanggu*)."[89] Similarly, maps and charts of China's existing or planned railroads illustrated the concentration of lines in China's east, central, and northeastern parts and served as a verification of the different levels of development between coastal and inland areas.[90] (See Fig. 2.9.) Further, use of statistics to document what portion of national production a given province or area generated dramatized differences in interregional economic development.[91]

At the same time, geography textbooks signaled to students in various ways China's overall lack of development vis-à-vis Euro-

---

88. For explicit references to "backwardness," see Wang and Zhou, 4: 5, 112.

89. Ge Suicheng, 1 (1933): 88.

90. Li Tinghan, 1 (1914): 84–86; Liu Huru, 1: 161–67; Wang Zhongqi, 1 (1924): 166–75; Wang and Zhou, 4: 152–54.

91. Wang and Zhou, *passim*.

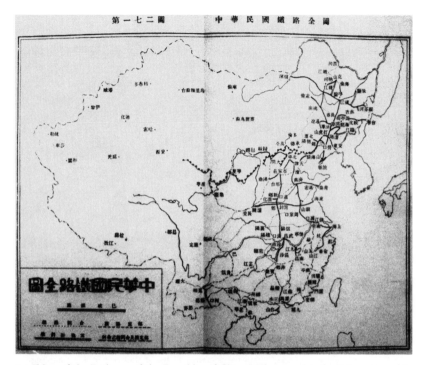

2.9 "Map of the Railways of the Republic of China." The completed rail lines, traced in the dark, solid lines, cover only a small part of the country and cluster in the eastern third of the territory. Even the projected lines, which are differentiated in type and stage of planning by the kinds of lines marking them, crowd eastern and southern areas, leaving untouched most of the West and Northwest. This spatial distribution of rail lines captured the relative levels of development between coast and interior, revealed the nation's lack of territorial integration, and suggested China's overall lack of economic development. Wang and Zhou, *Benguo dili*, 4: 152–53.

American countries and Japan. For instance, many textbooks pointed out that China imported manufactured goods while exporting primarily raw materials.[92] Wang and Zhou further noted that the length of China's rail lines in 1935 amounted to just one-fortieth of the rail lines in the United States.[93] Such comparisons in geography textbooks between China and "developed" countries reinforced the message that the Euro-American powers had developed first, setting a trend for the rest of the world.

---

92. Li Tinghan, 1 (1914): 79; Liu Huru, 1: 144–47, 157–58; Wang Zhongqi, 1 (1924): 206–13; Xie Guan, *Benguo dili*, 2: 170–78.

93. Wang and Zhou, 4: 152.

2.10 "Map of the Party Leader's Plan for 100,000 Miles of Railroad." This cartographic visualization of Sun Yat-sen's visionary plan of extensive infrastructure development contrasted starkly with the previous map of existing rail lines and those in the planning stages (Fig. 2.9). This map projected into the future a national geo-body that was to be both technologically modernized and spatially integrated in unprecedented ways. Wang and Zhou, *Benguo dili*, 4: 156–57.

The hierarchy of relative levels of civilization outlined in textbooks fed into an imperative for economic development that was articulated in terms of taking advantage of China's unused economic potential. Textbooks detailed the untapped resources in each kind of natural environment. The authors then described how the potential of these areas would be "developed" (*fazhan, kaipi, tuozhi*) as a threefold process of extending modern transportation into those areas, opening them to colonization by (Han) migrants from the nation's overpopulated areas, and using new techniques to exploit the resources available.[94] Some Nanjing-decade textbooks, such as the one by Wang and Zhou, outlined projects that were based directly

---

94. Ge Suicheng, 1 (1933): 18; Li Tinghan, 1 (1914): 23–25, 52–57; Liu Huru, 1: 133–47; Wang Zhongqi, 2 (1925): 83–86; Xie Guan, *Benguo dili*, 1: 77–88, 138–39.

on Sun Yat-sen's visionary plans for development. These focused specifically on transportation infrastructure such as seaports and railways.[95] (See Fig. 2.10.) The developmental perspective of all these textbooks was captured in Wang and Zhou's closing comment on China's Western frontier areas: "In sum, the Western Region today is still a great piece of undeveloped, virgin land. The vast wilderness can accommodate migrants. Rich minerals anxiously await development. Fertile land is sufficient for farming and pastoralism. Now, is not the development of the Western Region something for which all of us should strive?"[96] In making such a strong case for China's need for economic growth, geography textbooks encouraged students to think of themselves, and all other citizens, as potential contributors to national development.

V. IMAGINING CHINA

History and geography textbooks and instruction all taught national consciousness, but we have seen that they did so in diverse, and sometimes contradictory, ways, variously using myths of racial origin, narratives of cultural assimilation, and genealogies of territorial cohesion as the basis for portrayals of national unity. The importance in history and geography textbooks of territorial or geopolitical national images highlights a neglected area in the existing literature on nationalism in late Qing and Republican China, much of which has focused on racial and/or cultural definitions of national community.[97]

---

95. Wang and Zhou, 2: 76–78; 3: 6–8, 93; 4: 3–4. Ge Suicheng's 1933–34 textbook similarly pointed out the need for development or sketched out brief development schemas for nearly every region of the country (2 [1933]: 4–5, 29; 3 [1934]: 27, 61, 81, 83).

96. Wang and Zhou, 4: 61.

97. E.g., Dikötter, *The Discourse of Race*; Duara, *Rescuing History*; Gladney, *Muslim Chinese*, 82–87. But see also James Townsend, who in "Chinese Nationalism" describes a tension between what he calls ethnic and state nationalisms in twentieth-century China. In distinguishing between territorial and racial or cultural definitions of the nation, I mean to suggest how they *may* operate independently as discrete techniques for making the nation imaginable. But they can also be articulated together to give a definite cultural content to a territorially constituted geo-body, as in the previous section, when Republican geography textbooks' calls to consolidate border regions entailed culturally Sinicizing them to make them "civilized."

Race and culture were undeniably important markers of national identity in Republican China and deserve the attention they have received, but the textbooks reviewed in this chapter suggest that the image of the geo-body was equally important. In fact, because a territorial conception of the nation could accommodate or elide post-imperial China's undeniable ethnic, cultural, and linguistic diversity, it might have been rhetorically the most powerful and conceptually the most stable of national visions for twentieth-century China.

When juxtaposed in textbooks and in the classroom, these varied portrayals of the national community provided students with a menu of options from which they could develop their own images of the nation. Previous scholarship has established that nationalism was part of the collective worldview of many modern Chinese students, who persistently joined in nationalist protest movements throughout the Republican period.[98] At the same time, it is difficult to grasp exactly how educated Chinese youths were imagining the nation, even within the relatively circumscribed boundaries of the lower Yangzi region. Writings drawn from student-run journals suggest that they invoked different tropes to represent the nation in different contexts and moments but that territorial conceptions of the nation were consistently used by at least some students throughout this period.

During times of national crisis, some students certainly portrayed the nation in ethno-cultural terms. For instance, in May 1919 a student at Jiangsu Second Girls' Normal School argued that Qingdao had to be protected from being taken over by the Japanese, in part, because "Shandong is our culture's birthplace (*wo wenhua faxiang zhi di*). It is the place where Confucius' and Mencius' graves are, and *the spirits of all the nation's citizens (quan guomin jingshen) are attached to it.*"[99] The nation's citizens, here, were all assumed to share in a common and continuous Han culture. Similarly, a student at Shanghai Municipal Jingye Middle School, writing on the very eve of the Marco Polo Bridge Incident in 1937, characterized traitors as those who

---

98. E.g., Lang, *Chinese Family*, 269–323. Olga Lang's sample included students attending, in 1936, 22 colleges and eight high schools in Beiping, Tianjin, Taiyuan, Ji'nan, Shanghai, Hankou, Changsha, Fuzhou, and Canton. Cf. Israel, *Student Nationalism*; Wasserstrom, *Student Protests*.

99. Zhu Yunyu, "Qingdao cunwang yu quanguo zhi guanxi." Emphasis added. Significantly, Zhu also stressed Shandong's economic and geopolitical importance as part of China's territory.

undermined the unified body of the national people and denied their Han identity by "praising other peoples (*yizu*) as emperor" and adopting another culture.[100] This student author constituted the nation as an internally unified ethno-cultural community by drawing a sharp dichotomy between "Han" and "barbarian."

However, other essays drawn from student-run journals reveal that territorial nationalism was an important part of lower Yangzi region students' repertoire of nationalist discourse. These writings suggest that in Republican China the concept of the geo-body functioned as a dynamic alternative to racial and cultural approaches to nationalism. Significantly, the territorial conception of the nation presented in Republican-period student writings often incorporated images and phrasing characteristic of standard textbook formulations of the nation as geo-body. For instance, Qiang Yuanben, a student at Jingye Middle School in 1937, encapsulated the idea of the territorial nation vividly at the start of an essay on concession areas. Qiang began with the familiar image of a national map, brightly colored except for the discolored corner of the Northeast and the concession areas that he described as holes "nibbled away" from "mulberry leaf–shaped China."[101] China, here, was presented in its iconic and unified geopolitical form, a single territory threatened at its margins. Later in the essay Qiang's further evocation of the nation as a unified body made clear that he saw "the integrity (*wanzheng*) of our sovereign territory" as fundamental to his imagining of the nation. Qiang's focus on the nation as a territorial unit was echoed in numerous student essays from the 1910s through the 1930s, as youths outlined the threat of imperialist incursions.[102] In all of these accounts, students identified the nation, implicitly or explicitly, as a territorial unit that was clearly bounded and cohesive. Further, they argued that various frontier areas, though they might seem distant to their fellow students in the lower Yangzi region, were essential to the survival of the nation as a whole.

---

100. Liang, "Hanjian lun."

101. Qiang Yuanben, "Zujie zai zhongguo."

102. E.g., Chu Jinke, "Yu youren lun zhong ri jiaoshe shu," 3–5; Ni Qikun, "Huabei de weiji"; Wang Ti'an, "Zhong ri tixi sheng zhong de huabei wenti," 1–4; Yu Hetong, "Huabei teshuhua yu zhong ri gongtong fanggong liang da wenti"; [Zhang] Yuanxin, "Nei Meng wenti"; Yuan Yi, "Zhong ri dili shang zhi jiaoshe."

As if echoing the content of history and geography textbooks and lectures, students in these essays vividly portrayed the dangers of imperialist territorial encroachment and moved directly to calls for defense and development, especially in border areas. This rhetoric of mobilization for national defense was captured pithily in the words of one of Qiang's fellow students at Jingye Middle: "Japan's invasion of North China is a simultaneous political, economic, and military advance. The status of North China in reality is already 'specialized' (*teshuhua*), [but] today Japan still is calling for 'making North China special,' [demanding] change after change with the ambition not to stop until North China has become 'Manchukuoized' (*Manzhouhua*). In order to defend North China, we must have 400 million people unified as one and cooperating with the government to declare war on the enemy."[103] Significantly, Ni Qikun, here, cast the nation as an undifferentiated community of 400 million people, glossing over the ethno-cultural differences portrayed in geography textbooks. Such eliding of cultural diversity within the national population was facilitated by a focus on the nation as a cohesive and bounded territorial entity.[104]

The idea of nation as geo-body, then, was one way in which lower Yangzi region students could imagine China, and these essays, though limited in number and scope, suggest that at least some students were using it in their nationalist discourse. Student adoption of a territorial conception of the nation should perhaps not surprise us, since Japanese incursions throughout the Republican period continually reinforced the sense that the national geo-body was under threat. Japan's claim to German concessions in Shandong during the Paris Peace Conference after World War I, its occupation of Manchuria in the fall of 1931, and its forays into North China during the rest of the 1930s repeatedly brought to the fore issues of fixed borders and secure territory. Japan's invasion of Shanghai in the winter and spring of 1932 made the foreign threat to Chinese territory seem very immediate to lower Yangzi region students, many of whom took part in relief efforts and saw classes suspended for weeks or

---

103. Ni Qikun, 32.

104. Cf. Wang Ti'an; [Zhang] Yuanxin, 18. Both essays describe Inner Mongolia as integral to China's national territory, but they downplay the ethno-cultural distinctiveness of the Mongol people.

even months.[105] As teachers throughout the region repeatedly called attention to these foreign incursions they strengthened students' consciousness of the nation as a territorial body. In writing about these imperialist actions, the student authors discussed here referred back to their geography textbooks' images and maps, with all their associations with the Qing empire's historical legacy, to ground territorially based conceptions of the nation.

VI. CONCLUSION

Whether Republican-period students focused on the racial, ethnocultural, or territorial images of the nation in their history and geography textbooks, they were presented with a vision of the nation as threatened and incomplete. Geography textbooks carefully documented the many linguistic, cultural, and phenotypic differences among China's ethnic groups, creating a picture of ethnic and cultural pluralism that they argued could only be transformed into a single coherent community by efforts at "national unification" (*minzu tuanjie*) on the foundation of a standardized Han culture. At the same time, geography and history textbooks that portrayed the nation as a bounded, unified, and sovereign geo-body also described military and economic incursions from imperialism that threatened China's territorial integrity. Further, representations of internal socioeconomic diversity and the ranking of regions along a global hierarchy of development demonstrated that large parts of China stood below the standards of "modernity." Through these contrasting representations, geography and history textbooks and teachers suggested that China's viability as a modern nation-state depended on defense to secure the borders and development to achieve economic modernization. Even as history and geography textbooks offered ideal images of a unified national community and a bounded, stable national territory, they also delineated threats to the coherence of the national people and geo-body that served as the basis of rallying cries for action by citizens to realize the nation.

In this way, Republican-period history and geography textbooks' narratives and descriptions not only worked to cultivate nationalist

---

105. E.g., "Ben xiao dashiji," *Songjiang nüzhong xiaokan*, no. 28 (May 25, 1932): 2–3; Jiangsu shengli Suzhou zhongxue, ed., *Xuesheng zizhihui gaikuang*, 3; *Minli zhongxue sanshizhou jiniankan*, 20.

consciousness by presenting seamless representations of the Chinese national community that attempted to cover over internal tensions or gaps. In addition, these texts incorporated a form of nationalist discourse in which heightened emphasis on unevenness and contradiction became the basis for calls to struggle and civic action. In ways parallel to Chinese nationalist discourse at the turn of the twentieth century, the threat of becoming a "fallen country" (*wang-guo*) underpinned popular mobilization for an activist politics that figured "the people" as a nascent political agent.[106] At the same time, by un-suturing—or at the very least problematizing—the ethnically constituted Han national subject that had been formulated by revolutionaries in the years just prior to the 1911 Revolution, Republican-period nationalist education created the imperative for "the people" to be constituted through political action.[107] Rather than being primordial and given, the nation's people, history and geography textbooks suggested, had to be made through the efforts of its citizens, students among them. This discourse of China as a nation-in-the-making raised the pressing question of what kinds of action would produce the national community and thus characterize the true modern Chinese citizen.

---

106. Karl, *Staging the World*, 15–17, 35–38, 79–80, 90–93, 99–101.

107. Rebecca Karl demonstrates how late Qing revolutionaries' focus on the Han ethnic subject fostered the anti-Manchuism that underpinned the anti-Qing revolutionary project. But she also argues that the "narrowed recentering of 'Chinese-ness' around ethnicity, which allowed for 'the people' to be specified not through either political consciousness or revolutionary activism but through an essentialized ethno-racial designation . . . relieved Chinese revolutionaries of the responsibility for politically mobilizing 'the people' because they could claim a naturalized anti-Manchuism as the basis for revolution" (ibid., 118). The Republican-period de-centering of the ethnic subject that I have documented here created an imperative to mobilize nascent citizens to create a "people" that was not yet imagined to exist as a continuous and cohesive community.

3

Practicing Politics:

Student Self-government and

Civic Republicanism

After the 1911 Revolution, our nation's autocratic political system changed into a republican political system, and the monarchic national polity (*junzhu guoti*) became a democratic polity. However, democratic republican nations completely take the people to be their basis. If we look at our nation's current situation—whether in terms of politics, law, or society—as always it is a minority that controls a majority; the majority of citizens still occupy a position of subjugation. The [people's] title of our nation's basis was long ago stripped away without leaving a trace. How do you think this situation of having a democratic republic "in form but not substance" was created? Needless to say, it is because our citizens have no ability for self-government. . . . This is . . . [a] reason justifying why we should have a [student] self-government organization from the aspect of our political system and national polity.[1]

Secondary-level history and geography instruction encouraged Republican-period students to think of themselves as active citizens who would contribute to creating a modern nation. But as this quote from students at Hangzhou's Zhejiang Provincial First Normal School suggests, many students and intellectuals felt that the new republican institutions formed after 1911 had failed to generate true participatory government. A small group of corrupt elites controlled China's government, and the vast majority of citizens were not

---

1. Zhejiang shengli diyi shifan xuexiao xuesheng zizhi hui, *Zizhi hui chengli jiniance*, Xuanyan: 2–3.

politically involved. In this passage, Zhejiang First Normal's students identified Chinese citizens' incapacity for self-government (*zizhi*) as the root cause of the failure of the Republic. In response, they viewed the organization of self-government associations in their local school as one way to make China a true democratic republic by instituting democracy at the local level and training themselves for participatory politics as adult citizens. Correspondingly, in November 1919, they formed a student self-government association at their school and began enacting their own version of democratic politics.[2] They were not alone. From the late 1910s through the early 1920s, students at schools across the lower Yangzi region formed student self-government associations with the goal of preparing themselves to become republican citizens. These organizations remained a central part of secondary school life throughout the 1920s and 1930s, constituting a prime arena for students to practice politics.

Students in self-government organizations throughout these two decades performed an active mode of civic republican politics. This republicanism entailed direct participation in the management of their communities' affairs and cultivation of civic virtues that emphasized individual self-control and sacrifice for community welfare. Disparate intellectual influences and the living model of local elite management in lower Yangzi cities and towns all encouraged students and educators to adopt a civic republican rather than a liberal, Anglo-American model of democratic practice for self-government in their schools.

Within the overarching continuity of civic republican modes of student self-government during the Republican period, both schoolhouse politics and broader political trends caused shifts in the organization and aims of student government groups. Influenced by the mass nationalist politics of the late 1910s and early 1920s, some students and educators sought to integrate radical forms of direct democracy into self-government organizations. Moreover, student groups at a number of schools claimed for their organizations a great degree of autonomy from school authorities. After 1927, by contrast, the Nationalist Party sought to co-opt student self-government associations to further the party's goals of managed popular mobilization for national construction (*jianguo*). It imposed centralized

2. Shen Ziqiang et al., eds., *Zhejiang yishi fengchao*, 118–20, 171.

oversight on student associations and dictated their aims, activities, and organizational structure, claiming that the party was the sole source of republican virtue and the arbiter of the general will.

I. THE RES PUBLICA AND
THE PUBLIC SPHERE

Zhejiang First Normal students' call to build Chinese democracy by training people in self-government merged smoothly with the mainstream of cultural reform discourse in China during the 1910s. New Culture Movement thinkers such as Chen Duxiu promoted a broad conception of democracy that encompassed political participation and popular rights as well as social and economic democracy.[3] Reacting like Zhejiang First Normal's students to the post-1911 collapse of republican institutions that had been imposed by constitutional fiat on an unprepared Chinese populace, Chen, in an essay of December 1919, called for building democracy in China from the roots of local society up. "The foundation [for democracy]," he declared, "must consist of small units built by the people themselves."[4]

By building up from "small units," Chen meant that local communities and professional organizations could serve as sites to develop institutions and patterns of practice that would form the basis for democratic social, economic, and political systems in the nation at large. Local organizations would form the "cells" (*xibao*) and "organs" (*qiguan*) of the democratic political "organism" (*shengwuti*) of the nation.[5] The trade associations (*tongye lianhe*) and local self-government groups (organized at the village or city-block level, not at the county, city, or provincial level) that Chen envisioned would operate by a process of direct democracy. All members would have a direct say in decisions, officers would be popularly elected and rotated regularly, and the groups would deal with questions related to their members' practical concerns. Invoking John Dewey, Chen asserted that the practice of participatory politics at this level would be an education for democratic politics at higher levels.[6]

---

3. Chen Duxiu, "Shixing minzhi de jichu," 2: 28–39; Chow, *The May 4th Movement*, 230–32.

4. Quoted in Chow, *The May 4th Movement*, 231.

5. Chen Duxiu, 2: 32.

6. Ibid., 32–37.

Echoing Chen was a generation of educators who believed that student self-government associations, as democratic polities in miniature, would give students valuable opportunities to practice being active republican citizens. A guide to student self-government based in part on an organizational outline by the influential Zhejiang and Jiangsu Provincial Education Associations, represented in capsule form the feelings of many educators.

Today's students are tomorrow's citizens. The knowledge required by tomorrow's citizens should be nurtured in today's schools. Now the citizens of an autocratic nation need only the habits of being ruled (*bei zhi zhi xiguan*), but the citizens of a republic need the ability for self-government. Since our nation is a republic, it should have citizens with the capacity to rule themselves. And, the production of citizens with self-government ability must start with students' practicing self-government.[7]

A spectrum of educational reformers viewed self-government associations as mechanisms for training the new citizens of Republican China.[8] Likewise, many lower Yangzi region students themselves, in the late 1910s and early 1920s, called for new modes of political participation and training in their schools in preparation for being Republican citizens.[9] In the words of the student self-government organizers at Hangzhou's Zhejiang Provincial First Middle School:

Any republic requires that its citizens have the ability to govern themselves, so how is it that our Republic still wants to nurture citizens who are ruled by others? . . . Since we must be future citizens with the capacity for self-government, then we must exercise our ability for communal self-government . . . [and] practice it during the time we are students, otherwise we fear that in the future we will not be complete republican citizens.[10]

Because they viewed self-government organizations as training grounds for a new democratic politics in China, educators often imagined student self-government organizations as institutions of

---

7. Rui Jiarui, *Xuesheng zizhi xuzhi*, 1.
8. Muxin, "Jiaoyu yu demokelaxi," 1; Tao Zhixing, "Xuesheng zizhi wenti zhi yanjiu"; Tianmin, trans., "Demokelaxi yu xuexiao guanli," 2–4; Yang Xianjiang, "Xuesheng zizhi heyi biyao."
9. Chu Jinke, "Shu benxiao zizhihui zhi dazhi," 5–7; Zhejiang shengli diyi shifan xuexiao xuesheng zizhi hui, *Zizhi hui chengli jiniance*, Xuanyan: 2–3.
10. "Zhejiang diyi zhongxue xuesheng zizhihui chengli xuanyan." Cf. "*Zhejiang diyi zhongxuexiao xuesheng zizhihui banyuekan* fakanci," 86–88.

democratic government in miniature. One author took U.S. schools as a model for self-government and described them, in Dewey-like terms, as being microcosmic societies with institutions for legislation along with police and judges to enforce the laws.[11] Others described self-government organizations premised on the separation of legislative, executive, and judicial powers, or viewed local municipal and township self-government organizations in China as viable institutional models.[12]

Within these organizations, educators envisioned students managing daily life within their own communities and regulating themselves. Liberal educator Jiang Menglin identified self-government, in part, with the "responsibility for public service."[13] Yang Xianjiang, who later joined the Chinese Communist Party, characterized self-government work in the following terms: "Because student self-government necessarily includes all kinds of self-government enterprises and people to take them up, and [because] all rules and responsibilities are decided and carried out by the students themselves, it will certainly deeply affect their ideas about taking on duties, and we can hope that it will slowly transform their self-involved thoughts (*dandu weiji de sixiang*)."[14] By "taking on duties" Yang meant both undertaking practical tasks of management and enforcing and following the rules that ordered the student community.

Educators believed that students, like republican citizens anywhere, would effectively manage their communities only if they embodied and enacted civic mores. Yet many educators also viewed self-government as a way to cultivate civic morality through civic action. In the words of Rui Jiarui, "the definition of student self-government is one of ethics and feelings (*ganqing*) not of law; it is communal (*gongtong de*) and not individual; it is of duties, not of rights. Student self-government is not unrestricted activity (*ziyou xingdong*) but rather collective governance (*gongtong zhili*); it is not abandoning rules but rather establishing and following the laws oneself."[15] The orderly management of the community, in other words,

---

11. Muxin, 4.

12. Rui Jiarui, *Xuesheng zizhi xuzhi*, "Bianji dayi," 13; [Wu] Yanyin, "Zhuchi zhongdeng jiaoyu zhe jinhou zhi juewu," 4.

13. Jiang Menglin, "Xuesheng zizhi," 119. Cf. Tianmin, 2–3.

14. Yang Xianjiang. Also see Muxin, 4; Rui Jiarui, 4; [Wu] Yanyin, 4.

15. Rui Jiarui, 2–3.

was seen to depend on students' adopting the proper moral attitude of placing the welfare of the community first and fulfilling their duties to the group instead of asserting their rights and individual freedoms. At the same time, educators such as the Columbia University–trained Tao Xingzhi believed that active participation in self-government would reinforce and inscribe in students' thoughts and actions moral patterns that otherwise would remain abstract. In particular, he argued that participation in self-government would encourage students to emphasize the public welfare, take on group duties, and judge right and wrong in a public context.[16]

Leading educators believed that ethical commitment to the public welfare and the willingness to follow the rules that maintained collective order could be cultivated through intensive enforcement of rules, laws, and moral expectations by the students themselves. Tao asserted, for instance, that rules would be more effective if students made them themselves, because then the rules would be "even more internalized" (*gengjia shenru renxin*) and "always before [students'] hearts and eyes" (*xin mu zhong dou you ta zai*). Similarly, Tao argued that a minority of teachers and administrators could not possibly keep the whole mass of students under constant surveillance and manage student behavior, but that if students had collective responsibility they would themselves follow and enforce the rules.[17] Through management of their own self-regulated community, students would learn that collective order and individual and group freedom were dependent on self-discipline and regard for the rules. In the words of Yang Xianjiang, "students who undertake self-government are truly not in an easy position, [for they] must discipline (*molian*) and exert themselves, and only then can the name and substance [of self-government] correspond."[18]

Theorizing Civic Republicanism

The patterns of political participation these educators outlined in 1919 and the early 1920s paralleled in many ways modern Western notions of civic republicanism that were expressed first by Machia-

---

16. Tao Zhixing. Cf. Jiang Menglin, "Xuesheng zizhi," 120; Tianmin; Yang Xianjiang.
17. Tao Zhixing.
18. Yang Xianjiang.

velli and later elaborated by Rousseau, Tocqueville, and Anglo-American political theorists.[19] Civic republicanism stresses the citizen's full and active participation in the political community. Machiavelli used the example of citizens participating in a militia to defend their city as a literal and figurative example of how the citizen should take a direct role in the functions necessary to sustain community life. Citizenship in this mode is a form of active practice through direct participation in tasks vital to the community that rests on the moral commitment of the individual to regulate himself and choose to serve the community, placing the collective interest above his own. At the same time, only active involvement in civic affairs could generate individuals who were oriented toward the welfare of the community as a whole by cultivating in them the habits and outlook associated with group life.

Historically, civic republicanism, as a mode of political participation, has provided an alternative version of democratic politics to the practice of classical liberalism.[20] Chantal Mouffe, for instance, juxtaposes liberalism, in which "citizens are seen as using their rights to promote their self-interest within certain constraints imposed by the exigency to respect the rights of others," with "the civic republican view of politics . . . [that] puts a strong emphasis on the notion of a public good, prior to and independent of individual desires and interests."[21] The contrast between these two dynamics is sustained when one considers the normative formulation of bourgeois liberal democracy articulated by Habermas in his history of the public sphere of civil society. Habermas defines the citizen as an autonomous individual with private interests and perspectives that grow out of his, or potentially her, control and management of property. He argues that the free expression of these interests and perspectives in the public sphere of civil society, which in Europe and the United States became formalized in representative political institutions, can

---

19. On civic republicanism, see Miller, *Citizenship and National Identity*; Mouffe, ed., *Dimensions of Radical Democracy*; Oldfield, *Citizenship and Community*; Pocock, *The Machiavellian Moment*; idem, "The Ideal of Citizenship."

20. Pocock, *The Machiavellian Moment*, chap. 14. Cf. Pocock, "The Ideal of Citizenship."

21. Mouffe, "Democratic Citizenship," 226. Cf. Oldfield, chaps. 1, 2.

generate rational decisions about public affairs while preserving the freedom of the individual.[22]

In the Western Sinological literature, much research on the changing form of Chinese politics in the nineteenth and twentieth centuries has centered on the extent to which Chinese intellectuals and social elites theorized and practiced forms of democratic politics associated with Anglo-American liberalism. For instance, some have attempted to identify urban elite forms of voluntary association as an emergent public sphere.[23] Further, studies of "progressive" intellectuals such as Hu Shi have focused on the extent to which their thought faithfully reproduced the concern for autonomous individualism and personal rights that was characteristic of the Anglo-American liberal tradition.[24]

By shifting focus to the modes of political participation that elite intellectuals and educators taught and that students enacted in schools, we can see instead the emergence of a Chinese vision of democratic government organized according to a civic republican rather than a classical liberal model. This vision of participatory government stressed active and direct involvement by all citizens in community affairs. It also asserted that moral regulation and active self-discipline, rather than individual freedom and the expression of self-interest, were fundamental to republican politics. Active participation in the management and regulation of the student community along with self-discipline and moral training remained central to the practice of student self-government throughout the 1920s and into the 1930s.[25]

Republican-period intellectuals' elaboration of a mode of democratic politics organized around civic participation and moral regulation grew out of historical and theoretical precedents within China

---

22. Habermas, *Structural Transformation*. Cf. Alejandro, *Hermeneutics, Citizenship, and the Public Sphere*; Walzer, "The Civil Society Argument."

23. See, for instance, Rowe, "The Public Sphere"; Strand, *Rickshaw Beijing*; the sources in note 27 below; *Modern China* 19, no. 3 (April 1993).

24. Grieder, *Hu Shih*. Cf. Fung, *In Search of Chinese Democracy*.

25. Democratic education geared toward preparation for participation in a liberal democracy, in which discourse in the public sphere is the dominant paradigm, by contrast, would focus on rehearsing patterns of public debate, pluralistic discussion, and protest. See Westheimer and Kahne, "Education for Action."

itself, as well as from the influence of Dewey's vision of democracy and anarchist approaches to grassroots political organizations. At a conceptual level, late Qing reformer Liang Qichao's formulation of political participation, which was the basis for much later Chinese social and political theory, shared many elements with civic republicanism. Chang Hao argues persuasively that Liang's formulation of modern politics stressed active participation by the people in civic affairs so that they could contribute to national strength and development. In addition, Liang hoped that such civic participation would be conditioned by moral qualities that would encourage positive and constructive action by each person on behalf of society as a whole.[26] In all these dimensions, Liang's views closely paralleled civic republicanism.

In addition, the patterns of elite civic management that proliferated during the final decades of the Qing and continued into the Republican period at times more closely resembled civic republicanism than they did the growth of liberal democratic political institutions. Starting in the 1870s, urban and core-area elites, especially in the lower Yangzi region, engaged in various forms of public philanthropy and civic service, building and managing orphanages, relief agencies, and schools, as well as undertaking infrastructure projects for their local communities.[27] In these organizations, practical concerns of community welfare predominated. Although the periodical press began to serve as a forum for debate about public matters after the turn of the century, voluntary associations often operated more as sites for public management than as arenas for the kinds of open debate that Habermas associates with the public sphere.[28] Such forms of practical activity resonated clearly with a mode of civic republicanism that stressed active participation in the day-to-day affairs of local communities and public service over private interests. As lower Yangzi region educators and students sought models of political action for their self-government groups during the early

---

26. Chang, *Liang Ch'i-ch'ao*, chap. 6. Cf. Nathan, *Chinese Democracy*, chap. 3.

27. Bastid, *Educational Reform*; Culp, "Elite Association"; Rankin, *Elite Activism*; Rowe, *Hankow*; Schoppa, *Chinese Elites*.

28. Joan Judge also argues convincingly that rather than constituting a public sphere, the late Qing press gave life to a "middle realm" of reformist intellectuals who mediated between state and society. See Judge, *Print and Politics*.

Republic, they were surrounded by examples of elite organizations that assumed a civic republican form.

Further, models of democratic government that China's reformist and radical intellectuals each found attractive included many elements that resonated with civic republicanism. For instance, Dewey's vision of democracy and civic action placed great emphasis on the individual's membership in a sociopolitical community, and he equated freedom with the individual's full development and expression of his or her capacities through productive social work. Robert Westbrook, in his impressive study of Dewey and American democracy, argues that Dewey's conception of democracy centered on cooperative action and a commitment to the social good. Democratic citizens, for Dewey, realized themselves as individuals through cooperative activities in public life.[29] Dewey related these views quite clearly in his influential lectures in China during 1919 and 1920. There, he associated full human development with active sociopolitical participation and service to society. For instance, in a lecture at Beijing University concerning democratic institutions, he noted, "The best way to develop intellectually is to take the social welfare as our major concern, and think about ways of developing our abilities so that we can serve the common good."[30] He further described the social order as dependent on the individual's moral qualities that were rooted in the awareness of each person being a component of the social whole. Because of Dewey's tremendous popularity, especially among China's progressive educators and advocates of social reform, his views on democracy contributed to the trend toward civic republican approaches to political participation that found expression in calls for student self-government.

More radical political currents also encouraged a civic republican approach to self-government. Several versions of anarchism, which was immensely popular among radical intellectuals and lower Yangzi region students during the late 1910s and early 1920s, favored a political system based on small-scale communities that would be self-regulating and composed of people morally committed to the com-

---

29. Westbrook, *John Dewey*, 164.
30. Dewey, *Lectures in China*, 158. Cf. ibid., 221–22.

munity's harmony and welfare.[31] Such anarchist ideas about socio-political organization were reaching students in local schools. For instance, a lecture by Beijing University professor Yang Donglin at Zhejiang First Middle in 1924 related to students a vision of politics in which "guilds" (*gui'ertu*) at the local level would undertake all economic, social, and political functions, displacing existing state formations.[32] Anarchist models of political organization, too, stressed direct civic participation in grassroots communities. In this way, mutually reinforcing intellectual currents and existing examples of local voluntary association and gentry management all encouraged adoption of a civic republican model of political participation.

The similarities among civic republicanism, Dewey's progressivism, and certain forms of anarchism give some indication of how civic republicanism's main arguments and concerns have remained a vibrant part of Western political theory. This point is important. For by drawing this parallel between emerging modes of Chinese political participation and civic republicanism, we can see how Chinese intellectuals began to institute a dynamic form of modern democratic polity but one quite distinct from the Anglo-American liberalism to which most European and American scholars are most attuned. In fact, in constructing a mode of participatory politics that resembled civic republicanism, Chinese intellectuals opened the door to a potentially radical form of democratic politics. In contrast to rights-based and representative liberal democracy, civic republicanism invited direct civic action by all members of any community and was organized around a conception of a common good separate from the personal or sectional interests of any given social group, such as a dominant social elite.

However, the radical potential in civic republicanism is highly dependent on whether or not the people participating in that community have the opportunity to determine its boundaries and dynamics and to define the common concerns that will govern it. The danger in civic republicanism, as Mouffe diagnoses it, is the potential emergence of a communitarian approach "organized around a single sub-

---

31. E.g., Dirlik, *Anarchism*, 188–96, 234; Zarrow, *Anarchism*, 96–98.
32. Yang Donglin, "Wode lishiguan."

stantive idea of the common good" that would displace individual liberties and allow citizens little say in defining that good or their community.[33] As we reconstruct the practice and transformations of Republican-period student self-government we must ask whether students defined the aims and organization of their own communities or whether they operated under a "single substantive idea of the common good" and institutional forms that were assumed or imposed.

II. ACTION POLITICS

In many lower Yangzi region secondary schools, students and teachers quickly adopted student self-government as a mechanism for student training starting in 1919 and 1920. During the last decade of the Qing, many of China's secondary schools and colleges had formed student self-government organizations in order to promote democracy during the movement for constitutional government and to strengthen students' political organizations during the many protests, school disturbances, and boycotts of that period.[34] However, these early organizations generally seem to have lapsed after 1911. Between 1912 and 1919, few lower Yangzi region secondary schools organized student self-government groups, and those that existed sometimes operated primarily as a system of room monitors to maintain discipline in dorm rooms and study halls.[35] By contrast, in the years immediately following 1919, most secondary schools in the lower Yangzi region developed student self-government organizations, oftentimes with students playing an active role in initiating the

---

33. Mouffe, "Democratic Citizenship." Also see Young, "Polity and Group Difference."

34. Sang Bing, *Wanqing xuetang xuesheng*, 313–28. Cf. Weston, *Power of Position*, 71–72.

35. For an example of the room-monitor system, see Jiangsu shengli diyi shifan xuexiao, ed., *Ziyang xiang zizhi zhi*, 1. Beginning in 1915 Jiangsu Provincial Fifth Middle School in Changzhou had a student self-government association that operated on more democratic principles. See Chu Jinke, "Shu benxiao zizhihui zhi dazhi," 5–7; "Jiangsu shengli diwu zhongxuexiao xuesheng zizhi guiyue shixing xize"; "Jiangsu shengli diwu zhongxuexiao zizhi guiyue"; "Jiangsu shengli diwu zhongxuexiao zizhihui yishibu yizhi xize." Some Christian colleges also had student self-government organizations during the early Republic. See Lutz, *China and the Christian Colleges*, 173.

process.[36] Schools in other areas of the country developed student self-government organizations as well, but they were especially common in the lower Yangzi region, becoming a standard feature of secondary schooling there by the early 1920s.[37]

A powerful confluence of May Fourth activism and foreign educational models spurred students' and teachers' renewed interest in student self-government. The intersection of these factors comes through clearly in the following account from Zhejiang First Normal.

After the May Fourth Movement the atmosphere of the whole nation's world of thought was completely remade. It was felt that after this students' responsibility was increasingly great; because of the demands of the times [the nation] more than ever could not be without creative ability. During this time our self-government association was already in a period of gestation (*yunniang qizhong*) but had not yet been declared! Subsequently, when Professor Dewey and Professor Jiang Menglin came in turn to Hangzhou, one lecturing at the Provincial Education Association and one at this school, their contribution to promoting formation of this association was significant.[38]

---

36. Besides the schools that will serve as the main foci of the discussion below, lower Yangzi region schools that formed student self-government associations during the late 1910s and early 1920s included, but were not limited to, Jiangsu Provincial First Middle School (Zhang Daoren, "Jiangsu shengli diyi zhongxuexiao"), Shaoxing's Zhejiang Provincial Fifth Middle School (Er er shu bao she tong ren, "Er er shu bao she yan'ge"), Ningbo's Zhejiang Provincial Fourth Middle School (*MGRB* June 30, 1920), Jiaxing's Zhejiang Provincial Second Middle School (*MGRB* June 21, 1920), Shanghai's Minli Middle School (*MGRB* December 18, 1921, June 19, 1922), Yangzhou's Jiangsu Provincial Eighth Middle School (*MGRB* December 2, 1923), Huai'an's Jiangsu Provincial Ninth Middle School (*MGRB* December 17, 1923), and private Pudong Middle School (*Pudong zhongxuexiao nian zhou jiniankan*).

37. For a general discussion of the nationwide emergence of student self-government associations in middle schools and universities in the wake of the May Fourth Movement, see Lü Fangshang, *Cong xuesheng yundong*, 57–60. In a nationwide survey conducted in 1923, Zhang Nianzu found that fourteen of sixteen responding schools had student self-government organizations ("Zhongdeng xuexiao xunyu zhi yanjiu," 10). Zhili Girls' Middle School in Tianjin also established a student self-government organization, in 1918. See McElroy, "Forging a New Role," 360; "Transforming China," 134.

38. Zhejiang shengli diyi shifan xuexiao xuesheng zizhi hui, *Jilu*: 1. For other examples linking formation of student self-government to the May Fourth

Zhejiang First Normal's students had been active participants in the May Fourth protests, engaging in new forms of collective political action, as was characteristic of students in many schools throughout the region. At the same time, progressive teachers at the school and reformist principal Jing Ziyuan actively supported student self-government as a new mode of civic training.[39]

This environment of intense student activism and creative educational reform was highly conducive to the formation of student self-government associations, and student self-government groups experimented with a variety of forms. For instance, Jiangsu Provincial First Normal School initially modeled its self-government group on the patterns of local township government, whereas Zhejiang First Middle adopted a presidential model that placed great authority in the hands of an executive officer.[40] After a short trial period, though, a committee system became common at many schools. A general assembly of students would meet periodically to elect officers to fill positions on committees dedicated to particular tasks.[41] These committees then carried out the day-to-day operations of the self-government organizations.

At many schools, students were concerned to ensure that a maximum number of students would have opportunities to assume leadership roles in student government, in a pattern characteristic of civic republicanism. At Zhejiang First Normal, for instance, students stressed that positions should rotate frequently, so that all members of the association would be able to assume administrative roles, and they required that all students do some kind of physical work within the self-government area.[42] Similarly, at Zhejiang First Middle, rules that kept students from being elected to consecutive terms or serving simultaneously in multiple posts seem to have been designed to

---

Movement, see Jiangsu shengli diyi shifan xuexiao, *Ziyang xiang zizhi zhi*, 1; [Wu] Yanyin, 4.

39. For the reformist environment at the school during this time, see Cao Juren, *Wo yu wode shijie*, 1: 156–57; Yeh, *Provincial Passages*, 147–73, 187–89.

40. Jiang Weixian, "Ben xiao xuesheng zizhi hui de yange," 1–4; Jiangsu shengli diyi shifan xuexiao, *Ziyang xiang*, 1–2.

41. Jiangsu shengli diyi shifan xuexiao, *Ziyang xiang*, 1–3; Zhejiang shengli diyi shifan xuexiao xuesheng zizhi hui, Zhangcheng: 20–21.

42. Zhejiang shengli diyi shifan xuexiao xuesheng zizhi hui, Xu: 3–4; Zhangcheng: 13–14.

maximize the number of students taking active part in the daily management of affairs within the school community.[43]

Also in line with a civic republican model, most student self-government activity focused on managing the daily life of the student community and organizing practical activities in the local community. For instance, at Jiangsu First Normal there were self-government committees for discipline, hygiene, entertainment, cooking (i.e., supervision of the kitchen), gardening, and management of a store and bank as well as a book and newspaper reading room.[44] Zhejiang First Normal's self-government association also undertook a wide range of student activities while overseeing practical aspects of student life, such as managing the cafeteria.[45]

Further, at all these schools, various forms of self-regulation were an important part of self-government management activities. As expressed by student government organizers at Zhejiang First Middle, students saw the need to train themselves in patterns of self-discipline in order to prepare to be good citizens in the future.

Our organizing this self-government association is not a movement for self-liberation (*zifang de yundong*) or a movement to oppose the teachers and staff, and it is also not a mechanical organization (*jixiexing de zuzhi*). Rather, it is a movement appropriate for students in a republic; it is a movement of lifelong self-management (*ziji guanli ziji*); and it is also a movement to prepare to be citizens in the future. Speaking specifically, it is students uniting into a group, determining the rules, and doing for themselves any matters relating to student [life]. Students will control their own speech and conduct in preparation for collective self-government by citizens in the future.[46]

For these students an essential feature of self-government was students' formulating and enforcing rules of behavior, which they saw as necessary to being good citizens in a modern republic.

To this end, most self-government associations organized committees to monitor student behavior and hygiene to ensure that

---

43. "Zhejiang shengli diyi zhongxuexiao xuesheng zizhi hui huizhang," 5.

44. Jiangsu shengli diyi shifan xuexiao, *Ziyang xiang*, 1–13.

45. He Bingsong et al., eds., *Zhejiang shengli diyi shifan xuexiao du'an jishi*, Yu-lun: 9, 14, 25–26; Shen Ziqiang et al., 120; Zhejiang shengli diyi shifan xuexiao xuesheng zizhi hui, Zhangcheng: 11–15.

46. "Zhejiang diyi zhongxue xuesheng zizhihui chengli xuanyan."

students would follow modern standards of bearing and conduct. At Zhejiang First Normal, for instance, the inspection department enforced association rules that governed the most minute forms of student behavior in study halls, dorms, cafeterias, and playing fields, regulated the students' comings and goings, and even officiated their behavior outside the school.[47] The inspection department used continual surveillance carried out by a large number of student inspectors and Boy Scouts to ensure student compliance. In this way, political participation in the form of student self-government became a disciplinary mechanism for generating a modern form of cultural citizenship in ways parallel to the disciplinary mechanisms that Foucault describes proliferating in schools and other modern institutions in eighteenth- and nineteenth-century Europe and the United States.[48] The norms of disciplined behavior that student organizations enforced in their communities encouraged "modern" standards of civility and hygiene that were intended to make students acculturated modern citizens.

Students' choice to exercise self-discipline through the rules of self-government rather than to depend on teachers and administrators to impose outside controls is one indication of the level of independence that they had in organizing and managing their communities during the early 1920s. Students expressed their autonomous self-determination in several ways: they initiated the process of organizing and reorganizing their self-government associations themselves; they demarcated a sphere of student life that was outside the purview of the school administration; and within that sphere they made and enforced their own rules. At some schools, such as Zhejiang First Normal, students' autonomous organization was sanctioned and supported by the teachers and administrators. There, student independence in forming and running their own self-government organization was underwritten by a careful division of labor between the student association and the school administration that was delineated in a "Self-government Outline" (*zizhi dagang*) written in a meeting of teachers and staff before the establishment

---

47. Zhejiang shengli diyi shifan xuexiao xuesheng zizhi hui, Zhangcheng: 15–16, 19–20. Cf. Jiangsu shengli di yi shifan xuexiao, *Ziyang xiang*, 7–13.

48. Foucault, *Discipline and Punish*, 170–94.

of the association.[49] Further, when Zhejiang First Normal students formed their association, they independently organized a committee of representatives from different classes to draw up a charter and write the rules for the association.[50] By controlling the organization of their self-government associations, and determining collectively their communities' goals or "common good," they were acting out the most progressive side of civic republicanism.[51]

Whereas students at Zhejiang First Normal had teacher and administrator support in organizing their own self-government association, at nearby Zhejiang First Middle the association's capacity for self-definition and self-management was earned only through struggle with school administrators and provincial educational authorities. During 1920 and 1921 the association organized protests and petition drives directed against the head of the Zhejiang Provincial Department of Education, Xia Jingguan, and a new principal, Guo Weicheng. Guo, students claimed, sought to interfere in the association and ignored student proposals.[52] At stake in this struggle was the students' ability to define the aims, organizational form, and activities of their own community rather than operate under a conception of "the common good" and in accord with patterns of government imposed from outside. Zhejiang First Middle's students ultimately won their battle for self-determination when they ensured that Guo was replaced by the more cooperative Huang Renwang.

In self-government groups during the early 1920s, then, students embraced and enacted a civic republican conception of participatory politics organized around positive civic action and public-mindedness. Through self-government participation, they performed locally in their schools the kinds of active engagement with community life that their history and geography textbooks asserted all citizens needed to fulfill in order to build the Chinese nation. Further, many students sought to construct autonomous communities

---

49. Shen Ziqiang et al., 131–32; Zhejiang shengli diyi shifan xuexiao xuesheng zizhi hui, Jilu: 2–4.

50. Shen Ziqiang et al., 118–19; Zhejiang shengli diyi shifan xuexiao xuesheng zizhi hui, Jilu: 2–5.

51. At Jiangsu First Normal, any student could express dissatisfaction with committee management of student affairs at general assembly meetings (Jiangsu shengli diyi shifan xuexiao, *Ziyang xiang*, 3).

52. Jiang Weixian, 1–4.

wherein they would be able to determine their own aims and organizations and to impose on themselves the forms of discipline that would allow them to maintain an orderly self-regulating community. These student groups gave practical expression to a form of populist radical republicanism that would resurface in China during the Great Leap Forward (1958–61) and the Cultural Revolution (1966–76), when direct political action by self-regulating local communities was affirmed in opposition to centralizing, bureaucratic tendencies.[53] For May Fourth–era theorists such as Chen Duxiu, student self-government groups and other grassroots organizations were to provide an institutional infrastructure for a democratic Chinese state and a training ground for China's new citizens. As students initiated and ran self-government associations during the early 1920s they grew into politically active citizens who were accustomed to practicing a civic republican form of participatory government.

III. SELF-GOVERNMENT AS CONSTRUCTION

Between 1924 and 1927, the Chinese Nationalist Party and the Chinese Communist Party together sought to mobilize students for mass political action. This period between the Nationalist Party's reorganization and the culmination of its military campaign to reunify the country, the Northern Expedition (1926–28), witnessed intensive effort by the two parties to infiltrate student organizations and to draw students into their parties or allied organizations of various kinds.[54] Lower Yangzi region secondary students and recent graduates were often active participants in these party-led or party-affiliated movements for social reform and political revolution. However, in the wake of the Nationalist Party's anticommunist purge in April 1927 and the subsequent end of the Northern Expedition, fears of communist infiltration of student groups and concern over student-led political unrest caused Nationalist Party leaders to reassess the policy of mass student mobilization.

---

53. See, for instance, Maurice Meisner's discussion of Maoist approaches to communal government that were based on the model of the Paris Commune (*Marxism, Maoism, and Utopianism*, 147–48).

54. Lü Fangshang, 247–326.

After a long and involved negotiation about the party's approach to youth mobilization between 1928 and 1930, party leaders formulated new policies on student self-government that defined the parameters of student politics throughout the Nanjing decade. In 1928, two competing factions centered, respectively, in the Committee for Training the Masses (Minzhong xunlian weiyuanhui) and the Central Training Department (Zhongyang xunlianbu) offered a series of competing proposals for how the party should manage students' political action. Finally, in 1930 the party's Central Executive Committee (CEC) made a series of pronouncements that designated student self-government groups as the primary locus of student political action. Party leaders sought in this way to confine student activism within schools and to avoid the kinds of disruptive mass student protests by multischool student associations (*xueshenghui*) that had been common during the 1920s.[55]

To justify confining student activities to school-based self-government associations, Nationalist Party propaganda argued that, in the aftermath of the Northern Expedition, the needs of the times had changed. Now, during the period of party tutelage (*xunzheng shiqi*), the work of construction (*jianshe*) had replaced that of destruction.[56] Consequently, party ideologues asserted that there was no longer any need for large-scale student organizations that operated like political parties to engage in mass politics outside the confines of their schools. To "recklessly participate" (*lancan*) in politics, especially engaging in the destructive activity characteristic of the early 1920s and the Northern Expedition, was no longer appropriate.[57] Instead, students were to use school-based self-government organizations to focus on a variety of in-school activities.[58]

---

55. Huang, *The Politics of Depoliticization*, 33–75; Israel, *Student Nationalism*, 17–37.

56. Zhongguo Guomindang zhongyang zhixing weiyuanhui xuanchuanbu, *Xuesheng funü wenhua de tuanti*, 2–14.

57. The Third Party Congress formalized the transition from "destructive" to "constructive" work in party rhetoric regarding mass organizations. See Fewsmith, *Party, State, and Local Elites*, 154–56.

58. Ironically, the Nationalist Party's effort to replace political activism with constructive activity paralleled the attempt by missionary educators during the 1920s to steer their students into social service and away from nationalist protests, which the Nationalist Party had promoted during the Northern Expedition. See Graham, *Gender, Culture, and Christianity*, 120–37.

Previous discussions of the institution of student self-government groups under Nationalist Party rule have emphasized how they served to constrain student activism by confining students to campus-based activities.[59] Although political control was undoubtedly one goal of the new policies, there were other ways that the Nationalist Party could have limited or eliminated student activism. The Nationalist Party's continued promotion and encouragement of student self-government organizations suggests that party leaders and educators also viewed these groups as having some positive value in terms of youth training. That value, I argue, was to teach students forms of political participation and civic action that would contribute to "construction." The kinds of constructive work the party promoted through student self-government consisted of managing the practical affairs of student life and encouraging student self-discipline, continuing the civic republican approach to political participation that teachers and students had initiated during the previous decade, while now placing such activities under party supervision.

Sun Yat-sen and Student Self-government

When Nationalist Party leaders drafted party policies for student self-government, they built on a wide range of writings on local self-government by the party's founder and ideological beacon Sun Yat-sen. In his 1920 essay titled "First Steps of Local Self-government" ("Difang zizhi kaishi shixingfa"), Sun had identified the primary tasks of local self-government as the practical work of census taking, assessing land value, repairing roads, reclaiming wasteland, and establishing schools. Sun had argued that residents of the community should contribute one to two months of their labor, or the monetary equivalent, to the local self-government body each year. This stipulation made active involvement in constructive local enterprises the main idiom of self-government participation. Subsequent self-government work, suggested Sun, should center on establishing various cooperatives and bureaus for transport and trade with outside

---

59. See Huang, *The Politics of Depoliticization*, 44–82; Israel, *Student Nationalism*, 35–37.

areas. In this way, Sun defined self-government as the execution of practical local activity that would contribute to the community and to "national construction" (*jianguo*).[60]

Nationalist Party leaders carried this stress on practical activity in Sun's characterization of self-government over into their policies on student self-government associations. The Nationalist Party concretized in student self-government rules the imperative for students to engage in productive activities at the school level by mandating the formation of departments for academic research (*xueshu*), physical education, and recreation (*youyi*), and by allowing for the formation of consumer cooperatives.[61] Managing these basic elements of student life would prepare students for the kind of practical management of local community affairs that Sun identified with local self-government, teaching students the forms of political participation the Nationalist Party envisioned for their new polity. This stress on practical activity signaled the persistence of a civic republican model of political citizenship as concretized in student self-government.

Consequently, as had been true during the 1920s, student self-government associations in the lower Yangzi region managed aspects of school life that were of central importance to students. A number of schools continued to have committees or departments for managing or overseeing how the kitchen staff ran the cafeteria, and some used their hygiene departments to monitor kitchen cleanliness and food safety.[62] At Hangzhou Lower Middle School, students on the committee for managing the cafeteria actually rotated in setting the menu for student meals.[63] In addition, a number of schools organized consumer cooperatives (*xiaofei hezuo she*) or commissaries (*fanmaibu*), which carried items for students' daily needs.[64] Jiangsu Provincial

60. Sun, *Fundamentals of National Reconstruction*, 61–75.

61. Zhongguo Guomindang zhongyang zhixing weiyuanhui, "Xuesheng zizhihui zuzhi dagang," 1: 124–25.

62. "Dishijie shi zhengfu jinxing fangzhen," 15–16; Jiangsu shengli Suzhou zhongxue, ed., *Xuesheng zizhihui gaikuang*, 51–53, 82; "Shangzhongshi zuzhi," 43–45.

63. Jiang Lansun, "Bannian lai de xuesheng zizhihui," 110, 112.

64. Jiang Lansun, 112–13, 116; Jiangsu shengli Suzhou zhongxue, ed., *Xuesheng zizhihui gaikuang*, 10; *Songjiang nüzhong xiaokan*, no. 11 (May 20, 1930), 25; no. 17 (May 6, 1931), 15; no. 21 (February 30, 1931), 11; nos. 33–34 (November 15, 1932), 108.

Songjiang Girls' Middle School also had an office that coordinated students' work on the school garden.[65]

Also among the practical tasks carried out by student self-government associations were regulation of student behavior and rule enforcement. Echoing educators of the 1910s and 1920s, party leaders and educators championed self-government in part as a way to instill "order and discipline" in students and cultivate civic virtues and a group orientation.[66] As during the 1920s, student officials in public security departments (*gong'an ju*) and/or hygiene or public health departments (*weisheng ju*) carried out surveillance over their classmates. The self-government associations of many schools in the lower Yangzi region had at least one of these departments, both of which were primarily concerned with monitoring student behavior at the most basic level.[67] For instance, the Public Security Bureau of the Shanghai Middle Municipality (Shangzhongshi zhengfu), the student self-government organization at Jiangsu Provincial Shanghai Middle School, oversaw all aspects of student clothing, demeanor, and behavior, including everything from untidiness and absence from group meetings to infractions of rules of conduct.[68] Shanghai Middle

---

65. *Songjiang nüzhong xiaokan*, no. 4 (April 10, 1929), 15; no. 7 (November 15, 1929), 14; no. 8 (December 1, 1929), 13; no. 11 (May 20, 1930), 25; no. 12 (June 20, 1930), 6–7; no. 17 (May 6, 1931), 15; no. 21 (February 30, 1931), 11; nos. 33–34 (November 15, 1932), 108.

66. Chen Zhenbai, "Zhonghua minzu de shengming xian," 3–4; Qi Shuzu, "Dui zhongxuesheng jiang xunlian yu zizhi"; Xu Peihuang, "Zhongxuesheng kehou de xiuyang," 8–9; Zhongguo Guomindang zhongyang zhixing weiyuanhui xuanchuanbu, *Xuesheng funü wenhua de tuanti*, 6–7.

67. See, for instance, Chen Jiashan, "Xuesheng zizhihui gongzuo baogao," 173–74; "Jiangsu shengli Songjiang nüzi zhongxuexiao xuesheng zizhihui zhangcheng"; Jiangsu shengli Suzhou zhongxue, ed., *Xuesheng zizhihui gaikuang*, 15, 81–82; Suzhou Zhenhua nü xuexiao, *Zhenhua shenghuo*, 26; Suzhou Zhenhua nü xuexiao, ed., *Zhenhua nü xuexiao sanshinian jiniankan*, 86; "Zhejiang shengli Ningbo zhongxue xuesheng zizhi zuzhi dagang" and "Chunhui zhongxue xuesheng zizhihui zuzhi dagang" in "Zhejiang gexian xuexiao xuesheng zizhihui zhangcheng, mingce."

68. "Dishijie shi zhengfu jinxing fangzhen," 13; Jiangsu shengli Shanghai zhongxue chuzhongbu Shangzhongshi zhengfu, ed., *Shangzhongshi gaikuang*, 31–32, 37. Cf. *Songjiang nüzhong xiaokan*, no. 3 (February 5, 1929), 13; no. 4 (April 10, 1929), 15; no. 6 (July 1, 1929), 14; no. 7 (November 15, 1929), 13; no. 8 (December 1, 1929), 13–14; no. 11 (May 20, 1930), 25; no. 12 (June 20, 1930), 8–10; nos. 33–34 (November 15, 1932), 108.

Municipality's hygiene bureau similarly monitored the cleanliness of student areas of activity, seeing that dorms, study halls, and public spaces were kept clean, as well as regulating individual student cleanliness and health.[69] Officers of the bureau identified sick students, encouraged them to go to the nursing room (*yangbingshi*), made daily inquiries into all sick students' health, and reported on their status to the school doctor.[70] In general, an expressed concern for cleanliness and health also made possible continual monitoring of the students' most minute physical behavior, down to the level of individual students' clothing and bathing.[71]

Self-government associations were also instrumental in managing the academic, arts, athletics, and recreational activities that were the focus of student life outside of the classroom. Students writing at the time identified coordination of extracurricular activities as one of the central functions of student self-government associations.[72] Alumni from a number of regional schools also described the organization of extracurricular activities as one of the most important roles of student self-government. They listed extensive extracurricular activities that were central to student life at many schools and were organized either by the student self-government association as a whole or by class associations of various kinds.[73] Self-government associations coordinated a wide range of research and study societies. They oversaw arts, music, theater, and home economics clubs. And, they organized intramural and interschool sports competitions that were central to building students' collective identities.[74]

---

69. "Dishijie shi zhengfu jinxing fangzhen," 13.

70. "Shangzhongshi zuzhi," 38.

71. Jiangsu shengli Shanghai zhongxue chuzhongbu Shangzhongshi zhengfu, 25. Cf. Jiangsu shengli Suzhou zhongxue, *Xuesheng zizhihui gaikuang*, 81–83, 89–91; *Songjiang nüzhong xiaokan*, no. 6 (July 1, 1929), 14.

72. Ma Zhengjun, "Shimin shenghuo gaikuang," 19; Xu Shouyuan, "Guanyu Shangzhongshi zhengfu shijie jinian de hua," 17.

73. Interviews with alumni of Zhejiang Provincial Hangzhou High School (multiple subjects), Shanghai's Minli Middle, Jiangsu Provincial Suzhou Middle School, Shanghai Municipal Wuben Middle School, Jiangsu Provincial Yangzhou Middle School, and Suzhou's Zhenhua Girls' School.

74. Chen Jiashan; Jiang Lansun; Jiangsu shengli Suzhou zhongxue, ed., *Xuesheng zizhihui gaikuang*, 88; *Songjiang nüzhong xiaokan*, no. 5 (May 15, 1929), 9–10; no. 7 (November 15, 1929), 13; no. 25 (December 10, 1931), 6; nos. 33–34 (November 15, 1932), 108; "Songjiang nüzhong Nuli tuan biye jiniankan"; Suzhou Zhen-

These activities paralleled at the level of the school community the forms of constructive work that Sun Yat-sen had seen as essential to the development of self-government in China, and they provided a mechanism for the kind of active community involvement by all citizens that was central to civic republicanism. Thus, the party mandated student self-government associations not only to control student activists but also to train educated youths in a specific mode of political participation. Significantly, self-government was promoted for and practiced at boys' and girls' schools alike, suggesting that all students, regardless of gender, were being prepared for active civic life.

IV. DEFINING THE GENERAL WILL

By promoting student participation in practical management of school activities and collectively oriented modes of civic morality, student self-government organizations during the Nanjing decade continued to encourage in schools the civic republican patterns of political participation that students and educators had initiated during the May Fourth period. However, whereas student self-government groups during the early 1920s had been relatively autonomous and self-regulating, the Nationalist Party now exerted supervisory control over these student groups. Party directives defined the objectives for all student self-government groups and imposed a standard institutional structure on the disparate organizations that had operated in local schools during the previous decade.[75] The Party enforced implementation of these rules through multiple levels of guidance and oversight by school administrators, the local party branch, and county or municipal educational authorities.[76] The imposition of party oversight definitively marked the end of the radical republicanism initiated in student self-government groups during the early 1920s, for students were no longer able to define freely the goals and determine the institutional forms of their communities.

---

hua nü xuexiao, ed., *Zhenhua nü xuexiao sanshinian jiniankan*, 86; Suzhou Zhenhua nü xuexiao, *Zhenhua shenghuo*; Zhou Jianwen, "Wode huigu," 38–39.

75. Zhongguo Guomindang zhongyang zhixing weiyuanhui, "Xuesheng zizhihui zuzhi dagang."

76. See Zhongguo Guomindang zhongyang zhixing weiyuanhui, "Xuesheng zizhihui zuzhi dagang shixing xize."

Limited external supervision and control of student self-government associations had actually started before the advent of Nationalist Party rule. During the mid-1920s, teachers and administrators reacted to what they viewed as mismanagement of student affairs by reorganizing self-government groups and intervening in their activities. After several years of highly autonomous student self-government operations at a number of lower Yangzi region schools during the early 1920s, many teachers and administrators had become frustrated by uneven student participation in self-government groups, concerned about disorder in areas under student jurisdiction, and threatened by student challenges to school authorities.[77]

Undoubtedly the most dramatic case related to student self-government, which many educators read as a symptom of excessive student autonomy, was a tragic poisoning incident at Zhejiang First Normal in the spring of 1923 that was connected to the school's well publicized self-government association. In March, 22 students and two staff members died after eating food that two workers had laced with arsenic at the school's cafeteria, which was run by the student self-government association. Implicated in the plot was the student Yu Erheng, who had had a falling out with students in the self-government association. Relations between Yu and his classmates had soured because of the loss of over 100 yuan when Yu served as self-government association accountant.[78]

Because of Yu's apparent mishandling of association funds and the association's management of the cafeteria where the incident occurred, educators and other observers linked the tragedy to student self-government. Many used it as an opportunity to criticize excessive student autonomy and called for elimination, or at least increased administrative oversight, of student government. The words of a commentator writing in the *Education News* (*Jiaoyu xinwen*) are telling:

It is said that Zhejiang First Normal's poisoning case [occurred] because the school advocated student self-government, and in order to act cautiously in

---

77. E.g., Zhang Nianzu, 11–13. For a more extended discussion of these issues and the poisoning case at Zhejiang First Normal, see Culp, "Self-determination or Self-discipline?"

78. For an overview of the incident, see Hu Zhide, "Du'an jilüe." For day-to-day coverage of the case's unfolding, see MGRB, March–November 1923.

light of this, [some say] just eliminating student self-government will be sufficient. [These critics] do not know that this is like talk of 'refusing to eat for fear of choking' (*yin ye fei shi*). Promoting student self-government cannot be slighted for cultivating talent to manage affairs. For instance, American schools of the intermediate level and above especially emphasize this. But America's [student self-government associations] do not reach the point of our country's where there are no restrictions at all. Everything done by their students is still guided by administrative staff.[79]

In a characteristic reappraisal, the editorial used an authoritative American example to call not for the elimination of student self-government but increased administrative control over it by school authorities.

The doubts about student autonomy inspired by this tragic incident reflected a reassessment of student self-government by educators that was under way across the region during the mid-1920s, leading to increased teacher and administrator supervision over student organizations. Changes at Southeast University Affiliated Middle School, which nationally was considered a leading institution for educational reform, exemplify this shift. There, student government groups at all levels came to be subject to supervision by teachers. They were required to submit their rules to the school administration for approval and were watched closely by the school's teachers and staff to make sure they did not "exceed the scope of their authority, transgress school rules, or obstruct public welfare [*gongyi*]."[80]

The move by teachers and administrators during the 1920s to impose their supervision on student self-government groups set a precedent for outside control on which the Nationalist Party built. By promulgating uniform rules and establishing mechanisms for the oversight of student government, the Nationalist Party established itself as the central source for the supervision of student activities in

---

79. He Bingsong et al., eds., *Zhejiang shengli diyi shifan xuexiao du'an jishi*, Yulun: 48.

80. Liao Shicheng et al., 37–46; Tai Shuangqiu, "Xunyu shishi de yizhong jieguo." For other examples of greater administrator oversight, see *Jiangsu shengli diyi shifan xuexiao xuesheng ruxue xuzhi*, 12–18, for Jiangsu First Normal and the following sources for Zhejiang First Middle: "Xunyu tanhuahui ji jiaozhiyuan xuesheng lianxi huiyi jishi"; "Zhejiang shengli diyi zhongxue gaojibu tongxuehui jianzhang"; "Gaozhong tongxuehui weiyuanhui xize."

self-government associations. In doing so, it effectively extended the chain of hierarchical observation so that party and government representatives oversaw school administrators who were monitoring student officers. The increasingly intrusive role party representatives played in approving and overseeing student organizations further threatened student self-government associations' status as self-regulating communities.

At the organizational level, the party was quite effective at exercising direct oversight in lower Yangzi region schools. The immediacy of local government and party oversight is captured in the teacher Li Jinlin's account of Jiangsu Provincial Suzhou Middle School's attempts in 1930 and 1931 to get approval for its self-government association.

[In the fall of 1930] the school authorities directed students to begin planning a new student self-government association. The third-year students Wang Kaiji and four others formally applied to the Wu county party headquarters (Wuxian xian dangbu) to approve the organization. . . . Subsequently, in mid-October, they received a directive from the county party headquarters approving the organization [and also] transmitting a certificate of approval, so the [students] began drafting an association charter. After it was drafted, first I took it to the county party branch to ask their opinion. However, it had two points that the county party branch said it could not accommodate. One was that the institution with the highest authority [in the charter] was the representative assembly. The second was that the school was going to invite several teachers to organize a steering committee to take responsibility for directing the student self-government association. At that time the county party branch thought that the first point was fundamentally in conflict with the organizational outline for the student self-government association promulgated by the central government and that the matter defied the central government's regulations. With regard to the second point, although the school had the authority to supervise the student self-government association, [how it was to do so] was clearly stipulated in the charter (*huizhang*), so [establishing a steering committee] was redundant.[81]

Ultimately, the party branch decided to accept the organization of a steering committee but not the substitution of a representative assembly for a student general assembly.

---

81. Jiangsu shengli Suzhou zhongxue, ed., *Xuesheng zizhihui gaikuang*, 5–6.

Li's account demonstrates how the local party headquarters had decisive input at every stage of the self-government association organization process. Moreover, the party's rejection of certain elements of the original charter suggests that approval entailed more than a rubber stamp. Subsequently, the school's self-government organization, which formed only at the start of the fall 1931 semester, submitted its charter and requisite lists of members and officers to the authorities. Representatives of the Wu county party headquarters and local government officials also attended the opening general assembly meeting and the first representative assembly meeting, where association officers were sworn in.[82]

The process described here, which was replicated in other cases throughout the region, suggests that party and state oversight was rigorous and basically followed the guidelines delineated in the laws outlining student self-government organization and practice.[83] Local and provincial governments and party branches checked to ensure that the charter and organizational form of student self-government groups followed the prescribed models.[84] In addition, party and government representatives attended elections of new self-government association officers, participated in officers' swearing-in ceremonies, and kept records of self-government associations' officers.[85] Whether the schools were public or private, in cities or towns, or for boys or girls, student self-government associations in Jiangsu's and Zhejiang's secondary schools were subject to close party and government supervision.

Careful rule enforcement by local and provincial governments led to remarkable uniformity in the structure and stated aims of Nanjing-decade lower Yangzi region self-government organizations, a situation that contrasted sharply with the diversity of self-

---

82. Ibid., 6–7.

83. Cf. Suzhou Zhenhua nü xuexiao, ed., *Zhenhua nü xuexiao sanshinian jiniankan*, 83; see also SMA Q235-2-179, pp. 1–59, for Wuben Girls' Middle, and SMA Q235-2-1798, pp. 1–9, for Pudong Middle.

84. "Zhejiang gexian xuexiao xuesheng zizhihui zhangcheng, mingce."

85. Ibid.; SMA Q235-2-179, pp. 1–59; "Xuesheng zizhihui xiaoxi" (Student self-government news), *Songjiang nüzhong xiaokan*, no. 14 (December 31, 1930), 11; no. 16 (April 15, 1931), 15; no. 21 (September 30, 1931), 11–12; no. 24 (November 25, 1931), 5; no. 28 (May 25, 1932), 7; no. 33–34 (November 15, 1932), 107–8; no. 49 (December 1, 1933), 6–7; no. 59 (June 1, 1934), 10.

government models used by students during the 1920s.[86] In most student self-government groups in the lower Yangzi region, the body with the highest authority was the general assembly (Quanti huiyuan dahui), which was a combined meeting of all students. The general assembly usually met once or twice each semester and was responsible for establishing and revising the self-government charter, determining the major goals of the association, and overseeing subsidiary self-government bodies. At some schools self-government executive officers were chosen by the general assembly. Under it was a representative assembly (Daibiao dahui) composed of one or two representatives chosen by each class (*ji*). At some schools it, not the general assembly, elected the executive staff. Representatives met periodically throughout the semester and were responsible for making fundamental decisions about self-government association policy, overseeing the executive committee, inspecting executive committee reports and association accounts, and making decisions regarding disciplining association members or cases of disagreements between individuals or groups. Subordinate to the representative assembly was an executive committee (Ganshihui) of five to ten members who ran the "bureaus" (*ju*) or "departments" (*gu*) that organized many of the activities that made up student extracurricular life.

The Nationalist Party's successful imposition of a standard self-government organizational form, and the corresponding limitation of self-government work to a predetermined range of academic, athletic, and recreational activities, indicates that the party seriously constrained students' ability to organize their communities on their own terms. The party further imposed a common set of objectives on the associations that kept them focused clearly on a narrow set of educational aims that were included in nearly every school's self-

---

86. This synthetic description of the basic organizational form for lower Yangzi region schools' student self-government associations is based on association charters drawn from the following sources: "Dishijie shi zhengfu jinxing fangzhen"; Jiangsu shengli Shanghai zhongxue chuzhongbu Shangzhongshi zhengfu, ed., *Shangzhongshi gaikuang*; "Jiangsu shengli Songjiang nüzi zhongxuexiao xuesheng zizhihui zhangcheng"; Jiangsu shengli Suzhou zhongxue, ed., *Xuesheng zizhihui gaikuang*; Suzhou Zhenhua nü xuexiao, *Zhenhua shenghuo*, 70–81; "Zhejiang gexian xuexiao xuesheng zizhihui zhangcheng, mingce." The student association at Shanghai's Labor University took the same form and experienced the same types of regulation. See Chan and Dirlik, *Schools*, 159–60.

government charter. "This association's aims are to use the spirit of the Three Principles of the People to cultivate students' self-governing life within the school, and to promote the development of their academic, moral, physical, and group education."[87] Thus, the party also tried to preempt student communities' capacity to formulate their own conception of a "common good" by providing a uniform definition of the objectives for all student groups.

V. SELF-GOVERNMENT AND STUDENT LIFE

Even with increased oversight and uniformity imposed by the Nationalist Party, student self-government during the 1930s continued to attract the high levels of student involvement aimed for under civic republicanism. Students were drawn by opportunities for self-distinction and self-creation through practical management and extracurricular activities. At many schools, a relatively small group of students monopolized the highest positions of the student self-government associations. These students built a sense of identity through their leadership in the school community. The experience of Zhou Jianwen illustrates this trend of prominent student leaders. For his last two years at Shanghai Middle, Zhou served as a self-government officer, first holding the position of group officer (*zuwusheng*; roughly equivalent to a room monitor) and then serving as the mayor of the municipal government for the following three semesters.[88] Some student leaders assumed the airs and adopted the high-handed practices of corrupt government officials, whereas others elaborated an identity built on the ideal of selfless public service.[89]

However, the diverse forms of community management clustered within student self-government associations meant that much larger numbers of students could assume some kind of staff or officer posi-

---

87. This statement, or one like it, was included in nearly every secondary school self-government charter of the period. For one example, see "Jiangsu shengli Songjiang nüzi zhongxuexiao xuesheng zizhihui zhangcheng," 14.

88. Zhou Jianwen, "Wode huigu." In interviews, alumni of Songjiang Middle and Hangzhou High asserted that only a small minority of the best students tended to monopolize the highest positions in the self-government associations. Also see Zhang Lunqing, "Wode song dao," 7–8.

89. Gu Ruwen, "Xuesheng fuwu zizhi tuanti."

tion within the self-government framework. At Hangzhou Lower Middle, for instance, nearly one-third of the student body served as at least a staff member in one department or committee within the student self-government association.[90] Such widespread participation encouraged students to view themselves as active citizens of a self-governing student community.

Even students who did not participate in self-government associations as officers, departmental staff, or literacy-school teachers joined in community life through active involvement in the sports teams, study societies, and arts clubs that formed the infrastructure for extracurricular life in these schools. When they participated in these activities, students entered into the "constructive" mode of civic action sanctioned by the Nationalist Party, but they often did so for their own reasons and on their own terms. Many alumni interviewees stressed that students were drawn to art and theater activities and that they formed extracurricular clubs on their own initiative within the framework of the self-government organization to pursue their interests.[91] One former student from Zhenhua Girls' Middle recounted spending much of her free time practicing photography, using a darkroom provided by the school. Other students, she said, were engaged in music, writing, dance, and theater.[92] Student interest and intensive participation in groups for music, theater, and art indicate that student self-government associations provided extracurricular activities that offered students opportunities for individual development and identity formation through creative expression.

Similarly popular at both boys' and girls' schools were the wide spectrum of individual and team sports that were organized by

---

90. In 1935, the combined number of students serving in the representative assembly, the executive committee, and in the twelve departments as staff members was 208 (Jiang Lansun, 116–18). In the spring semester of that year, there were 632 students at the school overall (Zhejiang shengli Hangzhou chuji zhongxue bianji weiyuanhui, ed., *Zhejiang shengli Hangzhou chuji zhongxue liuzhou jiniance*).

91. Interviews with alumni from Zhenhua Girls' School, Yangzhou Middle, Suzhou Middle, and Hangzhou High. Statistics on student activity participation from Suzhou Middle suggest there was a great deal of student interest in music, theater, and arts clubs ("Disanjie xuesheng zizhihui ge yanjiu hui she renshu tongji biao").

92. Interview with alumna from Zhenhua Girls' School.

student self-government organizations.[93] With a focus on track and field and competitive sports such as tennis, soccer, ping-pong, basketball, and volleyball, student extracurricular activities continued the "liberal democratic *tiyu* [physical culture]" of the 1910s and 1920s. These forms of athletics, as Andrew Morris has persuasively argued, encouraged both "free will and self-restraint," allowing individual self-expression and achievement that would contribute to building the nation as a whole.[94] Students reporting on their participation in competitive sport took pride in their achievements and those of their classmates.[95] Association-run athletics provided students with opportunities for individual and collective distinction.[96]

Significantly, when reflecting on self-government, some Nanjing-decade students did not only focus on the many opportunities for self-expression, competition, and individual distinction that such groups provided. They also stressed the need for self-government groups to instill self-discipline and proper hygiene, for they maintained that self-regulation was the precondition to avoid being controlled by others.[97] For instance, Xu Fuzeng, a frequent self-government association officer at Songjiang Girls' Middle, argued that self-government was essential in order to avoid being controlled by others. "If one does things without a spirit of self-government, then s/he will be governed by others, and being governed by others is improper. The common people under a despotic monarchy are governed by others. Who is willing to be a person who has lost his/her freedom and is constrained (*shufu*) by others?" Then Xu went on to identify good self-government with adherence to group discipline and dedication to group wel-

---

93. "Dishijie shi zhengfu jinxing fangzhen," 14, 16–17; Jiang Lansun, 114; Jiangsu shengli Suzhou zhongxue, *Xuesheng zizhihui gaikuang*, 2, 55, 88, 91; *Songjiang nüzhong xiaokan*, no. 18 (May 20, 1931), 5; Suzhou Zhenhua nü xuexiao, *Zhenhua shenghuo*, 27, 57–59, 67.

94. Morris, *Marrow*, 47–52, 137–40.

95. "Songjiang nüzhong Nuli tuan biye jiniankan," 2–5; Suzhou Zhenhua nü xuexiao, *Zhenhua shenghuo*, 27, 67.

96. Many student athletic, arts, and academic activities were conducted through subordinate self-government bodies, such as "class associations" (*jihui*) or "dorm associations" (*quhui*) (interviews with alumni of Minli Middle, Yangzhou Middle, and Hangzhou High). See also Feng Yuanhuai, "Ji xun de yiyi"; "Songjiang nüzhong Nuli tuan biye jiniankan."

97. Cheng Liying, "Guanyu shangzhongshi shizhengfu," 14; Xu Yaoliang, "Guanyu shangzhongshi zhengfu de hua," 23.

fare.[98] Students like Xu accepted that the culture of citizenship required good hygiene and behavioral self-control that corresponded to teacher and Nationalist Party objectives for student self-government. But they did so in order to claim some degree of autonomy for their groups, despite the fact that both school administrators and party and government officials supervised self-government associations.

VI. CONCLUSION

Student self-government associations in Republican-period lower Yangzi region schools taught students political participation in a civic republican mode. In their schools, they practiced forms of civic action centered on community management, constructive activity, and self-regulation to instill public-spirited civic morality. Theoretical influences from Liang Qichao, John Dewey, and European anarchism intersected with ongoing patterns of gentry management that were prominent throughout the lower Yangzi region to encourage adoption of this civic republican approach to political participation. These multiple, long-term influences help to explain why the content of civic action continued to be defined in civic republican terms, despite the shift in political climate from the open, contestatory May Fourth period to the period of party tutelage after 1927. Student self-government contributed to making community-oriented civic action, rather than protection of individual rights and expression of self-interest in public debate, fundamental to twentieth-century Chinese definitions of democracy. This action-based approach to political participation resonated with the calls for citizens' activism in contemporary history and geography textbooks.

Though the content of civic action remained largely constant, its dynamics and institutional structure changed dramatically over the 1920s and 1930s. During the late 1910s and early 1920s, students at a number of regional schools experimented with radical forms of direct democracy at the local level by establishing self-government groups that were autonomous and self-regulating and that demanded high levels of participation by student-citizens. Nationalist Party policies on student government, by contrast, imposed close party

---

98. Xu Fuzeng, "Tuanti shenghuo de yaosu."

and government supervision on student self-government associations and dictated the goals and activities of those groups, even while continuing to encourage active civic engagement.

By adopting this approach to political action, Nationalist Party leaders took their place in a long line of revolutionary dictators that included Robespierre and Lenin. These parties and leaders claimed for themselves the power and authority to define the "general will" for the community as a whole while encouraging citizens' political activism and civic morality. Benjamin Schwartz captured this tension in the following terms in writing about Maoism: "[The] masses are not necessarily as they are but as they 'ought to be' and there can be no doubt of the leader's aspiration to make them what they ought to be. They are to be made public spirited and their virtue is no longer to be passive and negative, but active and dynamic. It is to be moral energy consolidated in the service of the nation."[99] The Mao-era Chinese Communist Party, in ways parallel to the Nanjing-decade Nationalist Party, promoted politics that incorporated active participation, civic morality, and determination of the goal of civic action by the party and/or leader. This continuity suggests that by cultivating party-supervised civic republicanism, the Nationalist Party may have contributed to cultivating habits of state-led civic participation that later flourished under the Chinese Communist Party.[100]

---

99. Schwartz, "The Reign of Virtue," 169.

100. But Maoism also contained within it the countervailing tendency, discussed earlier in this chapter, to focus on "the masses" as the source of political energy and virtue. See Meisner, *Marxism, Maoism, and Utopianism.*

4

Cultural Revolution and
the Social Organism

Cai Yuanpei, in his early Republican moral cultivation textbook, linked families and the nation. He described the family either as a foundational element of the nation (*guojia zhi jiben*) or as a miniature version of the nation (*guo zhi xiaozhe*), and he portrayed the nation as a family writ large.[1] Fulfilling duties within the family, asserted Cai, could serve as the basis for civic virtues that citizens would then enact in society. Further, maintaining well-ordered and harmonious families would contribute to building a peaceful and orderly nation.

Significantly, Cai suggested that conventional Confucian family roles and duties would encourage family harmony and serve as a model for more generalized civic duties in contemporary society. For instance, he asserted that filial piety was a natural result of the relationship between children and parents, and that children had a duty to repay their parents' benevolence in nurturing and raising their children. Patriarchal control and children's fulfilling filial obligations created a situation of order within the family microcosm that, by extension, would contribute to the ordering of society and the nation. Cai further considered filial piety to be the root of moral behavior in other social contexts.[2] Cai similarly described family harmony as grounded in hierarchical relations between husband and wife.[3]

Women's knowledge and ability is generally speaking inferior to men's, and because they just handle household matters, their experience in society is

---

1. Cai Yuanpei, *Dingzheng zhongxue xiushen jiaokeshu*, 38–39, 89–90.
2. Ibid., 38–48.
3. Ibid., 52–54.

more superficial than men's. So wives, unless they receive improper orders, must obey their husbands, and their basic duties are to be unshakably virtuous (*zhengu bu yu*) and share worries and joys. The husband's leading and the wife's following are the natural ethics of human relations. The husband is the master of the family and the wife assists him. Only when direction and assistance are harmonious is household management orderly.[4]

Elsewhere in his influential textbook, as we will see, Cai prescribed other behaviors for the modern citizen and portrayed alternative dynamics for a modern society. But in these passages, by grounding social harmony and national welfare in late imperial family patterns, Cai made the age and gender hierarchies of Confucian ethics bases for the behavior that would constitute national society.

By the end of the 1910s, New Culture Movement reformers, anarchists, and socialists, Cai among them, were challenging the basic assumptions underlying the hierarchical view of society presented in portions of this early Republican moral cultivation textbook. Many of these challenging new ideas about social roles and social order penetrated lower Yangzi region local schools, first through informal channels such as lectures and study societies, and then through mass-publication civics and language textbooks that repackaged or directly reprinted writings on social reform. These media, in part, promoted social equality and portrayed each individual, regardless of gender, age, or class, as a valued member of society. Textbooks and other media frequently also sanctioned radical reform, and occasionally even revolution, to realize these social ideals. These media contributed to eroding the doxa, or self-evidence, of late imperial Neo-Confucian social norms and helped to expand the "universe of discourse" about social roles and organization during the Republican period.[5] After 1927, the ruling Nationalist Party deployed censorship,

---

4. Ibid., 53–54.
5. Bourdieu defines doxa as a shared, implicit understanding of the patterns of social life that "goes without saying because it comes without saying." A crisis that disturbs patterns of social practice, by contrast, can reveal the arbitrariness of the social order and destroy doxa's self-evidence. When this happens, social categories, roles, and relations that were once assumed become part of the "universe of discourse," that is, open to review, question, critique, and reformulation by actors with different interests and agendas (*Outline*, 166–69). China's late imperial Neo-Confucian doxa can be seen as having been undermined by the internal and external crises of the nineteenth century and the "decanonization" of

surveillance, and new textbook standards to limit messages of social radicalism, effectively narrowing the boundaries of that universe. Instead, party-sanctioned civics and language textbooks highlighted party authority, commitment to collective development, and a return of neo-traditional social norms.

Across these dramatic fluctuations in what students learned about social roles, relations, and processes of change, civics textbooks and other media consistently presented students with an image, rooted in late nineteenth-century social theory, of society as an organism composed of cellular citizens. By casting all citizens as dynamically interdependent, the organic image of society linked each student's behavior to the welfare of society as a whole, reinforcing the messages about self-regulation and active civic participation that students encountered in other dimensions of their schooling.

I. NETWORKS OF CHANGE: STUDY SOCIETIES AND THE LECTURE CIRCUIT

During the early Republic, study societies and public lectures served as major conduits outside the formal curriculum for transmitting new ideas about society and citizenship to students in lower Yangzi region schools. Both techniques for transmitting ideas had historical precedents in the region that contributed to their efficacy. Study groups and fledgling political societies had become an increasingly prominent part of Chinese social life starting in the mid-nineteenth century. "Societies" (*she*) and "associations" (*hui*) proliferated especially in Southeast China, where the roots of gentry activism were deep.[6] Intellectual associations and political societies had also blossomed among Chinese students in Japan during the first two decades of the twentieth century and in many late Qing schools, where students circulated reformist and/or revolutionary writings.[7] These "sprouts" of student reading societies grew into a veritable forest

---

the Confucian classics at the start of the twentieth century (Chang, *Chinese Intellectuals*; Elman, *Cultural History*, chap. 11).

6. Rankin, *Elite Activism*; Schoppa, *Chinese Elites*, 8–9, 34–39, 72–77, 100; idem, *Blood Road*, 38–39. Cf. Averill, "The Cultural Politics of Local Education."

7. Harrell, *Sowing the Seeds*, 99–100, 103–6, 145–46; Rankin, *Early Chinese Revolutionaries*, 61–69.

during the New Culture Movement. By the late 1910s, students and faculty at prominent schools such as Beijing University formed associations of various political stripes.[8] Through publications, social activism, and personal contacts, these groups were well known in educational circles throughout China, serving as models for local forms of organization.

Teachers and school administrators who had participated in or had knowledge of these archetypal study societies often adopted a supportive attitude toward students who formed study groups and created publications to express their ideas. Such tolerance coupled with numerous ready models led to the proliferation of study societies in lower Yangzi region secondary schools during the 1910s and 1920s. For instance, in the spring of 1923, with the help of reformist educator Liu Dabai, a group of eight students from the 1922 incoming class at Shaoxing's Zhejiang Provincial Fifth Middle School formed a group that they called, after their entering year, the Twenty-two Reading Society (Er er shu bao she). They formed the new reading society out of dissatisfaction with the school's existing National Learning Research Society (Guoxue yanjiu she), which was likely more conservative. Members paid dues to purchase books and magazines and collected readings contributed by members and outsiders, lending from their growing library of books and journals to society members and fellow students. The heart of its collection was seven volumes (*juan*) of the New Culture Movement journal *New Youth* along with three volumes of the reformist education journal *New Education* donated by Principal Shen Suwen.[9] Judging from this base collection, the group's research focused on contemporary issues of social and cultural reform.

The mention in the foregoing account of the National Learning Research Society at Zhejiang Fifth Middle suggests that not all of the region's students were primarily concerned with a New Culture approach to social and cultural reform. In fact, several other area schools had more culturally conservative national learning groups.[10]

---

8. Chow, *The May 4th Movement*, 41–77; Schwarcz, *The Chinese Enlightenment*, 55–144; Van de Ven, *From Friend to Comrade*, 38–50; Weston, 134–39.

9. Er er shu bao she tongren, "Er er shu bao she yan'ge," 107–21.

10. Zhenhua Girls' Middle School provides one example (Zhang Quanping, "Guoxue shangdui hui gaikuang").

At Jiangsu Provincial Second Normal School in Shanghai, for example, the National Learning Research Society formed in 1922 with the help of a teacher and had 150 members. The group's aims were to "cultivate ethics and discuss learning" (*xiude jiangxue*) and to "learn new things by reviewing the old" (*wengu zhixin*) through study of classical texts.[11] But the group was not narrowly traditionalist. Society members argued that "national learning is like gold in the sand: it has many good properties, but it also has lots of dross. But people who hope to research national learning take out their methods to scoop up sand and extract gold, organizing what benefits the present and makes a contribution to society."[12] Their goal was a rational reformulation of classical culture that would cull its best parts, using them in combination with elements of Western learning, to reform (*gaizao*) Chinese society.[13]

In stark contrast to these reading groups, in which students explored moderate or gradual approaches to social and cultural change, were radical study societies such as those formed under Jing Ziyuan at Ningbo's Zhejiang Provincial Fourth Middle School in the early 1920s. Young, reform-minded teachers hired by Jing, including Xia Mianzun and Zhu Ziqing, fostered an open environment of experimentation in which teachers and students organized societies to research new ideas and to publish journals and wall-poster newspapers expressing their views.[14] The Snow Flower Society (Xuehuashe), was organized by teachers and students from the former Zhejiang Provincial Fourth Normal School, which had been combined with Zhejiang Fourth Middle in 1923. Under the guidance of CCP member Wang Ziwang[15] and others, the Snow Flower Society edited a journal that "publicized progressive thought and attacked the evil

---

11. Song Xuewen, "Qing lai taolun guoxue yanjiushe de jige wenti," 25.

12. Ibid., 22.

13. Ibid., 23.

14. Dong Qijun, "Jing Hengyi yu Zhejiang shengli disi zhongxue." Both Jing Ziyuan and Xia Mianzun had been active in Japan during the late Qing, with Jing joining the Revolutionary Alliance and Cai Yuanpei's Chinese Educational Association. Zhu Ziqing graduated from Beijing University in 1920 and was a member of the New Tide Society (Chow, *The May 4th Movement*, 55–57; Yeh, *Provincial Passages*, 80–82; *Zhejiang renwu jianzhi*, 3: 79–80, 154–55, 238–39).

15. *Zhejiang renwu jianzhi*, 3: 267–68.

power of local feudalism."[16] Students from the school also organized a Social Science Research Association (Shehui kexue yanjiuhui), whose pamphlet, "besides analyzing the situation inside and outside the country and introducing the basic theories of Marxist-Leninism (such as class struggle, surplus value, prices, and profit), also introduced Marxist-Leninism's rudimentary works."[17] Parallel to these groups, the Huoyao Society, which was reputedly a front organization for the school's underground Communist Youth Group, also published several journals.[18] These three societies for research and publication gave students arenas in which to experiment with new ideas, and Marxist thought clearly was a significant part of the mix.[19]

Apart from formally organized study societies, students also engaged with new ideas about society through their own extracurricular readings. For instance, during the early 1920s "publications such as *New Youth*, *The Guide*, *Chinese Youth*, and the *Communist Manifesto*, Bukharin's *ABC's of Communism*, *The Guide Collection* (*congshu*), *China's Customs Tariff Problem* (*Zhongguo guanshui wenti*), *The Unequal Treaties*, and *The Anti-Christian Movement*, as well as many other Marxist-Leninist books and publications published by Shanghai companies, all flooded into [Ningbo's] Fourth Middle."[20] These publications freely circulated among teachers and students for extracurricular reading. Similar kinds of radical publications and New Culture Movement periodicals also circulated at other regional schools, such as Zhejiang Provincial First Normal School and Yangzhou's Jiangsu Provincial Eighth Middle School, and in students' summer reading societies.[21] At Jiangsu Eighth Middle, Principal Li Gengsheng actively promoted students' reading journals to expose them to "new thought."[22] Besides the informal circulation of books and journals, students also set up bookstores and book-buying cooperatives that purchased and circulated left-wing journals and publications for so-

---

16. Dong Qijun, 80.

17. Ibid., 81.

18. Ibid., 80–81.

19. Many students also formed study societies during summer vacations and other breaks. See Chen Heting, "Linhai geming qingnian de yaolan."

20. Dong Qijun, 82.

21. Chen Heting; Dong Shulin, "'Zhe yishi xuechao' de yingxiang," 4; Wang Shouhua, "Riji liangze," 196.

22. Xiao Zhizhi, "Li Gengsheng zhuan."

cial and cultural reform.[23] In addition, school libraries subscribed to journals and purchased books that reflected a wide spectrum of thought.[24] Through all these channels, students were exposed to ideas that ranged far beyond their textbooks, and many experimented with approaches that questioned, challenged, or reconceptualized the existing social and cultural orders.

Along with study societies and new publications, public lectures became a powerful technique for disseminating new ideas in China during the late 1910s and early 1920s. Public oratory had been an important part of earlier political movements, especially in the last decade of the Qing and the early Republic.[25] But lectures gained a new prominence during the cultural reform movements of the late 1910s and early 1920s. Lecture tours by John Dewey, Bertrand Russell, and others served as key points of departure for some of the scholarly debates that formed the core of the New Culture Movement, introducing ideas about social change, the relationship between education and society, and the philosophical perspectives of rationalism and pragmatism.[26] Lecture tours by prominent world intellectuals established the formal lecture as a major idiom for presenting new ideas at leading academic institutions such as Beijing University and Nanjing Higher Normal School.

Secondary schools throughout the lower Yangzi region followed this model and by the late 1910s were inviting prominent scholars, visiting dignitaries, and noteworthy local intellectuals to give lectures at their schools. The lower Yangzi region's dense transportation networks made regional schools very accessible to lecturers traveling to and from Shanghai. High-profile lectures could generate publicity for schools just as they did for local communities, marking them as local or even regional centers of intellectual life. But lectures also complemented study societies and extracurricular readings as a way

---

23. "Hangzhou yizhong shubao fanmai qishe shi yi"; Yeh, *Provincial Passages*, 158.

24. For library acquisitions at Zhejiang Provincial First Middle School, see "Tushuguan qishi" and "Tushuguan qishi" (2); "Tushuguan tonggao" and "Tushuguan tonggao" (2); "Tushuguan goudao xinshu."

25. Shao, *Culturing Modernity*, 169–73; Strand, "Citizens in the Audience"; Wang, *In Search of Justice*, 113–14.

26. Chow, *The May 4th Movement*, 191–93; Keenan, *The Dewey Experiment in China*, 30–34; Shao, *Culturing Modernity*, 174–75.

for new ideas to reach students.[27] As elite intellectuals of widely varying political stripes went to lower Yangzi regional schools, complex debates about China's trajectory of social and political change were played out through their presentations.

Some school administrators, such as Ren Mengxian at Jiangsu Provincial Fifth Normal School in Yangzhou, embraced pluralism as a value and purposely invited speakers with different perspectives to their schools.[28] But other schools focused on exposing students to a particular perspective on the period's debates over social politics. In Ningbo, for instance, intellectuals with competing agendas concentrated their efforts on specific schools. Jing Ziyuan invited prominent Nationalist Party intellectuals including Hu Hanmin and Dai Jitao to Zhejiang Fourth Middle to speak while also welcoming anarchists Wu Zhihui and Shen Zhongjiu and CCP activists Yun Daiying, Yang Xianjiang, Shi Cuntong, and Chen Wangdao. In 1925, the more conservative "nationalist clique" (*guojia zhuyi pai*), worried about communist student radicalism then brewing in Ningbo, sent Chen Qitian and Zhang Zhizhu to private Xiaoshi Middle School where they lectured, respectively, on nationalism (*guojia zhuyi*) and "The World's New Tide" (*shijie xinchao*).[29]

As these examples from Ningbo suggest, the spectrum of views on social reform presented in school lectures was just as broad as the readings that students encountered through their study societies and extracurricular readings.[30] At some schools, such as Jiangsu Provincial Second Middle School in Shanghai, the focus could be on cultural revival and national learning.[31] At others, like Zhejiang Provincial First Middle School, the lecture might present an anarchist

---

27. Qin Shao demonstrates how celebrity lectures raised publicity for Zhang Jian's many reform ventures in Nantong (*Culturing Modernity*, 169–76). She also emphasizes how local elites used lectures to control students and reinforce their own power. By contrast, this section demonstrates some of the ways that lectures could open local communities to new ideas that challenged the existing social order.

28. Shen and Xiao, eds., *Jiangsu sheng Yangzhou zhongxue*, 45–46.

29. Dong Qijun, 78, 90.

30. E.g., Pudong zhongxue xiaoshi bianxie zu, 209–10; Wu Sihong, "Dageming shiqi de Shaoxing xianli nüzi shifan," 229; Zhejiang sheng zhengxie wenshi ziliao weiyuanhui, *Zhejiang jindai zhuming xuexiao he jiaoyujia*, 207.

31. Hu Pu'an, "Guoxue zhi yanjiu," 16–17.

approach to social change.[32] At still others, such as Shaoxing Girls' Normal School, lectures might relate ideas about women's liberation and social revolution.[33]

Yet lectures differed from study societies, extracurricular readings, and textbook readings in one important way. The personal charisma of a visiting speaker could add persuasive power to his or her message and dramatically sway students' thought and action. The potent mix of message and charisma is illustrated by a lecture that socialist youth movement leader Yun Daiying gave at Jiangsu Eighth Middle and Jiangsu Fifth Normal in Yangzhou during the fall of 1925. Yun, a reputedly powerful speaker, lectured on the topic "livelihood problems for normal school students" (*shifansheng he fanwan wenti*). Eyewitnesses claimed the lecture had a great impact on students: "for us it was a revelation, making us realize that study was not for our own livelihood (*bushi wei geren de fanwan*) but for the revolution (*shi wei geming de daoli*)."[34] After the speech Yun introduced a student who had been active in the May Thirtieth protests into the Chinese Communist Party and helped to establish a party cell at Jiangsu Eighth Middle, highlighting further the potency of the lecture medium.

II. REPRINTING RADICALISM AND AUTHORIZING REVOLUTION

Study groups, extracurricular readings, and lectures offered informal and ad hoc avenues through which students encountered new ideas about social organization and civic roles. By the early 1920s ideas of social and cultural reform were finding their way into mass-publication national language and civics textbooks produced by China's main publishing companies. By including literature written in the vernacular, national language textbooks published during the 1920s often exposed secondary students to a wide range of social reform writings. In addition, civics textbooks presented liberal and socialist ideas to students in a systematic way, giving them all the implicit and explicit sanction that goes with a formally published and approved textbook.

---

32. Yang Donglin, "Wode lishiguan."
33. Wu Sihong, "Dageming shiqi," 229.
34. As quoted in Fan and Su, "'Wusi' yundong zai Yangzhou," 45–46.

Republican-era writers advocating social change often adopted a vernacular style both as a political gesture and in an effort to address a broader audience. Because the new national language textbooks published after the promulgation of the New School System in 1922 collected a wide range of the most current vernacular literature, they often included pieces on social reform and worked in tandem with study societies and reading groups to spread reformist and revolutionary ideas.[35] Significantly, even the two leading secondary-level national language textbooks produced by the dominant mainstream publishers, Commercial Press and Zhonghua Book Company, included many writings critical of Chinese social and cultural conventions and introduced new conceptions of social roles and organization.[36] Zhou Zuoren's descriptions of utopian socialism, Cai Yuanpei's anarchist writings on integrating work and study, Hu Shi's calls for individual autonomy, and empirical analyses of social inequalities were all included.

In addition to these textbooks from the two major publishing houses, authors and publishers with particular political agendas used national language textbooks to promote their social visions. Perhaps the clearest example of this strategy was a lower middle school national language textbook compiled by two anarchist teachers from Wusong Middle School, Sun Lianggong (see Appendix B) and Shen Zhongjiu,[37] and published by the Nationalist Party–supported Popular Wisdom Book Company (Minzhi shuju).[38] Drawing heavily from

---

35. Some schools, such as Shanghai's Minli Middle School, continued to use selections by their own teachers as the readings for national language classes (*Shanghai Minli zhongxuexiao yichou nian zhangcheng*, 45–49). However, others turned to the language textbooks produced by mainstream publishing companies (e.g., Liao Shicheng et al., 77–81; Wang Ruyu, "Ni xinzhi zhongxue shiyong guowen jiaoxue cao'an").

36. Gu, Fan, and Ye, *Guoyu jiaokeshu*; Shen Xingyi, *Chuji guoyu duben*.

37. The other author, listed only as "Zhongjiu," was Shen Zhongjiu. See Wang Zicheng, "Huiyi Minzhi shuju," 113. Shen was a progressive educator with anarchist leanings who had been active in Hangzhou educational circles around the time of the May Fourth Movement. For Shen's activities during the 1910s and 1920s, see Chan and Dirlik, *Schools, passim*; Yeh, *Provincial Passages*, 129–36.

38. Sun and [Shen], *Chuji zhongxue guoyuwen duben*. Popular Wisdom Book Company opened in 1922 in Shanghai with funds collected by the Shanghai branch of the Nationalist Party. Popular Wisdom focused primarily on publish-

reformist journals such as *New Youth*, *Awakening*, *Weekly Review*, *Young China* (*Shaonian zhongguo*), *Eastern Miscellany*, and *Reconstruction* (*Jianshe*), this textbook focused its readers' attention on the seminal issues of "feudal customs," oppressed youth, women's social status, labor exploitation, and rural reform.

Even from these brief examples, one can see that vernacular language textbooks during the 1920s exposed students to diverse writings on social reform that paralleled the wide array of ideas they encountered in study societies, reading groups, and lectures. Writings in vernacular language textbooks critically reassessed Chinese social institutions such as the patriarchal family and challenged the oppressive conditions experienced by workers, peasants, women, and the young in the hierarchies of late imperial society and in modern industrial capitalism. At the same time, many of these writings advocated an egalitarian social order and called for a broadening of full social membership to China's most oppressed and deprived groups.

As with many language textbooks, New Culture Movement influence on the civics textbooks written according to the new curriculum of 1923 was direct and powerful.[39] Because the New School System curriculum standards incorporated New Culture Movement agendas, and reformist intellectuals wrote influential textbooks themselves, many of this period's textbooks actively promoted social and cultural reform. They often portrayed as an ideal an egalitarian society wherein individuals and social groups would be freed from the constraints of traditional hierarchies and the oppression of modern industrial society, both of which were seen to be harmful to "human dignity" (*renge*).

Tao Menghe (see Appendix B), one of China's leading young sociologists, in the following passage of his high school *Social Issues* (*Shehui wenti*) textbook, summarized many of the ideas about modern society that were contained in 1920s civics textbooks:

---

ing Sun Yat-sen's writings during the period between the company's founding and 1925. See Bian Chunguang et al., eds., *Chuban cidian*, 540. The press also was a vehicle for a number of leftist writings on literary reform, social reform, and women's rights during the First United Front (Wang Zicheng, 113–14).

39. As a convenient shorthand, I include high school "social issues" texts within the "civics" rubric.

The movement for freedom and liberation slowly followed changes in mode of production. One aspect is that workers gradually extended their rights: Not willing to be machines for production and not willing to be exploited by capitalists, they knew to use the power of unification to improve their position. After reading the history of the labor movement, we know that labor's struggle has been protracted but has most always attained success. Another aspect [of the movement for freedom and liberation] is that women have also demanded their rights and the position they should have in society. Within the household, occupationally, and politically, they have in succession attained equal status with men. Moreover, in regards to children, most people have discovered their inherent rights (*guyou de quanli*) and that they are not accessories of their parents or slaves of the family but rather budding elements of society (*shehuili hanbao weifa de fenzi*). Children should receive the best education that society can arrange and should have opportunities to develop. This movement for liberation and freedom is unstoppable, [and] following industrial development it steadily advances.[40]

Most civics textbooks of the 1920s echoed Tao, describing and promoting increasing equality for oppressed social groups, liberation of the individual, and recognition of all people's innate human dignity.

In line with New Culture discourse, civics textbooks criticized the traditional, large, patriarchal family for encouraging dependence, conflict, and laziness.[41] They called for egalitarian social relations within the family that would preserve the human dignity of all its members.[42] Many textbooks also advocated for free choice in marriage and families that would cultivate children to be independent and capable adults.[43] In short, they rejected the hierarchies and unequal power relations in the late imperial Chinese family, stressing instead equality and the human dignity of all family members.

Some textbook authors, such as reformist intellectual Shu Xincheng (see Appendix B), in his innovative textbook from Zhonghua

---

40. Tao Menghe, *Shehui wenti*, 108–9.

41. Gu and Pan, *Gongmin xuzhi*, 2: 40–42; Shu Xincheng, *Chuji gongmin keben*, 1: 15–16; Xiang Jutan, *Gongminxue*, 26–27. Xiang was a teacher at Huai'an's Jiangsu Provincial Ninth Middle School, so his textbook can be seen as reflecting the perspectives and concerns of local educators in contrast to textbooks by recognized intellectuals such as Shu Xincheng and Tao Menghe.

42. Gu and Pan, 2: 35; Shu Xincheng, *Chuji gongmin keben*, 1: 13–14; Xiang Jutan, 31.

43. Gu and Pan, 2: 36–40; Shu Xincheng, *Chuji gongmin keben*, 1: 13–14.

Book Company,[44] further offered extensive discussions of the labor and women's movements, supporting those groups' claims for greater social equality and emphasizing their humanity. In the case of workers, he used their claim to basic human dignity as citizens to justify their rights to organize labor unions and engage in collective bargaining and strikes to obtain better working conditions and standards of living.[45] Shu similarly used women's status as human beings to support claims of equality with men: "Men and women are both human. This statement is acknowledged by most people. Since men and women both belong to humankind, logically they should have the same social status, and in terms of human dignity (*renge*) are naturally equal."[46] Shu went on to connect women's natural rights as human beings with their political rights as full citizens and the ability to have their work valued in the same way as men's.[47] In his discussions of women's and workers' rights, Shu challenged the inequalities inherent in patriarchal society and modern industrial capitalism.

Implicit in these critiques of the social inequality and oppression characteristic of past and existing social institutions was a new view of the person as an individual with innate human dignity and personal rights. Civics textbooks elaborated this perspective in part through their discussions of the basis of ethical judgment and in their portrayals of individual rights. In terms of ethics, the textbooks called for self-conscious and independent actors to make moral judgments based on logical reasoning about their circumstances.[48] Shu Xincheng, for instance, described an evolutionary process of ethical development whereby instinctive (*benneng de*) and customary (*sushang de*) ethics were replaced by a more advanced form of "self-conscious" ethics (*zijue daode*). In the practice of self-conscious ethics, Shu asserted, individuals came to ethical decisions through logic and free choice, taking into account both individual development and social welfare.[49] Some civics textbooks also

---

44. With Ministry of Education approval, Shu's text went through 14 printings in five years and continued to be published, with some revisions, into the 1930s (Peake, *Nationalism and Education*, 175–76).

45. Shu Xincheng, *Chuji gongmin keben*, 2: 47–52. Cf. Tao Menghe, 99–109.

46. Shu Xincheng, *Chuji gongmin keben*, 2: 55–56.

47. Ibid., 2 (1923): 54–62. Cf. Tao Menghe, 107–9.

48. Gu and Pan, 2: 2–4; Xiang Jutan, 25–26.

49. Shu Xincheng, *Chuji gongmin keben*, 3: 54–60.

outlined in great detail the rights and freedoms that individuals were assumed to have in democratic societies.[50]

Through this stress on the moral autonomy and rights of the individual, many civics textbooks of the 1920s seemed to be promoting a form of liberal individualism. In this framework, all people, regardless of their social or economic status, were considered free and equal members of society based upon their intrinsic value as human beings. According to these textbooks, the citizens of an ideal modern social order needed to be freed from the various forms of social constraint and obligation that characterized both late imperial Chinese society and industrial capitalism so that they could pursue their own interests and develop their own capacities. They would become free to function as autonomous people with innate human dignity. From this perspective the social order of modern China was to be quite distinct from the hierarchical and role-centered social order of the late imperial period.

III. NARROWING HORIZONS, PARTY TUTELAGE, AND CONFUCIAN REVIVALISM

The tumultuous and dynamic political environment of the 1910s and 1920s facilitated the spread of social and cultural reform ideas to local schools. Provincial, let alone national, governments intervened little in local schools, allowing a wide spectrum of views to be explored in study societies and lectures. Limited government censorship coupled with reformist intellectuals' input into the formulation of curricula and textbooks meant that language and civics textbooks of the 1920s could relate radical ideas about social and cultural change.

This open, permissive environment changed dramatically after the founding of the Nationalist government in 1927. The Nationalist Party aspired to far more extensive centralized control than had earlier postimperial regimes. After Chiang Kai-shek's purge of CCP members in April 1927, the government vigorously sought to eradicate any signs of communist influence in lower Yangzi region schools. Moreover, the Nationalist Party leadership was ambivalent about New Culture Movement agendas for social and cultural

---

50. E.g., Zhou Gengsheng, *Gongmin jiaokeshu*, 1: 50–76.

reform and promoted its own program for social change based on Sun Yat-sen's writings and political legacy. The new Nationalist government sought to both control and redeploy the media and techniques that had developed to spread new social ideas during the 1910s and 1920s. The result was a significant shift in the messages about social change and modern citizenship that students received in lower Yangzi region secondary schools.

Party and government leaders made a major stride toward controlling student study societies and reading groups in 1930 simply by declaring that all such groups had to be organized under the auspices of the heavily monitored student self-government associations. Party-mandated teacher and administrator supervision over all reading groups and study societies, combined with strict oversight of self-government organizations by local party and government officials, ensured that study societies focused on Marxism and anarchism would be a thing of the past or would operate entirely underground. In fact, student study societies in lower Yangzi region secondary schools increasingly followed the contours of the curriculum or concentrated on less politically sensitive topics.[51]

In addition, operating under a new political mandate, provincial government officials and school authorities during the Nanjing decade also closely investigated secondary students' extracurricular readings for politically sensitive materials. For instance, in 1933 the Jiangsu Department of Education conducted a survey of student extracurricular readings in all the province's middle schools, with a concern for their effect on students' "character and thought." Schools were directed to provide the name, publisher, and author or editor for the journals, reference books, and fiction students were reading.[52] The Jiangsu Department of Education's 1936 ban of and calls for vigilance against a long list of supposed CCP-related publications reveals the authorities' concern that students might be reading leftist materials that were published under cover of the National Salvation Movement.[53]

---

51. E.g., Jiangsu shengli Suzhou zhongxue, ed., *Xuesheng zizhihui gaikuang*, "Disanjie xuesheng zizhihui ge yanjiu hui she renshu tongji biao."
52. Wu xian jiaoyuju xunling, no. 349, SMA2 Jia5.422.
53. Wu xian jiaoyuju xunling, no. 18, SMA2 Jia5.431.

Individual schools correspondingly exhibited increased scrutiny of students' extracurricular readings. At Jiangsu Provincial Yangzhou Middle School, for instance, teachers were expected to "direct students to read carefully books that are beneficial to their bodies and minds."[54] Likewise, teachers and administrators at Zhejiang Provincial Jinhua Middle School checked student extracurricular readings, confiscated banned books, compiled statistics on student readings, and subscribed to government-sponsored publications such as *Zhejiang Youth* in an effort to manage students' reading.[55] Government and school authorities closely monitored print media, constraining students' opportunities to experiment with radical approaches to social change.

The spectrum of views students encountered through lectures narrowed after 1927 as well, though not only due to an intentional policy orchestrated by the Nationalist government. Rather, the changed political environment seems to have made party and government leaders, as well as party-aligned intellectuals, desirable speakers at many schools. A striking example of the change in intellectual and political atmosphere comes from the private Zhenhua Girls' Middle School in Suzhou. During March and April of 1927, contemporaneous with the Nationalist Party's occupation of the lower Yangzi region, Zhenhua hosted a series of lectures about Sun Yat-sen's ideology, the Three Principles of the People. In commenting on this situation, the editors of the record of school events noted, "Since the [Wu County] Education Department implemented party-centered education (*danghua jiaoyu*), this school has invited famous people to lecture in hopes of according with the trends of the new tide."[56] After the Japanese invasion of Manchuria in 1931, lectures on international relations and China's military crisis became common at many regional schools.[57] In general, lectures on party principles and

---

54. Zhou Houshu, "Yangzhou zhongxue jiaoxun heyi hou zhi chubu shishi," 62.

55. Zhejiang shengli Jinhua zhongxue chuban weiyuanhui, 24–25.

56. Chen and Ji, "Benxiao ben xuenian dashiji." After 1934, the New Life Movement similarly became a focal topic for lectures on Nationalist Party ideology (*Minli xunkan*, no. 7 [May 20, 1936]).

57. *Zhejiang shengli gaoji zhongxue xiaokan*, 64–65. Cf. *Minli xunkan*, nos. 2–3 (March 20–April 10, 1936); *Songjiang nüzhong xiaokan*, no. 26 (December 30, 1931): 4–7; no. 27 (January 20, 1932): 7–9.

nationalist themes became a central part of school life during the Nanjing decade.

This is not to say that lectures about social issues ceased altogether. In fact, a review of lectures given at Jiangsu Provincial Songjiang Girls' Middle School between 1929 and 1934 reveals lectures on many topics that had engaged social reformers during the 1920s, including women's education, social responsibility, and social roles, as well as lectures on family problems and superstition.[58] However, during the Nanjing decade there was a much greater possibility that these lectures would present students with a more conservative and party-centered perspective on modern society than they might have encountered during the previous decade. For instance, at Songjiang Girls' Middle in 1929, lectures on family problems, women in society, and women's education were presented by China's most famous eugenicist, Pan Guangdan. In these talks, Pan called for compromise between China's large family and the Western small family, and he criticized the women's movement in the West for not recognizing intrinsic physical and psychological differences between men and women that made full gender equality impossible.[59] Such lectures were a far cry from the messages about gender politics students had heard at regional schools just a few years before, when, for example, CCP member Chen Wangdao had lectured at Shaoxing Girls' Normal about "women's liberation and revolution."[60] Radical perspectives on social change and sociopolitical order, such as anarchism or Marxism, and even to some extent liberal individualism, were now eliminated from local schools' lecture schedules. The "universe of discourse" about society narrowed considerably under Nationalist Party restrictions.

State, Family, and Society

The Nationalist government, through the dual pressures of its curriculum standards and censorship apparatus, sought to align the messages of language and civics textbooks more closely with its ideology

---

58. *Songjiang nüzhong xiaokan*, no. 3 (February 5, 1929): 16–20; no. 6 (July 1, 1929): 22–30; no. 12 (June 20, 1930): 28–30; no. 18 (May 20, 1931): 6–11; nos. 19–20 (June 15, 1931): 25–27; no. 21 (September 30, 1931): 13–14; no. 43 (May 15, 1933): 8–11.

59. *Songjiang nüzhong xiaokan*, no. 6 (July 1, 1929): 22–30.

60. Wu Sihong, "Dageming shiqi," 229.

and policies. In language textbooks, the range of readings students encountered shifted somewhat from the precedents of the 1920s. Mass-produced language textbooks and large collections of selected readings meant to allow teachers to compile their own textbooks still included many readings on rationalizing Chinese culture and creating a more equal society.[61] However, these texts included few readings advocating complete reformulation of social institutions or radical anarchist and Marxist approaches to correcting social inequalities. In their place were readings that celebrated the Nationalist Party, called for party and state guidance in gradual processes of social leveling and reform, and promoted an ideal of young people's dedicating themselves to national development and social service.

In ways similar to language textbooks, many state-sanctioned party-doctrine and civics textbooks of the Nanjing decade paralleled civics textbooks of the 1920s in identifying the liberation and equality of women, workers, and the poor as important goals of social reform. However, Nationalist-era textbooks privileged the role of the state in engineering social equality by mediating between competing social groups and improving social welfare by guiding national development. Further, in many cases, the textbooks described service to national society as a whole, not struggle for a particular group's social and economic interests, as the proper role of true citizens. Together these formulations led to a vision of all social sectors working equally under party leadership for the benefit of all, a vision that resonated clearly with Sun Yat-sen's approach to development and nation building.

The party-doctrine and Three Principles of the People textbooks published between 1927 and 1934 confronted issues of poverty and economic inequality through discussions of Sun Yat-sen's "people's livelihood" (*minsheng zhuyi*) lectures. Consequently, the central themes in those discussions were the fundamental unity of all classes in national society, the importance of cooperation over conflict in dealing with poverty and social inequality, and the central role of the party and state in planning for development and mediating potential class conflicts. One Three Principles of the People textbook author,

---

61. E.g., Zhu Jianmang, *Chuzhong guowen*; Zhu and Song, *Chuzhong guowen duben*; Fu Donghua, *Guowen*; *Kaiming huoye wenxuan zongmu*; Jiang and Zhao, eds., *Beixin wenxuan*.

Zou Zhuoli, highlighted these points in his discussion of people's livelihood. He drew heavily on Sun's ideas and wrote the section as an extended polemic against Marxist ideas of class struggle being the motive force of human evolution.[62] Zou instead stressed the unity and cohesiveness of human societies and how, rather than conflict, coordination and cooperation under state guidance generated social development and improved welfare for all social classes. Zou also denied that China had deep class fissures or major income distribution problems. He argued, "All Chinese people are poor. There is no wealthy privileged class (*da fu de teshu jieji*) but only general common poverty. The so-called unequal distribution of wealth is nothing more than distinguishing between the terribly poor (*dapin*) and slightly poor (*xiaopin*) within the impoverished class."[63] Emphasizing national poverty effectively made all China's social problems developmental problems, a question of how best to speed and coordinate China's economic growth while also mediating what were perceived to be minor differences of wealth between social classes.[64]

At one level, by emphasizing national welfare and downplaying inequalities within Chinese society, textbooks like Zou's obscured the intensity of the exploitation and poverty experienced by China's workers and peasants. At the same time, by placing primary emphasis on national development, these textbooks made a privileged place for the state as a specialized body that would assess overall social welfare from a transcendent position, plan a trajectory of development, and coordinate competing social interests.[65] Zou described the state as the primary agent promoting socioeconomic development by equalizing land rights, controlling (*jiezhi*) individual capital, and developing state capital in the areas of large industry, transport, banking, and communications.[66] This image of the party-state, which

---

62. Zou Zhuoli, *Sanmin zhuyi jiaoben*, 2 (1929): 60–72. Zou's critique of Marxism and his corollary emphasis on national development and interclass harmony and cooperation clearly resonated with Dai Jitao's writings. See Dai Jitao, "Sun Wen zhuyi zhi zhexue de jichu."

63. Zou Zhuoli, 2: 67.

64. Ibid., 66–71.

65. Chatterjee, *The Nation and Its Fragments*, 200–219.

66. Zou Zhuoli, 2 (1929): 68–71. For a discussion parallel to Zou's, see Su Yiri, *Minsheng zhuyi*, 54–70. For a somewhat different approach, see Wei and Xu, *Chuzhong dangyi*, 6: 11–21.

was rooted in Sun Yat-sen's writings, as the all-seeing, centralized planner of nationwide development reflected the "high modernist" aspiration of total control that James Scott associates with Leninism.[67] The state-centered approach to encouraging social equality and managing national socioeconomic development also worked rhetorically to legitimize Nationalist Party rule by casting it as the primary agent of positive socioeconomic change.[68]

After 1933 social inequalities and reform were discussed most fully in the "Social Issues" volumes of high school civics textbooks. Following brief discussions of the nature of society and the causes of social problems, these textbooks included a series of chapters focused on particular social issues, such as the reorganization of the family, population problems, the women's and labor movements, the decline of the village, poverty, and crime. "Social Issues" textbooks' discussions of the so-called labor problem (*laodong wenti*) provide a prime example of their approach to questions of social equality.

In terms of workers' social status and the labor movement, the textbooks detailed the inequities that came with an industrial economy and lamented the damage done to workers' livelihoods and families because of the exploitation of capitalism.[69] The textbooks then identified labor laws under state auspices as the primary way to protect workers and moderate the relations between labor and capital. One textbook argued that "if we relied on the natural trend of labor and management, and let them continue to dispute, the contention between labor and management would never come close to being resolved, and workers would always occupy a position of being oppressed and exploited. Thus, workers, in order to advance their welfare and maintain their livelihood, have to depend on legislative protection."[70] State guidance was seen as central to harmonizing relations between antagonistic classes. At the same time, textbooks commented on the backward state of Chinese industry and argued

---

67. Scott, *Seeing Like a State*, chap. 5.

68. Other textbooks explicitly identified the Nationalist Party as representative of the whole people's welfare and interests (Guo and Wei, *Gaozhong dangyi*, 1 [1932]: 4–10).

69. Du and Zhang, *Gaozhong gongmin*, 1: 33–44; Li Zhendong, *Shehui wenti*, 59–82; Sun Benwen, *Shehui wenti*, 69–92.

70. Du and Zhang, *Gaozhong gongmin*, 1: 41–42.

that the basic issue in China was mainly one of underdevelopment that both labor and capital had to work to redress.

Consequently, some Nanjing-decade "Social Issues" textbooks also described state-led coordination of labor and capital for national economic development as a concern more fundamental than the inequities between these two social classes. In the words of leading sociologist Sun Benwen (see Appendix B) in his textbook, "China's workers' present problem is not how to handle the oppression of this country's capitalists. Rather, it is how to cooperate with the capitalists and to help the government develop [China's] state capital (*guojia ziben*) and promote commerce and industry in order to eliminate foreign economic oppression."[71] These textbooks called for basic protections of the rights and livelihood of workers as full citizens to create the conditions whereby workers could contribute optimally to developing national society as a whole. In turn, contributing to national development was portrayed as the primary way for workers to fulfill their responsibilities as social members and also benefit themselves as national citizens. The connection between social membership and workers' rights was made explicitly when calls for protection of the workers themselves were justified by noting that things harmful to the workers, as national citizens, were by extension harmful to the nation as a whole.

Workers are constitutive elements (*goucheng fenzi*) of the state and nation, as well as essential factors in enriching the state and strengthening the military (*fuguo qiangbing*) and thus in reviving the nation. Consequently, long hours of work, meager wages, onerous treatment, and cruel use of child and female labor is not only a problem for the workers themselves, but is also a problem for the health of the citizenry (*guomin jiankang*) and national recovery (*minzu fuxing*).[72]

Workers' equal status and basic welfare as citizens were promoted in these textbooks but as means to maintain the health of national society and advance its development.[73]

---

71. Sun Benwen, 91. Cf. Li Zhendong, 81–82. This formulation reflects long-held Nationalist Party assumptions about classes being component units of a single social whole. See Tsin, *Nation*, 53–54.

72. Li Zhendong, 79. Cf. Wei and Xu, *Chuzhong dangyi*, 6: 16.

73. The "Social Issues" volumes of civics textbooks took similar approaches with discussions of rural poverty and women's rights. See Du and Zhang, *Gaozhong gongmin*, 1: 46–63; Li Zhendong, 101–6; Sun Benwen, 121–30, 133–50.

Nationalist Neotraditionalism

Even as Nanjing-decade textbooks called for the pursuit under party-state leadership of social equality within a modern, industrialized society, they also drew on reworked Confucian models of hierarchy and the family-based society to encourage unity and social order.[74] They did so in part to instill a sense of cultural distinctiveness in modern Chinese social organization and roles. But the effort to re-invent traditions of moral practice and hierarchical social roles ran into trouble at several different levels.

Many Nanjing-decade textbook authors portrayed the Chinese family as a hothouse for cultivating civic virtues, as a basis for the social order, and as an icon of Chinese distinctiveness. In doing so, they related family and nation in ways similar to the sections of Cai Yuanpei's textbook discussed at the start of this chapter. For instance, Du Weitao (see Appendix B) and Zhang Liuquan, in their high school civics textbook, portrayed filial piety and companionship (*you*) as the essential elements of Chinese culture and as fundamental relational patterns cultivated within the Chinese family.[75] The practice of such filial piety and companionship within the family nurtured a sense of altruism (*lita*) that could be extended out into society, forming the basis for social order and progress. Students were called upon to serve the social group, follow laws and rules, and develop their capabilities. Moreover, in this time of crisis for the nation, textbooks asked students to "control their desires and join with the group (*keji hequn*)," restricting their own freedoms and interests to sacrifice for the nation.[76] These patterns of "submergence of personal desires" and "selfless impartiality" were characteristic of the hierarchical norms associated with the Neo-Confucian family.[77] Du and Zhang also called for Chinese youths to extend their love of family into love of their ethnicity and nation, connecting the two levels as microcosm and macrocosm. The equation of family and nation in

---

74. For analysis of one textbook's discussion, based on Chen Lifu's writings, of social ethics in neotraditional, hierarchical terms, see Culp, "Setting the Sheet of Loose Sand," 70–72.

75. Du and Zhang, *Gaozhong gongmin*, 3: 170–73.

76. Ibid., 173–80.

77. Munro, "The Family Network," 264–69.

these textbooks encouraged students to foster a deep attachment to the nation by viewing it as an ascriptive community that they were connected to as they were to their own families—that is, by blood and feelings.

This attempt to root more generalized social mores in hierarchical family relations clashed with the New Culture critique of the Confucian family and problematized the connections earlier reformers had made between an emerging small, modern Chinese family and its Western counterpart. Textbooks of the 1930s still challenged the Chinese large family system in ways that were reminiscent of May Fourth–period critiques and consistent with the Nationalist government's family-reform policies, which encouraged gradual evolution from the patriarchal joint family to a smaller conjugal family.[78] Textbooks castigated extended, patriarchal families for encouraging conservatism, dependence, and parochialism. They charted a "natural" trend toward smaller families with industrialization. And, they criticized traditional inequalities in norms of chastity and the arranged marriage system that had given young people little say in their own matches.[79]

Yet some authors also celebrated filial piety as the central feature of Chinese family life and the basis for its natural affections, or they called for multigenerational coresidence to continue to be a central part of the modern Chinese family.[80] Du and Zhang further contrasted images of Western and Chinese families, valorizing the latter. They portrayed the Western family as a group devoid of feelings, a loose collection of individuals who were most interested in their own material interests and who were bound together by nothing more than the legal obligation for parents to raise their children. By contrast, they characterized the Chinese family as being tied by strong emotional bonds and selfless dedication by each member to the welfare of the family unit.[81] These Nanjing-decade civics textbook portrayals of the ideal family were consistent with Nationalist government family reform policies, which Susan Glosser argues "attempted

---

78. Glosser, *Chinese Visions*, chap. 2.

79. Du and Zhang, *Gaozhong gongmin*, 1: 10–16; Li Zhendong, 16; Sun Benwen, 9–29; Sun Bojian, *Gongmin*, 1 (1935): 49–50.

80. Li Zhendong, 18; Sun Benwen, 29–33; Sun Bojian, 1 (1935): 48–49, 52–53.

81. Du and Zhang, *Gaozhong gongmin*, 3: 170–72.

to preserve some of the ties of obligation, duty, and deference that characterized the traditional family ideal."[82]

When the family was taken as the fundamental unit of society, these characterizations of the distinctive Chinese family and contrasts in family styles represented the perceived differences between Western liberal society and the distinctive Chinese national society envisioned by textbook authors and many Nationalist Party leaders. Chinese society was to be bound in collective harmony by close emotional ties, with all of its members committed to selfless sacrifice for national welfare and progress, as well as obedience to the rules and norms determined by the state as patriarch—a reconstituted traditional family writ large. Because these family patterns were associated with Confucian norms, a family-based Chinese nation would be culturally distinctive. But such calls for family-based solidarity were complicated by two decades of sharp criticisms of the Confucian family, some of which appeared in these very textbooks. These critiques had eroded the family's self-evidence as a "natural" social unit organized according to hierarchies of age and gender.

IV. CITIZENS AND THE SOCIAL BODY

Republican-period civics textbooks and extracurricular readings presented multiple, competing visions of social order—hierarchical versus egalitarian, individualist versus collectivist—and social reform—for example, anarchist, socialist, Marxist, and statist. Underlying these dramatic variations in portrayals of social life and social change was a remarkably stable conception of modern society as a functionally integrated whole or organism. This view of social order was presented in normative terms in moral cultivation, civics, and party-doctrine textbooks and taken as an assumed point of departure for discussions of society and civic action in many textbooks and other contemporary texts and writings.

Tao Menghe concisely stated this widely held conception of society near the start of his high school *Social Issues* textbook from 1924. "Society really is a kind of organic group (*youji de tuanti*), because the countless social relations [that compose it] are very close. If

---

82. Glosser, *Chinese Visions*, 98.

there is slight change in one aspect of one kind of relationship, it can affect the whole body; a person's every movement and action has a power that can influence society."[83] Though not all textbooks explicitly used the metaphor of the organism, many of them developed the same formulation of society as a dense network of interrelations in which the individual was an enmeshed component part. In describing society, authors stressed the interdependence of individual and social group. Individuals relied on society for various kinds of material and cultural support, and the welfare and order of society as a whole grew from the contributions of individuals.[84] Civics textbooks of the 1920s characterized all social groups, ranging from the family, the school, the occupation group, and local society to the nation, as sharing this organismic, functionally interrelated structure.[85]

Both the moral cultivation textbooks of the 1910s and the party-doctrine textbooks of the Nanjing decade paralleled the neat formulation presented in the civics textbooks of the New School System period (1922–28). For instance, Cai Yuanpei, in his influential early Republican moral cultivation textbook, repeatedly related individual to society in a part-to-whole formulation. Characteristic of this approach is Cai's explanation of "self-restraint" (*zizhi*):

If one does not use clear reason and a firm will to control them [desires], then their damage is more than words can describe. And, it is not just a matter of one person. If a whole country's people are the slaves of desire, then the improvement of the political system and the advance of arts and letters cannot be obtained and expected, and the future of the country cannot be inquired about.[86]

In this passage, the actions of individual citizens in aggregate constitute society. Elsewhere, Cai argued that individuals were to work to benefit society, for it was society that nurtured and protected them.[87]

---

83. Tao Menghe, 3.

84. See Shu Xincheng, *Chuji gongmin keben*, 1 (1923): 1–9; Xiang Jutan, 63–64; Zhou Gengsheng, 1 (1926): 1–2.

85. National language textbooks of the 1920s also included essays that defined society in organic terms and stressed the interdependence between individual and society (e.g., Gu, Fan, and Ye, *Guoyu jiaokeshu*, 2: 7–9; 5 [1923]: 146–52; Shen Xingyi, 1: 42–47, 193–99; 3: 23–33, 228–36).

86. Cai Yuanpei, *Dingzheng zhongxue xiushen jiaokeshu*, 11.

87. Ibid., 61–63.

Similarly, party-doctrine and civics texts after 1927 portrayed individuals and society as being in a direct, part-to-whole relationship, and they described society as being composed of functional interrelations among people.[88] In the words of one Three Principles of the People textbook from 1928, "The individual is an element making up society, and society is formed out of the union (*jihe*) of many individuals. If an individual is not within society, he cannot provide for his survival (*buneng mou sheng*)."[89] Based on the assumption that individuals and the social group were functionally interrelated, textbooks reasoned that because individuals benefited from the development of the social group, they should rightfully be committed to contributing to society and adhering to norms of social order.[90] In the words of one lower middle school civics textbook, "we must respect social discipline and support public welfare, not misunderstanding freedom and overemphasizing private interest. We must know that the welfare and freedom of the individual must be pursued within the welfare and freedom of the group."[91]

The intellectuals, educators, and editors who wrote moral cultivation, civics, and party-doctrine textbooks during the Republican period had a common reference for this organic vision of society in the writings of China's first social theorists at the end of the Qing dynasty. Those social theorists, in turn, had been influenced by Western sociologists such as Herbert Spencer. Liang Qichao and Yan Fu, in particular, had emphasized the need for social groups (*qun*) ranging from the family to the nation-state to unify and become strong through the committed actions of the members who composed them in aggregate.[92] They portrayed a part-to-whole relationship between citizen and society because it seemed to encourage national unity and provide a justification for mobilizing citizens for national

---

88. Du and Zhang, *Chuzhong gongmin*, 1: 55–56; Hu Yuzhi, *Minzu zhuyi*, 1; Lou Tongsun, *Minquan zhuyi*, 42–43; Su Yiri, *Minsheng zhuyi*, 2; Sun Benwen, 1–2; Sun Bojian, 1 (1935): 4, 6–8; Wei and Xu, *Chuzhong dangyi*, 1: 1–3.

89. Su Yiri, *Minsheng zhuyi*, 2.

90. Ibid., 4; Sun Bojian, 1 (1935): 4, 6–8; Wei and Xu, *Chuzhong dangyi*, 6: 66–69.

91. Du and Zhang, *Chuzhong gongmin*, 1: 56–57. Emphasis in original.

92. Chang, *Liang Ch'i-ch'ao*, 95–100, 172–76, 204–5; Nathan, *Chinese Democracy*, chap. 3; Pusey, *China and Charles Darwin*, 62–71, 111–12, 289–307; Schwartz, *In Search of Wealth and Power*, 56–58.

service at a time of radical disintegration and foreign threat that they interpreted in a pseudo–social Darwinist framework of a collective struggle for survival. Significantly, Spencer himself resisted the conclusion that the organic relationship between individuals and the collective privileged the latter, and he defended the autonomy of individuals within society.[93] But in the atmosphere of national crisis during the late Qing and Republican periods, China's intellectuals and educators consistently stressed the priority of the social collective.[94]

The ability of the organic vision of society to link the two planes of collective solidarity and individual action made it a very powerful formulation. By relating the welfare of the social collective to individual action, Liang's and Yan's vision of society as an organism built on a Confucian structure of the harmonious state rooted in personal moral behavior that dated back to the foundational text the *Daxue* (Great learning).[95] At the same time, the horizontal interconnection of a broad-based community of citizens corresponded to the modern idea of the nation-state as distinct from the hierarchically ordered dynastic realm or the potentially universal cultural community.[96] For although textbooks applied the image of the social organism to all levels of society, they often identified the nation-state as the highest and most important level of social organization in the early twentieth-century context.[97] Like the image of the nation as geo-body, the idea of the social organism helped to make the abstraction of the nation imaginable to students by providing them with a concrete

---

93. E.g., Spencer, "The Social Organism," 276–77.

94. Benjamin Schwartz argues forcefully that Chinese intellectuals' priorities might have been more logically consistent than Spencer's, for the organic "analogy points to the ever-growing role of the state and subordination of all sub-organs of society to the goals of the state" (*In Search*, 76–77).

95. For the *Daxue*, see Chan, *A Source Book*, 84–94. For late imperial examples of the relationship between sociopolitical order and the actions of individuals, see de Bary, "Introduction," 11–22; Rowe, *Saving the World*, chaps. 9, 12; Wright, *The Last Stand*, 60–63.

96. On this contrast, see Anderson, *Imagined Communities*, 5–37.

97. E.g., Cai Yuanpei, *Dingzheng zhongxue xiushen jiaokeshu*, 60–61; Du and Zhang, *Chuzhong gongmin*, 1 (1935): 70–71; Hu Yuzhi, *Minzu zhuyi*, 1: 2–3; Shu Xincheng, *Chuji gongmin keben*, 1 (1923): 35–36; Sun Bojian, 1: 78; Zhou Gengsheng, 1 (1926): 11. Some textbooks acknowledged that an international society was possible, but they presented it as an ideal future goal, identifying the nation as the highest operative social body (e.g., Wei and Xu, *Chuzhong dangyi*, 1: 12–21).

metaphor for the shape and structure of the national community. Further, by describing the national community as a network of functionally interrelated and interdependent individuals, the civics and language textbooks suggested that the nation was constituted by bonds created through citizens' civic action rather than by ascribed qualities like race or culture.

The widespread use of the image of the social organism in educational materials and other writings of the Republican period made it a fundamental touchstone in debates about society and politics. Used as a basic assumption in civics and language textbooks, it had power as a grounding premise in the arguments of educators and students. In fact, throughout the Republican period, lower Yangzi region students' writings were replete with references to an organic society, especially during the 1920s and 1930s after this social model became common in educational materials. When students deployed the organic metaphor, they often adopted language that mirrored closely that of their textbooks and other writings of the period, and they presented that image of society as an unquestionable assumption from which they extended other arguments. Because of its ubiquity, the social organism metaphor provided students with an authoritative foundational trope from which to argue. It also conditioned how they portrayed individuals and their relationship to social groups.

For instance, Jiang Jiaxiang, a second-year high school student at Southeast University Affiliated Middle School, in writing an essay on student life at the school in 1926, used as premises for several of his arguments statements such as the following: "People are members of society and cannot live separately from society. Society is a group, so people cannot lack group spirit." "That students cannot live on their own is like people not being able to separate from society and live; how important is group life!"[98] Both statements could have come directly from the lower middle school civics textbook by Shu Xincheng, who taught at the school, or Tao Menghe's *Social Issues* textbook. This image of society was also reflected in other student writings of the 1920s. Students promoting a range of policies, perspectives, and political movements took as axiomatic the individual's

---

98. Jiang Jiaxiang, "Fuzhong xuesheng shenghuo," 382, 380.

integration within and dependence on society and, just as importantly, the power of individual action to influence society.[99]

A conception of society as an organic whole composed of cellular individuals was even more common in student writings during the 1930s, when again it often appeared as an unquestioned and undefended assumption. As stated concisely by a student from Jishan Middle School in Shaoxing, "People cannot separate from the masses (*qunzhong*) and live independently, because people fundamentally were produced from within the masses. I think this is a general law that nobody can deny."[100] The image of being embedded within society, in turn, underpinned the argument that individuals had a responsibility to commit themselves to society's welfare. Students sometimes expressed this sense of social duty with the set phrase "All people have responsibility for the flourishing or decline of the state" (*guojia xing wang, pifu you ze*) or Sun Yat-sen's assertion that "Human life takes service as its objective."[101] As Nanjing-decade lower Yangzi region students used the image of the social organism as a rhetorical device, they described themselves as being embedded within, dependent upon, and thus responsible to national society.[102]

V. CULTIVATING SOCIAL CELLS

As these student writings suggest, the direct relationship drawn between individual and community in the organic conception of society meant that an individual's actions were all fraught with social implications. In the words of Tao Menghe, "because sometimes people's actions are just for themselves without considering others, it is thought that these are 'individual actions' (*geren xingwei*), but actually individual actions all have social influence. Thus thinking that society and the individual are mutually opposed is mistaken. In the

---

99. See, for instance, Sheng Youxuan, "Wumen xuesheng de zeren"; Wang Weihua, "Yichan," 44–45; Wu Tao, "Xuesheng zhi tongbing"; "Zeren fakanci," 94–95.

100. He Qinghu, "Zizhu he huzhu," 2–3.

101. Wang Duanteng, "Qingnian yinggai renshi," 4–5; Wang Yongquan, "Gaozhong biyesheng de zeren," 12. Cf. [Wang] Binglong, "Rensheng de yiyi."

102. For similar formulations in other student writings, see Chengrong, "Qingnian he huanjing"; Gu Ruwen, "Xuesheng fuwu zizhi tuanti"; Li Yunfang, "Huzhu lun."

world there are only social people, not individuals."[103] Many moral cultivation, party-doctrine, and civics textbooks, because of the integral connection they drew between individual and society, argued that individuals' thoughts and actions should not be "for themselves" but rather should, to varying degrees, be oriented toward the general social welfare.[104]

These textbooks, based on the part-to-whole relationship they drew between citizen and society, mapped out a variety of socially productive personal behaviors and ethical orientations that they expected students, as future citizens, to embody. At the start of the Republic, Cai Yuanpei's moral cultivation textbook defined the public morality for the new state's citizens by listing a range of attitudes and practices that resonated with the "civic virtues" (*gongde*) outlined by Liang Qichao in *The New Citizen*. Cai called on young people to be diligent, brave, knowledgeable, trustworthy, and physically strong.[105] These virtues together defined a new kind of active, capable, and engaged citizen for China's modern social order and were all justified by and oriented toward the strength and welfare of the nation.

The process of delineating the qualities that would allow student-citizens to contribute optimally to the welfare of the social whole continued through the 1920s and 1930s. Writing in 1923, educators Gu Shusen (see Appendix B) and Pan Wenan offered an impressive list of civic virtues for the individual citizen to learn and perform. They included self-sufficiency (*zizhu*), struggle, and creativity, as well as honesty, obedience, self-control, steadfastness, tenacity, boldness, bravery, hard work, modesty, diligence, meticulousness, love, and happiness.[106] Other textbooks paralleled Gu and Pan's in calling for individuals to cultivate a range of civic virtues that would positively affect society as a whole.[107] Because an organic view of society dictated that "the interests of the individual reside within the welfare of

---

103. Tao Menghe, 4.

104. On reform of the individual to transform society, see Huang Jinlin, *Lishi, shenti, guojia*, chap. 2.

105. Cai Yuanpei, *Dingzheng zhongxue xiushen jiaokeshu*, 3–20.

106. Gu and Pan, 2: 10–26. They also discussed personal hygiene as a way to strengthen individual bodies to benefit society (1: 2).

107. See, in particular, Shu Xincheng, *Chuji gongmin keben*, 3: 71–76; Zhou Gengsheng, 1 (1926): 6–8.

the group," individual citizens were expected to shape their minds, bodies, and behaviors so as to benefit society.[108]

Nanjing-decade civics textbooks refined their characterizations of the different levels and kinds of civic virtues in an attempt to shape students' public behaviors. In their accounts, the ideal citizen had to be self-disciplined, obedient, and accord with public mores, but s/he also had to have a sense of moral inwardness and develop his or her own abilities and talents so s/he could be a productive contributor to society as a whole. Consequently, these 1930s textbooks encouraged the development of productive and creative individuals and also sought to shape their personal conduct to create cohesive and orderly social groups.[109] Du and Zhang's discussion of the moral demands that group life placed on each person illustrated a dialectical movement between individual refinement and dedication to social welfare. They described four kinds of moral requirements: first, to control selfish desires that wrongly encouraged people to put themselves above others and the family above the nation; second, to develop one's own capabilities so they might contribute to collective life; third, to cultivate moral character, which included keeping one's clothes neat and tidy, maintaining one's appearance, being polite to others, and moving with good posture; and fourth, to cooperate and participate in mutual aid.[110] Students, here, were isolated and marked out as individuals, but their refined individual behavior and personal achievement was meant to serve the good of the social whole.[111] Social membership carried with it moral requirements for both self-cultivation and social service.

By grounding the social order in the moral action of individuals, moral cultivation, civics, and party doctrine textbook authors followed the Confucian pattern of individual cultivation for the public welfare, but the new civic morality departed from Confucian precedents in several important ways. The organic conception of society placed individual and society in a direct, unmediated relationship,

---

108. Shu Xincheng, *Chuji gongmin keben*, 3: 76.
109. E.g., Wei and Xu, *Chuzhong dangyi*, 1: 4–12.
110. Du and Zhang, *Chuzhong gongmin*, 1: 58–60. Sun Bojian, 1 (1935): 89–91, lists a similar collection of attributes in his discussion of patriotism (*aiguo*). Also see Sun Bojian, 1 (1935): 22–34.
111. Du and Zhang, *Chuzhong gongmin*, 1: 7–29.

whereas Confucian views of society stressed dyadic personal relations.[112] Moreover, in the fully integrated social whole textbooks described, all members, regardless of their social status or roles, were required to perform the same duties, in the same ways, toward all members of the public, whereas in late imperial society, each social position demanded differentiated relation- and class-specific moral duties.[113] In addition, the content of the social person's moral duties differed from Confucian ethical precepts, especially insofar as the new ethics focused on physical decorum in public space, creative self-development, and qualities such as boldness and a readiness to struggle. The Confucian legacy helped to legitimate a definition of social membership that expected citizens to act ethically for the greater good, but the content, orientation, and agents of that moral practice were changed significantly by Republican-period textbook authors and educators.

Civics textbooks outlined in ideal terms the virtues of the socially responsible republican citizen. But those ideals were diverse, amorphous, sometimes in mutual tension, and presented in abstract form. The precise content of civic morality and the techniques used to encourage students to embody and internalize it were only concretized by local educators and government officials in schools' cultivation (*xiuyang*), training (*xunlian*), and character-development education (*xunyu*), to which we now turn.

---

112. See, for instance, Du and Zhang, *Chuzhong gongmin*, 1: 54–56.
113. Munro, *Images*, 138–44.

5

Tempering Bodies, Molding
Moral Characters: Training, Cultivation,
and Cultural Citizenship

It is known that schools do not only impart to students academic learning and inculcate knowledge. Transforming character, cultivating morals, and nurturing sound citizens are in truth equally important. Chairman Chiang's promotion of the New Life Movement lies in making citizens' lifestyles accord with propriety, righteousness, integrity, and a sense of shame. If each school uses this to train students, and if teachers and staff members personally practice it, then students' academic learning and moral education can hope to advance together.[1]

This statement from the Shanghai Municipal Bureau of Education in 1934, which introduced methods for implementing Chiang Kaishek's New Life Movement, portrayed social training and moral cultivation as goals of schooling that were just as important as the transmission of knowledge. Many have cast the New Life Movement as an unprecedented attempt to shape the behaviors and mores of China's citizens according to a uniform standard of public behavior. But in the context of secondary education, the emphasis on training and cultivation encouraged by the movement was consistent with several decades of character-development education (*xunyu*).[2] Those goals are echoed, for instance, in the words of educator Pan Wenan a decade earlier, "The objectives of secondary school character-

---

1. Shanghai shi jiaoyuju, "Shanghai shi zhong xiao xue shixing xin shenghuo banfa dagang" (May 19, 1934) (SMA Q235-1-323).

2. E.g., Huang Jinlin, *Lishi, shenti, guojia*, chap. 2.

development education are to mold students' dispositions, elevate students' aims, and cultivate their moral habits and ability for self-control, [so that] when placed in group life they can clearly understand the relationship between self and others and sacrifice their own personal interests (*xiaoji de sili*) in order to plan for public welfare."[3]

Projects to create and instill modern forms of civility, and competition among social groups over which group's behavior was most representative of civilization and national culture, have been a common dimension of the nation-building process throughout the world.[4] The resulting forms of dress, decorum, language, and symbolic expression that mark people as members of a community and organize their daily public actions can be characterized as "cultural citizenship."[5] Henrietta Harrison has introduced us to some of the new conventions of ritual, dress, and etiquette that characterized cultural citizenship in China during the early Republic and first years of Nationalist Party rule.[6] The exploration here of character-development education and related forms of civic training, such as Scouting and military drill, will extend Harrison's groundbreaking work by demonstrating the diverse and sometimes contradictory ways that lower Yangzi region students were taught to act as citizens in their daily lives. Even in what Harrison, following Charles Keyes, calls the "proposed world" of school training, the forms of cultural citizenship students encountered were varied and often in tension.

Schools taught plural idioms of cultural citizenship because educators and political leaders drew liberally from late imperial patterns of cultivation (*xiuyang* or *peiyang*) and various modern forms of training (*xunlian* or *duanlian*) borrowed from Japan and Euro-America to try to shape students' attitudes and behaviors. Educators used late imperial methods of exhortation, moral example, mentoring relationships, and guided self-reflection as techniques for the cultivation of students' moral character. Training in secondary schools, which

---

3. Pan Wenan, "Jinhou zhongdeng xuexiao zhi xunyu," 2. Cf. Liao Shicheng et al., 25.

4. E.g., Bourdieu, *Distinction*; Corrigan and Sayer, *The Great Arch*, chap. 6; Guha, *Dominance Without Hegemony*, chap. 2; Elias, *The Civilizing Process*; Kasson, *Rudeness and Civility*; Williams, *Stains on My Name*.

5. Ong, "Cultural Citizenship as Subject-Making." Cf. Rosaldo, "Cultural Citizenship."

6. Harrison, *The Making of the Republican Citizen*, esp. chaps. 2–3, 5.

found its fullest expression in Scouting, arts, athletics, and military drill, was the effort to transform students' thought and behavior through compartmentalized and repeated practice of actions that were justified by reference to an articulated sociopolitical ideology. The resulting mélange, which combined contradictory elements that could variously be characterized as liberal, Confucian, authoritarian, or fascist, provided students with multiple, sanctioned modes of performing citizenship in their daily lives.

I. SPACE, TIME, AND THE SCHOOL DAY

China's modern schools introduced students to the norms of civic life in a modern society, in part through daily routines within their walls, especially the organization of the school day and rules governing activities in the classroom, dormitory, gym, and assembly hall. The new practices students encountered in the structured environment of the school encouraged habits that prepared them for active lives in the archetypal spaces of modern society, such as parks, playing fields, offices, and factories. Yet these practices were also combined with patterns of interaction and modes of cultivation that were reminiscent of late imperial Confucian moral education.

The daily routines of late imperial academies and elementary-level schools provide a baseline for assessing continuities and departures in Republican-period character-development education.[7] Longmen Academy near Shanghai offers a particularly good example. It was established by local officials and elites in Jiangnan during the post-Taiping period in order to revive traditional forms of education, and it was characteristic of the kinds of academies that were converted to secondary schools at the start of the twentieth century.[8] At Longmen, the spatial boundaries of the school and the parameters of its student body were fluid and vague, with many students living away from the school. The main pedagogical tools were monthly essays, on which the student's status at the academy depended, and critical dialogue with the tutor that centered on the student's

---

7. Keenan, "Lung-men Academy," 506–9; Meskill, *Academies in Ming China*, 51–58, 96–107.

8. Longmen Academy was the predecessor of Jiangsu Provincial Second Normal School and, subsequently, Jiangsu Provincial Shanghai Middle School.

"scholarly journal." The order of the student's day followed a naturalized cycle of four time divisions, each of which had a generally outlined syllabus. Study was conducted entirely in private. Students' individual study was to stress equally academic learning and implementation of moral principles. Indeed, moral cultivation was a central concern of many late imperial academies. In Ming-era academies, especially those inspired by Wang Yangming, great emphasis had been placed on individual self-reflection and personalized instruction and guidance. In addition, ritual gatherings and choreographed modes of etiquette among different categories of people at the academies were both used as techniques for moral cultivation.

At lower Yangzi region clan and community schools during the Qing, the school day was broken into four major blocks—before and after breakfast, afternoon, and early evening—with no further discrete divisions. There was no fixed curriculum. All students were thrown together in common, not divided into distinct classes or spaces, and each memorized the Four Books and Five Classics at his own pace. Each student also studied morality books and was drilled in etiquette and terms of address for teachers and students of varying ages, all of which "was to give him an elementary idea of his social position and the basic and formal rules of daily social intercourse with his superiors and inferiors."[9]

Many of the daily routines of China's first modern secondary schools differed starkly from their late imperial predecessors. Starting in the late Qing and early Republic, schools across the country adopted central institutions of modern schooling: the standard timetable, the regular distribution and categorization of students in space, and new systems of rules and surveillance to enforce student behavior in those spaces.[10] Together they rehearsed students in new cultural forms regarding time, space, and semiritualized social interactions.

The schedules of early Republican schools blocked out students' activities in regular segments almost to the minute. Note, for instance, the schedule at Shanghai's Minli Middle School in 1925. "6:30 rise and wash; 7:30 breakfast; 9–12 attend class (ten-minute breaks each hour); 12:10 lunch; 1–4 attend class (ten-minute breaks each

---

9. Leung, "Elementary Education," 392–400.
10. E.g., Carter, *Creating a Chinese Harbin*, 61; McElroy, "Transforming China."

hour); 6:15 dinner; 7–9 preparation and review; . . . 9:45 students living in the dormitory go to bed; 10:00 lights out."[11] At many schools, the period between rising and breakfast was occupied by morning exercise (*zaocao*) or physical training.[12] Mandated extracurricular sports or other student activities organized by self-government associations often absorbed the period between class and dinner in the afternoon.[13] Such organization and control of time as a valued resource, and the coordination of simultaneous collective activities, reflected the shifting conceptions of time that have accompanied industrialization, with its commodification and synchronization of labor.[14] This industrialization of time was in full swing in the quickly developing lower Yangzi region during the first third of the twentieth century.[15]

The school's physical environment was likewise divided into regular compartments and spaces for discrete activities that were determined by a bell schedule. (See Fig. 5.1.) At Jiangsu Provincial First Normal School and Pudong Middle School, for example, rules dictated the particular function of each space in the school and the specific time when it should be used.[16] Dormitories, dining halls, classrooms, study rooms, music rooms, playing fields, gyms, libraries, and laboratories now all had their own special patterns of activity associated with them. (See Figs. 5.2, 5.3, and 5.4.) Such compartmentalization encouraged the idea that public spaces were dedicated to specific kinds of behavior. Further, these patterns of activity were often confined to particular times of the day that were marked off by bells or other signals. Such uniform patterning of activity fostered the sense that the school was a clearly defined community moving in prescribed ways through space and time—the school as an ordered social body. Arbitrary numbering and assignment of spaces—such as

---

11. *Shanghai Minli zhongxuexiao yichou nian zhangcheng*, 42–43. Cf. *Suzhou zhongxue gaikuang*, 30–31.

12. E.g., *Jiangsu shengli diyi shifan xuexiao xuesheng ruxue xuzhi*, 2, 4; Liao Shicheng et al., 283; Wang Yankang, "Zhongxuexiao jiji xunyu," 1–5.

13. Liao Shicheng et al., 283–84; *Suzhou zhongxue gaikuang*, 30–31.

14. Thompson, "Time, Work-Discipline, and Industrial Capitalism."

15. Honig, *Sisters and Strangers*, 140–48; Shao, *Culturing Modernity*, 88–99.

16. *Jiangsu shengli diyi shifan xuexiao xuesheng ruxue xuzhi*, 5–8; *Pudong zhongxuexiao nian zhou jiniankan*, 24–27.

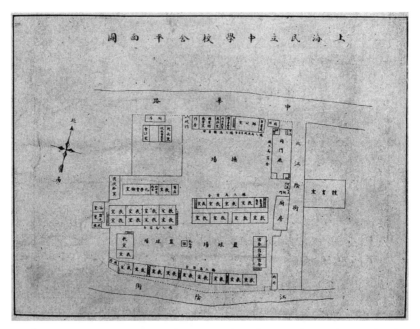

5.1 "Plane Diagram of the School Grounds of Shanghai's Minli Middle School." This imaginary bird's-eye view of the Minli Middle School grounds illustrates the compartmentalization of space that was fundamental to modern secondary schooling in Republican China. Classrooms, dormitories, offices, laboratories, playing fields, and many other dedicated spaces were blocked out for specific activities. *Shanghai Minli zhongxue sanshizhou jiniankan*, "Shanghai Minli zhongxue xiaoshe pingmian tu."

desks, seats in the cafeteria, and dorm rooms and beds—raised the image of the school as a modular system of unmarked cells into which homogeneous students, devoid of distinguishing markings, such as native place or kinship, wealth or age, were inserted.[17] This modularity was enhanced when schools, such as Southeast University Affiliated Middle School, organized students into set decimal groups that became basic units for all students' activities.[18] This abstract structure helped to make imaginable the generic self as an individual citizen in relation to the composite whole of a wider community.[19]

---

17. E.g., *Pudong zhongxuexiao nian zhou jiniankan*, 25–26; *Shanghai Minli zhongxuexiao yichou nian zhangcheng*, 44.

18. Liao Shicheng et al., 343–44.

19. See Mitchell, *Colonising Egypt*, 34–80.

理　化　實　驗　室

5.2 "Physics and Chemistry Laboratory." Spaces such as labs, libraries, and playing fields, because of their layout and content, became functionally dedicated, with prescribed behaviors associated with them and set times to be used. *Shanghai Minli zhongxue sanshizhou jiniankan*, "Lihua shiyan shi."

Over the course of the 1910s and 1920s, students and educators negotiated the practice of daily school life within the normative frameworks of time and space established by teachers and administrators. This process is captured in second-year high school student Jiang Jiaxiang's revealing discussion of student life in Nanjing's Southeast University Affiliated Middle, one of China's leading institutions for educational reform. [20] Jiang presents as absolute the minimum 25 credits' worth of academic work a week (which could amount to considerably more than 25 hours of class time), distributed across the days of the school week, and two hours of mandated evening study hall that anchored students' days. Besides these two large blocks of time each day, writes Jiang, "everything is freely managed by the students." Yet his discussion reveals that students, in fact, struggled to balance academics and extracurricular activities in

20. Jiang Jiaxiang, "Fuzhong xuesheng shenghuo," 376–83.

二　之　影　攝　館　書　圖

5.3 "Photograph of the Library (2)." *Shanghai Minli zhongxue sanshizhou jiniankan*, "Tushu-guan sheying zhi er."

their few hours of free time in the afternoons. "Among those espe-cially fond of sports, some do not fully prepare in their time outside of class; and, some of those especially focused on their studies may work as many as twelve hours a day." In an effort to further their studies while participating in other aspects of school life, students stretched the school day by studying early in the mornings or late into the night (*kai zao che, kai wan che*).

Students also challenged the organization of space through little battles over where to carry out extracurricular activities. At issue during Jiang's time at Southeast University Affiliated Middle was where students could practice their Chinese music (*guoyue*). Students enthusiastic about practicing had spilled over from the designated music room into the study halls, a practice against school rules that was leading to "minor conflicts" (*xiaoxiao chongtu*) between students and administrators. In general, Jiang's account suggests that schools' schedules were densely packed and spaces were highly structured, with students working at the margins to carve out times and places to pursue their own goals.

角　　　一　　　場　　　操

5.4 "A Corner of the Playing Fields." *Shanghai Minli zhongxue sanshizhou jiniankan*, "Cao-chang yi jiao."

Specific rules tailored for different spaces within the school de-lineated modern patterns of public decorum.[21] "One's posture when sitting in the classroom must be upright and [one's attitude] serious (*yansu*)." "Frequently organize and put away one's books, pens, and inkstone." "Do not pour out washing water on the ground at will." "When going to meals do not talk and laugh loudly." "One's seat must accord with the arranged number; one cannot switch seats ran-domly or fight [over seats]." These standards reflected commonly held elite ideals of decorum, physical hygiene, and behavior suited to specific spaces and times. At schools such as Southeast University Affiliated Middle they were enforced through surveillance by all of the school's teachers, who each submitted reports on individual stu-dents to the school administration's guidance department (*zhidaogu*)

---

21. Liao Shicheng et al., 25–28. Cf. *Jiangsu shengli diyi shifan xuexiao xuesheng ruxue xuzhi*, 3; Pan Wenan, "Jinhou zhongdeng xuexiao zhi xunyu," 2–3, 4–5; *Pu-dong zhongxuexiao nian zhou jiniankan*, 24–27; *Shanghai Minli zhongxuexiao yichou nian zhangcheng*, 43.

at the end of each semester. These reports provided the basis for a student's conduct grade. Students could also be co-opted into enforcing rules over themselves and their classmates through systems of room monitors and dorm captains or through their self-government organizations.[22] Some principals also took an active role in enforcing rules. At Shanghai's Minli Middle School, for instance, Principal Su Benyao regularly roamed the classrooms while class was in session to observe student behavior. Students he saw misbehaving would be called to his office during the lunch period for a personal interview and receive punishment of so many pages of calligraphy copying.[23]

During the Nanjing decade, the Nationalist Party sought to centralize standards for character development within secondary schools. But in practice party or government rules often corresponded closely to precepts that had been developed by educators during the previous two decades.[24] As central and local government agencies implemented programs such as the New Life Movement and the system of "uniting instruction and character-development education" (*jiaoxun heyi*) they built on training techniques pioneered during the 1920s and earlier forms of moral cultivation to expand the concern with character-development education in secondary schools.

At the private Zhenhua Girls' Middle School in Suzhou, for instance, teachers viewed the New Life Movement as a response to the general decline in civic virtue, etiquette, and hygiene that had plagued China throughout its modern history. This decline was described as a simultaneous loss of ancient virtues and a failure to adopt the Euro-American "spirit of self-strengthening, not standing idle, and being roused to action and capable of great achievements."[25] As part of the movement, the school instituted three graduated levels of moral and behavioral rules, which teachers and students alike were expected to follow. The primary level of regulations was composed mostly of rules for individual behavior, with a great stress on

---

22. *Jiangsu shengli diyi shifan xuexiao xuesheng ruxue xuzhi*, 4–5; *Pudong zhong-xuexiao nian zhou jiniankan*, 26; *Shanghai Minli zhongxuexiao yichou nian zhangcheng*, 43; Tai Shuangqiu, "Xunyu."

23. Separate interviews with two Minli Middle School alumni.

24. Zhongguo di'er lishi dang'anguan, ed., 2: 1031–35, 1063–65, 1274–76.

25. Suzhou Zhenhua nü xuexiao, ed., *Zhenhua nüxuexiao sanshi nian jinian-kan*, 44.

hygiene, cleanliness, timeliness, and propriety. The expectations for higher levels were increasingly focused on social mores, culminating in a model citizen who would have the following qualities: "wear and use national products; not be false or deceptive; cultivate the habits of being able to be active and quiet when appropriate; have a spirit of hard work and enduring hardship; treat people amiably and politely; obey the ideas of the majority; have a firm will, not fearing difficulty and retreating; be able to transform society and not be transformed by it; be able to take responsibility and persevere in service; [and] love society and the nation with zeal."[26] At all levels, Zhenhua's New Life Movement rules echoed standards of student conduct that had been common at many regional schools during the previous two decades.[27]

Even as they built on previous forms of character-development education, programs such as the New Life Movement were also paralleled and reinforced by other party policies for educational training. One example is the system of uniting instruction and character-development education, which was supposed to make moral education more efficient by distributing responsibility for it to all the teachers in the school.[28] Offices for character-development education and instruction (*jiaowu*) were to merge in a new guidance office (*jiaodaochu*). Full-time teachers (*zhuanren jiaoyuan*) would live together with students, providing constant supervision. Academic-year groups (*xueji*) would have advisers (*jiren daoshi*), and each academic-year group would be divided into class groups of 20 to 30 students—often coterminous with the group of students who took required classes together—that would have their own supervisory teachers. These teachers would constantly observe student conduct, whether in classrooms, dormitories, or public places. Advisers and supervisory teachers would have regular discussions with classes, groups, and individual students, assess and grade student conduct and attitudes, read students' notes or diaries, and guide extracurricular activities.

---

26. Ibid., 45–46.

27. E.g., Liao Shicheng et al., 25–28; *Pudong zhongxuexiao nian zhou jiniankan*, 3–6.

28. [Jiangsu jiaoyuting], "Yinianlai Jiangsu zhongdeng jiaoyu zhi huigu yu zhanwang," 74–76; Jiangsu jiaoyuting, comp., *Jiangsu sheng xianxing jiaoyu faling huibian*, 64–72.

The system of uniting instruction and character-development education was congruent with techniques of linking training and instruction that had become popular in education circles and implemented at schools such as Southeast University Affiliated Middle during the 1920s. As a result, educators at several prominent regional schools adopted the system, and it became common throughout Jiangsu during the mid-1930s.[29]

In school practice and government directives, school administrators, educational reformers, and government officials combined new forms of discipline with late imperial approaches to moral cultivation in eclectic and complex ways. Detailed rules were coupled with surveillance, examination, and grading in systems of conduct assessment (*caoxing kaocha*). As with the ledgers of merit and demerit of the seventeenth century, conduct evaluation spelled out normative patterns of social behavior.[30] But where the ledgers of merit and demerit had concentrated on proper performance in archetypal hierarchical social relations, modern conduct assessment in many cases focused on proper behavior in public spaces. Surveillance and systematic evaluation operated in tandem with more traditional approaches to cultivation, such as group and individual discussions or counseling, guided readings, and review of student journals, to inspect and shape student behavior.[31] Modern, disciplinary approaches to training coexisted with individuated and highly personalized moral cultivation characteristic of some late imperial academies.

In fact, at many regional schools close interpersonal relationships between teachers and students continued to serve as a primary basis for influencing students' attitudes and behaviors through the 1920s and 1930s. State directives on character-development education during the Nanjing decade recognized and encouraged such mentoring relationships and teaching by example. Teachers who lived with and related closely to students were expected by educators, students, and

---

29. Hu Huanyong, "Suzhou Zhongxue jiaoxun heyi tan," 49; Shen Yizhen, "Jiaoxun heyi"; Zhou Houshu, "Yangzhou zhongxue jiaoxun heyi."

30. Brokaw, *The Ledgers of Merit*, chap. 4.

31. For examples of these techniques, see Hu Huanyong, "Suzhou Zhongxue jiaoxun heyi tan"; Jiangsu shengli Shanghai zhongxue chuban weiyuanhui (1933), 84–86; Shen Yizhen, "Jiaoxun heyi"; Zhou Houshu, "Yangzhou zhongxue jiaoxun heyi."

officials to be moral exemplars and act as surrogate parents.[32] One student commented, "If students make a mistake, training instructors must use a sincere attitude like a father or brother, honestly telling the student what mistake he made and hoping that he will quickly rectify it."[33] The expectation at schools such as Jiangsu Provincial Yangzhou Middle School was that these familized relationships with teachers, who were to act as moral exemplars, would lead to "imperceptible influences (*mohua qianyi*) and changes of disposition" in the students and an internalization of civic mores.[34]

Such mentoring relationships exercised a subtle form of moral suasion. At the same time they reinforced personalized patterns of social order based on hierarchy and reciprocity that were quite different from the homogenizing patterns that organized other dimensions of school life and civic training. An alumnus from Wusong Middle School, north of Shanghai, recalled growing very close to several teachers who brought students to their homes or took them on outings. Teachers, he said, were extremely attentive to students' thought (*sixiang*), lifestyles (*shenghuo*), and ethical cultivation, triggering strong feelings that motivated students and shaped their behavior. He recalled giving gifts to his mentor at Lunar New Year, and the teacher gathered this student's friends to see him off when he left for college. These interactions, remembered vividly and fondly after 60 years, echoed late imperial Confucian models of reciprocity.[35] Sentiment helped to reinforce students' feelings of academic and moral responsibility, paralleling late imperial forms of moral cultivation. These relationships also described a form of social order organized by hierarchies based on age, experience, and moral authority and held together by strong emotional bonds that were parallel to conventional family relations and the ideal family as presented in Nanjing-decade civics textbooks.

---

32. Jiangsu jiaoyuting, 64; Pan Wenan, "Jinhou zhongdeng xuexiao zhi xunyu," 5; *Pudong zhongxuexiao nian zhou jiniankan*, 2; Zhou Houshu, "Yangzhou zhongxue jiaoxun heyi," 36–37.

33. Xingzhi, "Zhongdeng xuexiao xunyu zhi shangque," 19.

34. Shen and Xiao, eds., *Jiangsu sheng Yangzhou zhongxue*, 57.

35. Alumni from Hangzhou High School also recounted strong mentoring relationships that developed between teachers and students who lived in the same dormitories and spent extensive time together outside the classroom.

Other aspects of school life also stressed ritualized enactment of hierarchical status distinctions inherited from late imperial society. In the classroom at many schools, students were to stand at attention in an expression of respect when the teacher entered and left the room at the beginning and end of class, or when guests visited. Students were also to salute and assume the correct posture and tone whenever speaking with teachers, guests, or the principal.[36] These simple, repeated movements, though distinct from the bows of late imperial academies, enacted in similar ways the status difference between teachers and other social superiors, on one hand, and students, on the other. Such hierarchical relations based on ascribed status contrasted with the homogenizing tendencies of some modes of social organization within the school and forms of cultural citizenship that valued all citizens equally.

Digressions and Transgressions

Character-development education was intended to make students into morally upright and civilized citizens, but, not surprisingly, many students diverged from schools' moral and behavioral standards. Zhejiang Provincial First Middle School student Wu Tao listed many of his classmates' failings in a 1924 article in that school's journal.[37] Students, he said, were envious and jealous, pleased when others failed, and given to arguing and fighting. They also willfully resisted group discipline, ate and drank excessively, and gambled. Further, students loved luxury; "They wear a new hat, new clothes, plain glass spectacles, . . . preparing all the fashionable adornments as if they were good-looking beauties." Wu's complaints were echoed by educator Zhang Nianzu. He claimed that it was common for higher level students to "intentionally cause disturbances, scribble things on the walls, spit randomly, smoke, drink, and even visit prostitutes and gamble."[38]

---

36. *Jiangsu shengli diyi shifan xuexiao xuesheng ruxue xuzhi*, 6 [*Shanghai Minli zhongxuexiao yichou nian zhangcheng*, 42]; *Pudong zhongxuexiao nian zhou jiniankan*, 24. Similar patterns of authority were common at Changshan Middle School in Shandong as well. See Thøgerson, *A County of Culture*, 69.

37. Wu Tao, "Xuesheng zhi tongbing."

38 . Zhang Nianzu, "Zhongdeng xuexiao xunyu zhi yanjiu," 1. Cf. Tai Shuangqiu, "Xunyu." An alumnus who attended Suzhou Middle School during

Beyond typical individual transgressions of rules and morals, students also often stirred up "school disturbances" (*xuexiao fengchao* or *xuechao*) to challenge the structures of disciplinary authority that imposed and reinforced regulations and moral standards. These localized power struggles disrupted many Chinese schools from the late Qing through the Republican period.[39] Many of the school disturbances focused on issues regarding personnel or finances, but a large number erupted over disciplinary rules and school administrators' efforts to enforce them.[40]

Brief accounts of several early 1920s school disturbances in lower Yangzi region middle schools reveal how students challenged the rules, regulations, and surveillance that constituted character-development education.[41] At Zhejiang Provincial Eleventh Middle School in Lishui, tension over seating in the cafeteria led to conflict that embroiled the principal. At Zhejiang Provincial Fourth Middle School in Ningbo, an ill student was not allowed to go to his room to rest during a class period, leading to student protest over inflexible and unfair application of the rules. Students at Jiangsu Provincial Eleventh Middle School in Haizhou criticized the principal for lacking proper morals. At Pudong Middle, students were outraged that the principal had expelled as many as 70 students over the summer for disciplinary reasons, such as that a given student "did not fit with the school's lifestyle" or had "missed too many classes." In each of these cases, issues having to do with the organization of time and space within the school, basic standards of behavior, or administrators' role as moral guide and exemplar became a lightning rod for student criticism and challenge.

Despite students' many challenges to schools' systems of order and discipline, character-development education introduced these

---

the mid-1930s recalled that theft was a problem among students. An alumnus from Songjiang Middle School cited students who were expelled for getting locked outside the school gates for missing curfew, and a man who studied at Yangzhou Middle during the first years of the Nanjing decade remembered students' being expelled for excessive absence at morning exercise.

39. Sang Bing records 502 school disturbances between 1902 and 1911 (*Wanqing xuetang*, 5). Cf. Lü Fangshang, 67–88; Weston, 92–96.

40. Chang and Yu, *Xuexiao fengchao de yanjiu*, 17–19. The source reports lower Yangzi region secondary schools accounted for 34 out of a reported 106 disturbances nationwide in 1922. Ibid., 14.

41. Ibid., 53, 56, 57, 62, 64.

students to a repertoire of behaviors and attitudes that were closely associated with modern standards of timeliness and public decorum. At the same time, individual and group moral education also reinforced expectations about respect and reciprocity toward elders and exemplars, such as teachers, that had been inherited from the late imperial period. Even if students did not always fully internalize and embody these standards of behavior, they learned them, could use them strategically, and, as we will see, teach them to other social groups as a way of privileging themselves.

II. THE CHINESE SCOUTS AND HYBRID TRAINING

In addition to the daily routines, discipline, and mentoring relationships of school life, many lower Yangzi region educators and the Nationalist Party promoted Scouting (*tongzijun*) as a means of instilling in students other forms of civility and social action that they viewed as essential for modern citizens. Advocates favored Scouting because it provided students with opportunities to enact and practice many new forms of civility and civic action. Throughout the Republican period, Scout training combined instruction in forms of etiquette and hygiene rooted in Euro-American culture with skills training and the promotion of social service. Through these diverse kinds of training, Scouting encouraged forms of cultural citizenship that included self-control and refinement, independence and individual creativity, and performance of civic involvement through social service.

Origins and Organization

Chinese educators' adoption of Scouting as a technique for elite-led civic training was consistent with the founding principles of the Boy Scouts. In Great Britain in 1908, Sir Robert Baden-Powell created the Boy Scouts in reaction to what he and many others saw as a decline in the moral quality of British youths as a result of industrialization.[42] Scouting organizations quickly took root in China during the first years of the Republic, just years after their invention in Great Britain. Missionary schools formed the earliest Boy Scout

---

42. Rosenthal, *The Character Factory*.

troops in cities in Southeast and Central China, though accounts vary as to exactly where and by whom the first troop was formed.[43] By all accounts Scout training spread quickly among both Chinese-run and missionary schools during the 1910s, with the greatest density of interest in Scouting concentrated in Shanghai and the surrounding provinces of Jiangsu and Zhejiang. By 1917, 48 of Jiangsu's 60 counties were said to have Scouting organizations of some kind, and Shanghai alone was said to have as many as 600 Scouts organized into eleven troops.[44]

In November 1915, educators interested in Scouting formed the Chinese National Scouting Association (Zhonghua quanguo tongzijun xiehui) in Shanghai, with the aim of organizing troops (*dui*; composed of two or more patrols [*pai*] of six to twelve scouts) into a hierarchical, nationwide organization.[45] Centralization was undercut by the creation of competing managing organizations, tensions between Chinese and foreign proponents of Scouting, and the great degree of autonomy enjoyed by local troops and Scouting organizations.[46] But many educators in the lower Yangzi region remained enthusiastic about Scouting and promoted it actively.[47] A number of prominent regional middle schools formed Scout troops and made them a central part of their curricula in the decade between 1917 and 1927.[48]

---

43. One origin story describes the first troop being formed in Shanghai at Huatong Academy (Huatong gongxue) or the Gezhi Academy (Gezhi gongxue) in April 1913. But other versions, perhaps driven by nationalist motives, claimed instead that the Chinese educator Yan Jialin formed the first Chinese Boy Scout group at Wuchang's Wenhua Academy (Wenhua shuyuan) in February 1912. See Cheng Jimei, *Tongzijun zuzhifa*, 3; Fan Xiaoliu, *Xinbian tongzijun chuji kecheng*, 57–59; Kemp, "Boy Scouts Association of China" and "Tongzijun hui baogao"; Zheng Haozhang, *Shanghai tongzijun shi*, 1–3; Zhonghua quanguo tongzijun xiehui, ed., *Tongzijun guilü*, Yuanqi: 1.

44. Kemp, 1–2; Liu Chengqing, *Chuji tongzijun shiyan jiaoben*, 26.

45. Zhonghua quanguo tongzijun xiehui, ed., *Tongzijun guilü*, 1–11.

46. Zheng Haozhang, *Shanghai tongzijun shi*, 7–26.

47. Cheng Jimei, Xu: 1; "Tongzijun jiaoyu cujin hui jinxing zhuangkuang"; Zhang Junchou, "Tongzijun jiujing shi shenma?"; Zheng Haozhang, *Shanghai tongzijun shi*, 8–26.

48. Troops were formed at Ningbo's Zhejiang Fourth Middle, Shanghai's Jiangsu Second Normal, Nanjing's Southeast University Affiliated Middle, Pudong Middle, Hangzhou's Zhejiang First Middle, Jiaxing's Zhejiang Provincial Second Middle, and Jinhua's Zhejiang Provincial Seventh Middle Schools (*Zhe-*

Educators who embraced Scouting did so because they saw it as an active mode of education that incorporated the many attitudes, behaviors, and capacities that Chinese elites believed the citizen should embody while it used forms of active learning characteristic of American progressive education. For instance, Zhang Junchou, an educator from Wu County in Jiangsu, characterized Scouting as a form of experiential learning that built on students' innate love of activity (*hao dong de tianxing*) to improve the quality of China's children, who in the future would determine the quality of Chinese society. Zhang identified the objectives of Scouting education as being "to develop inherent talents and instincts, practice service to the group, cultivate sound individuals, and create a progressing society (*jinhua de shehui*)."[49]

A New Order. After 1927 Nationalist Party policies promoted the expansion of China's Scouts while consolidating Scouting organizations under party leadership through techniques characteristic of the party's approach to other dimensions of education management, such as group and individual registration, administrative centralization, and curricular control.[50] By building on existing Scouting groups, the party chose not to set up an entirely new and independent party youth division, as Mussolini and Hitler did contemporaneously in Italy and Germany.[51] This approach reflects again the Nationalist leadership's tendency to use co-optation and indirect management rather than state monopoly and direct control in state building and highlights a qualitative difference between the Nationalist regime during the Nanjing decade and Europe's fascist states.[52]

---

jiang shengli disi zhongxue yilan, 20–21; Liao Shicheng et al., 65, 340–41; *Jiangsu shengli di'er shifan xuexiao yilan*, 9–15; Zhu Junda, "Bannian lai zhi tongzijun," 3–4; *Pudong zhongxuexiao nian zhou jiniankan*, 8; "Tongzijun zhi zuzhi"; *Zhejiang shengli di'er zhongxue yilan*, 48; Zhejiang shengli Jinhua zhongxue chuban weiyuanhui, 28–29).

49. Zhang Junchou, "Tongzijun jiujing shi shenma?," 5. Cf. Cheng Jimei, 1–6; *Pudong zhongxuexiao nian zhou jiniankan*, 8.

50. For the process of local troop registration with party authorities, see Dai Jitao, *Dai Jitao xiansheng wencun*, 2: 806–9; Zheng Haozhang, *Shanghai tongzijun shi*, 27–37.

51. Koon, *Believe, Obey, Fight*; Stachura, *Nazi Youth in the Weimar Republic*.

52. The Three Principles of the People Youth Corps (Sanmin zhuyi qingniantuan), which was established during the Sino-Japanese War, more closely paral-

There were two major steps toward reorganization. The first came in 1928 and 1929, under the direction of the party's Central Training Department (Zhongyang xunlianbu), which organized a "central command" (*silingbu*). The second step came in 1934 with the formation of a general association (*zonghui*). In both cases, the Scouting organization was neatly hierarchical, with provincial and municipal- or county-level organizations operating under the Chinese Scout Central Command or General Association, depending on the period. Troops (*tuan*) were often organized on the level of one school and were composed of patrols (*xiaodui*) of six to nine youths that constituted the lowest level of organization.[53] Councils (*lishihui*), which oversaw troops and patrols, were organized at the county/ municipal, provincial, and national levels.[54]

The Nationalist Party consistently maintained close control over Scouting organizations at all levels. Councils at the county/municipal and provincial levels included as ex officio members one member of the local party branch's standing committee and the head of the local bureau (county) or department (provincial) of education.[55] The first National Council was appointed entirely by the Ministry of Education and approved by the party's Central Executive Commit-

---

leled both fascist and Soviet-style party youth divisions. See Huang, *The Politics of Depoliticization*, chaps. 5–7. The Zhejiang Provincial Youth Group (Zhejiang sheng qingniantuan), which was formed by the Zhejiang Department of Education in the summer of 1936, can be seen as a forerunner of the Three Principles of the People Youth Corps. But the Zhejiang Youth Group's training closely paralleled that of the Chinese Scouts, and Scouting continued to be the primary mechanism of youth civic training at the lower middle school level throughout the Nanjing decade. See "Zhejiang sheng qingniantuan dagang."

53. Chen Chaozong, "Zhongguo tongzijun de xianzai yu guoqu"; Zhongguo Guomindang zhongyang zhixing weiyuanhui xunlianbu, *Fagui huikan*, 1, 12–21; Zhongguo tongzijun zonghui choubeichu, ed., *Zhongguo tongzijun zonghui choubeichu gongzuo baogao*, "Zhongguo tongzijun zonghui choubeichu zuzhi jingguo," 1–2; 1–12, 33–38, 172–74. My translation of *tuan* as "troop," *zhongdui* as "group," and *xiaodui* as "patrol" is based on the standard translation of the Scouting rules in Zhongguo tongzijun zonghui choubeichu.

54. Initially, military terminology was used to mark out the Scouting hierarchy. County or municipal administrative bodies were designated as "divisions" (*shi*), and the provincial level organization was called an "army" (*jun*).

55. See Zheng Haozhang, *Shanghai tongzijun shi*, 27–49, for examples of the influence of the Shanghai party branch and Bureau of Education on Shanghai's Boy Scout Association and the Chinese Boy Scout Shanghai Council.

tee, and subsequent National Councils each had ten members appointed by that ministry.[56] Throughout the Nanjing decade, Chinese Scouting's members and personnel at all levels were held to strict standards for registration and training.[57] Further, under the 1928–29 regulations, the Central Training Department itself produced and distributed, as well as reviewed, Scout publications. Under the 1934 regulations, all Scouting materials were subject to approval by the General Association before publication.[58]

During and after the party's reorganization of the Chinese Scouts, important figures in the Nationalist Party leadership were closely involved with managing it. The position of commander (*siling*) during the late 1920s and early 1930s was filled by the head of the Central Training Department.[59] Major party politicos, ideologues, and educators who were involved in the reorganization of the Scouts from 1932 to 1934 subsequently became fixtures on the National Council that set Scouting policy after 1934.[60] Party leaders who were actively involved in reorganizing and managing China's Scouts constituted an eclectic group that included leading members of both the proto-fascist Society for Vigorous Practice (Lixingshe), such as Teng Jie and Liu Jianqun, and of the rival CC clique, most notably Chen Lifu and Zhu Jiahua.[61] In the planning stages for reorganizing the Scouts

---

56. Ministry appointees dominated the council, for as of 1934 only one provincial and five municipal councils (Jiangsu Province, Shanghai Municipality, Shantou, Ji'nan, and Hangzhou) were prepared to elect National Council members, and each was slated to elect only one member (Zhongguo tongzijun zonghui choubeichu, "Zhongguo tongzijun lishihui yilan biao," 167).

57. Zhongguo Guomindang zhongyang zhixing weiyuanhui xunlianbu, *Fagui huikan*, 17, 21; Zhongguo tongzijun zonghui choubeichu, 52–57, 60–61.

58. Zhongguo Guomindang zhongyang zhixing weiyuanhui xunlianbu, *Fagui huikan*, 7–11, 62–63; Zhongguo tongzijun zonghui choubeichu, 61–62.

59. Wu Yaolin, *Tongzijun quanshu*, 8; Zhongguo Guomindang zhongyang zhixing weiyuanhui xunlianbu, *Fagui huikan*, 1–2.

60. Wu Yaolin, 9; Zhongguo tongzijun zonghui choubeichu, 1–2.

61. The participation of CC clique members and other high-ranking Nationalist Party leaders who were not aligned with the Society for Vigorous Practice in the reorganization and management of the Scouts during the Nanjing decade suggests that the proto-fascist organization may not have dominated the central Scouting organization, as some have argued. See, for instance, Eastman, *The Abortive Revolution*, 65; Wakeman, "A Revisionist View," 164–65. Members of the Society for Vigorous Practice and of the CC clique seem to have competed for influence in the Chinese Scouts as in so many other contexts. For accounts

in 1933 and 1934, an especially pivotal figure was Dai Jitao, who was one of Chiang Kai-shek's closest advisers and one of the party's leading educational and cultural theorists. Dai had final say in many key decisions and took the lead in revising the Scouting rules and oath.[62]

Chinese Scouting expanded rapidly and consistently under party rule, but it remained dominated by organizations from the lower Yangzi region. On the eve of the Ministry of Education's 1934 policy directive making Scouting a required class in all lower middle schools, there were 86,536 registered Chinese scouts (including some in Japan and the Philippines), of which 55,098 (roughly 64 percent) were from Jiangsu, Zhejiang, Shanghai, and Nanjing.[63] The Nationalist Party successfully fostered growth of the Chinese Scouts in part because its curriculum continued to focus on comprehensive civic training that was congruent with the action-centered educational model favored by many educators, even while the party adapted Scout training to some of its own objectives. Local educators' receptivity is reflected in part by the fact that most lower Yangzi region secondary schools established and registered Scout troops and made them a central part of their physical education curriculum by the early 1930s, before Scouting became a required class in 1934.[64]

---

of that rivalry, see Huang, *The Politics of Depoliticization*, 100–177; Wakeman, *Spymaster*, esp. chap. 9.

62. Dai Jitao, *Dai Jitao xiansheng wencun*, 2: 817–20; Zhongguo tongzijun zonghui choubeichu, 5–12.

63. For the 1934 figures, see Zhongguo tongzijun zonghui choubeichu, 168–69. For the directive making Scouting a required course for lower middle school, see ibid., 146–47. As of 1927, 47 of Jiangsu's counties and Shanghai municipality had 346 troops and 13,387 scouts, and there were only 17,949 registered scouts nationwide in 1930 ("Jiangsu gexian tongzijun tuan tongji biao" [June 1927]; *Zhongguo tongzijun chuji kecheng*, 74).

64. *Hangzhou shizhong yilan*, 154; Jiangsu shengli Shanghai zhongxue chuban weiyuanhui (1930), 40, 228–36; *Shanghai Minli zhongxue sanshizhou jiniankan*, 100; *Suzhou zhongxue gaikuang*, 8; *Zhejiang shengli di'er zhongxue yilan*, 26; *Zhejiang shengli disi zhongxue yilan*, 20–22, 217; Zhejiang shengli Hangzhou chuji zhongxue bianji weiyuanhui, *Zhejiang shengli Hangzhou chuji zhongxue liuzhou jiniance*, 79–80; Zhejiang shengli Jinhua zhongxue chuban weiyuanhui, 28–29; *Zhenhua nüxuexiao sanshi nian jiniankan*, 65–66, 118; Zhongyang daxuequli Shanghai zhongxuexiao mishuchu chuban weiyuanhui, comp., *Zhongyang daxuequli Shanghai zhongxuexiao yilan*, 126; Zhongyang daxuequli Songjiang zhongxue yilan bianji weiyuanhui, ed., *Songjiang zhongxue yilan*, 38; Zhongyang daxuequli Taicang zhongxuexiao, ed.,

Active Civic Training

In their formulation of Chinese Scout training, Chinese educators built on the organization and techniques that had been used in Europe and the United States to erect a three-stage training process that was incredibly eclectic.[65] The Euro-American Scouting curriculum was already a diverse mix of moral precepts, practical skills training, athletics, and civic education.[66] With the goal of making Scouting a comprehensive and active mode of civic training, Chinese educators emphasized lessons on hygiene, etiquette, military-style drill, and social service. At the same time, Chinese Scouting included a wide range of skills training that allowed students opportunities for individual achievement and creative development. Together these various dimensions of Scout training encouraged new forms of cultural citizenship, but their very diversity granted students flexibility in defining how they would act as citizens.

Hygiene and Etiquette.   Hygiene and etiquette lessons, which were central parts of the Scout training curricula throughout the 1920s and 1930s, reflected elite Chinese educators' concern to instill their own vision of civility in China's young people by transforming their most basic physical actions and choreographing their social interactions. Hygiene lessons introduced standards of physical decorum and cleanliness that Chinese elites had learned through the painful process of being criticized by Westerners for supposedly being uncivilized and unhygienic. Etiquette lessons taught students hybrid patterns of polite behavior adopted from Treaty Port culture that integrated Chinese elite and Western bourgeois standards of behavior. This spatialized ranking of hygiene and etiquette standards paralleled the hierarchy of development that was outlined in students' history and geography textbooks and that provided an implicit rationale for adopting those practices.

Yilan, Jiaowu gaikuang: 4, 12; Zhongyang daxuequli Yangzhou zhongxue chuban weiyuanhui, *Yinianlai zhi Yangzhong,* Yinianlai zhi jiaowu: 3, 4–5.

65. Scouts progressed from elementary (*chuji*) to intermediate (*zhongji*) to advanced (*gaoji*) levels.

66. For examples from the United States, see Boy Scouts of America, *Handbook for Scout Masters; Revised Handbook for Boys.* For Britain, see Rosenthal, *The Character Factory.*

Instruction in basic hygiene was included within Scout training from the late 1910s through the 1920s.[67] Lessons justified good hygienic practice by explaining that it would produce strong bodies that could serve the nation. "Today's children in our Scouts are tomorrow's masters of the nation. The future is great and the burden is heavy, so we should train the body to take up difficult and great work in the future. Because of this, hygiene is esteemed."[68] Scouts were then taught to cut and clean their fingernails, brush their teeth, breathe through their noses to filter out bacteria, and to spit in the proper places.[69] Hygiene lessons were also standard parts of the elementary training curriculum throughout the Nanjing decade, and they varied little over that time, with continuing emphasis on strengthening the individual body in order to strengthen society.[70]

Through hygiene lessons, Chinese educators and Nationalist Party leaders were also concerned with cultivating in Chinese citizens a very Western conception of public decorum. Figures as diverse as local reformer Zhang Jian, writer Lu Xun, and party leader Sun Yat-sen all adopted a critical perspective on their countrymen's personal habits, often by way of reflecting on and responding to foreigners' criticisms.[71] Moreover, Ruth Rogaski demonstrates that Chinese elites in Tianjin adopted Western practices of hygienic modernity as a result of perceived deficiencies in indigenous practices of personal and public health.[72] Such concerns were implicit in Scouting handbooks' repeated injunctions against spitting in public, relieving oneself in undesignated areas, and sneezing or coughing loudly, not to mention in calls to eat from separate plates, all of which reflected Euro-American criteria of good manners and public decorum. Scouting organizers' concern with personal hygiene stemmed from a shared belief that the reordering of social life could only happen through the readjustment of personal behavior. Thus,

---

67. Cheng Jimei, 13; Zhongguo tongzijun xiehui, *Tongzijun chubu*, 13.

68. Cheng Jimei, 87.

69. Ibid., 87–88.

70. Er er wu tongzijun shubao bianyi she, *Tongzijun chuji kecheng*, 101–3; Fan Xiaoliu, *Xinbian tongzijun chuji kecheng*, 209–15; Liu Chengqing, 189–98; *Zhongguo tongzijun chuji kecheng*, 170–83.

71. Fitzgerald, *Awakening China*, 103–6; Liu, *Translingual Practice*, chap. 2; Shao, *Culturing Modernity*, 151–52, 176–77, 183.

72. Rogaski, *Hygienic Modernity*, esp. chap. 8.

lessons on hygiene in Scouting handbooks, which were often written by New Culture–influenced educators, directly paralleled many of the concerns of Chiang Kai-shek's New Life Movement.

Scouting lessons on etiquette reflected similar concern with re-ordering Chinese social life by imposing Western or elite Chinese standards of behavior. Between the late 1910s and 1937, the Chinese Scout curriculum gave scouts step-by-step instructions for decorous behavior in a range of social contexts. Youths were guided through how to visit relatives and friends, how to receive a guest, how to hold and attend dinner parties, and, in some cases, how to hold meetings.[73] The handbooks were extremely detailed in describing various social encounters, dealing with questions such as the following: When does one send a New Year's card instead of visiting? When is a sympathy visit appropriate? How should one bid farewell before a long trip? How does one greet people of different social status? When is it time to take one's leave when meeting with someone?

In many ways, handbooks portrayed what could only be described as Treaty Port etiquette. These rules of conduct demarcated a realm of "civilized" (*wenming*) practices that predominated in coastal cities such as Shanghai. There, self-defined social elites from the lower Yangzi region's economically and culturally advanced core—the people from "south of the river" (*jiangnan*)—were mingling with foreigners in the hybrid business and social environment of Shanghai's concessions.[74] In Scouting handbooks, Western bourgeois conduct and/or elite Chinese forms of etiquette provided standards for proper social behavior in a range of contexts. For instance, some handbooks printed seating charts to show where guests of different status would sit in either Western-style or traditional Chinese banquets and parlors, and they guided students in polite ways to eat with both Chinese and Western utensils.[75] An intermediate handbook

---

73. For the pre-1927 period, see Cheng Jimei, 16; Shen Baoqi, *Tongzijun benji kecheng*, 114–26; Zhongguo tongzijun xiehui, *Tongzijun chubu*, 83–90. For the Nanjing decade, see Hu Liren, *Zhongji tongzijun*, chap. 11; Zhao Bangheng et al., *Tongzijun zhongji keben*, chap. 10: Etiquette (*Liyi*).

74. Honig, *Creating Chinese Ethnicity*, 36–57; Yeh, *The Alienated Academy*, 49–88.

75. Hu Liren, 100–103; Shen Baoqi, 121–25; Zhao Bangheng et al., *Tongzijun zhongji keben*, 10: 9–11, 25–35. The Chinese National Scouting Association handbook discussed only Chinese-style patterns of polite seating and eating. See Zhongguo tongzijun xiehui, *Tongzijun chubu*, 85–86.

from 1936 acknowledged that either shaking hands or bowing might be appropriate in different contexts, recommending that students take their cue from the person they were greeting.[76] The urban elites writing Scouting handbooks and shaping the Scouting curriculum sought to define a new conception of civility by combining norms from Jiangnan elite culture and Western bourgeois culture. But the resulting handbooks reveal the many ways that those rules of conduct were still awkwardly joined, unstable, and in flux.

The hybrid Treaty Port etiquette the Scouting curriculum taught was also clearly an elite class culture. Scouting handbooks' examples of etiquette walked students through many contexts and encounters that would have matched the experience of only well-to-do urban students. Great attention was paid to formal banquets, how to deal with doormen, and when to use one's name cards. Handbooks also made unself-conscious references to employing servants for particular social tasks. On one hand, by focusing on urban elite contexts, handbooks taught skills that would be useful to many of their readers, for as we have seen Republican middle schools constituted a relatively high level of education that was most accessible to children of the urban commercial, official, and professional classes. Such youths *needed* to know where to sit at a Western-style banquet and how to interact with doormen. On the other hand, the introduction of elite modes of etiquette in these widely distributed handbooks suggests that these class-based forms of civility were being marked as representative of an emergent national culture and being taught to other social classes.[77]

Practical Skills and Self-production. Moral training, hygiene, and etiquette were important elements of the Scouting curriculum, but the vast majority of Scout training was dedicated to teaching young people a wide range of practical skills, such as knot tying, semaphore, first aid, washing, cooking, and cleaning. By instructing all scouts in

---

76. Zhao Bangheng et al., *Tongzijun zhongji keben*, 10: 5–6. Cf. Shen Baoqi, 117–18.

77. According to Norbert Elias, in Germany and France different definitions of civility, each associated with a rising national bourgeoisie, came to be identified with national definitions of, respectively, culture and civilization (24–30). Cf. Kasson, *Rudeness and Civility*.

these many different tasks and techniques, Scout training threatened to erode hierarchies based on class and gender. At the same time, Scouting's broad training curriculum, which included many kinds of arts, athletics, and technical knowledge, allowed students outlets for creativity, individual growth, independence, and achievement through the mastery and performance of specialized skills.

Scouting instructed elite, male scouts that cooking, washing, and cleaning were practical necessities that they needed to master in order to be self-sufficient and capable. In the words of one handbook lesson on washing clothes, "The scout's life is one of labor; no matter what he does, he must always do it himself. For example, the Scout curriculum has lessons in daily life, such as washing, cooking, and sewing."[78] By encouraging male scouts to take on what had always been considered female tasks, the Scouting curriculum aimed to de-gender such work. Thus, one Scouting handbook from the 1920s criticized the traditional view that treated cooking as "an internal housekeeping matter and a natural responsibility of women" and instead declared it a "daily necessity" that all scouts, both male and female, would have to master.[79] Although few handbooks so directly addressed the gender implications of teaching male scouts to cook and clean, they were implicit in all the portions of the Scouting curriculum that introduced male scouts to tasks that had always been considered women's work.

By teaching scouts to work with their hands, the Scouting curriculum also challenged class hierarchies and constructed a more populist conception of cultural citizenship. It did so by identifying techniques such as starting fires, cooking, sewing, cleaning, and building as essential survival skills that the lower Yangzi region's "young masters" (*shaoye*) needed to perfect.[80] These kinds of manual physical work in China had always been identified with low class status. Now Scouting's lessons related that all citizens were expected to be able to care for themselves, even if that meant doing work that had traditionally been reserved for nonelite groups.

---

78. Liu Chengqing, 185. Cf. Zhao Bangheng et al., *Tongzijun zhongji keben*, chap. 9: 1.

79. Cheng Jimei, 91.

80. Zhao Qikun, "Xu" (Preface). Cf. Zhao Bangheng et al., *Tongzijun zhongji keben*, chap. 9: 1.

From the opposite perspective, as Girl Scout troops formed in lower Yangzi region schools during the 1930s,[81] female students were learning many of the same kinds of skills as their male counterparts, even when those skills were pseudo-militarized and were considered masculine in China. Girl Scout handbooks identified "women warriors" (*nüyongshi*) as the historical predecessors of the modern Girl Scouts.[82] In retelling the stories of women who had served as scouts and soldiers, they encouraged young women to claim for themselves militarized, masculine patterns of civic action.[83] In general, the skills training presented to female scouts within these handbooks closely followed that given to male scouts, including such martial and "outdoor" skills as orienteering, semaphore, hiking, and drill.[84]

This is not to say that there were no gender-based differences between Girl Scout and Boy Scout training. Girl scouts' physical training exercises were noticeably different from those of boy scouts, including more stretching and different types of games.[85] Further, the intermediate Girl Scout handbook's lesson on service used Florence Nightingale as an exemplar, suggesting that nursing was a privileged form of service for women.[86] However, the bulk of the training curriculum included the same kinds of practical skills and outdoor activities that were central to Boy Scout training. At Zhenhua Girls' School in Suzhou and at Zhejiang Provincial Hangzhou Girls' Middle School, for instance, female scouts received training in different

---

81. Several efforts were made to establish and sustain Girl Scout troops during the 1910s and 1920s. See Fan Xiaoliu, *Xinbian nü tongzijun chuji kecheng*, 31–32; Zheng Haozhang, *Shanghai tongzijun shi*, 8, 13, 19, 25, 32, 34, 46. But active and sustained Girl Scout training in lower Yangzi region secondary schools seems only to have become common under Nationalist Party rule during the 1930s. The prominent Zhenhua Girls' Middle School and Zhejiang Provincial Hangzhou Girls' Middle School, for instance, only had troops beginning during the Nanjing decade. See below.

82. As Kam Louie and Louise Edwards note, martial prowess (*wu*) was sometimes associated with women in legend, but "the public corporealisation of [masculine qualities] *wen* and *wu* necessarily occurred in the male sex, the scholar and soldier being the primary signifiers of a masculine world. . . . Women, however, . . . remain outside this classificatory system" ("Chinese Masculinity," 140–41).

83. Fan Xiaoliu, *Xinbian nü tongzijun chuji kecheng*, 23–28.

84. Ibid., *passim*; Fan Xiaoliu, *Xinbian nü tongzijun zhongji kecheng*.

85. Fan Xiaoliu, *Xinbian nü tongzijun zhongji kecheng*, 125–37.

86. Ibid., 19–21.

combinations of first aid, semaphore, wireless, telegram, rowing and boating, bicycling, shooting, tracking, knot tying, hiking and camping, military-style drill, mountain climbing, measurement, and map making.[87]

Thus, Scout training introduced male and female scouts to forms of activity that clashed with long-established patterns of inequality within Chinese society, implicitly challenging a gendered division of social capacities and encouraging privileged youths to master practical skills that demanded physical labor. Through cross-class and cross-gender forms of training, Scouting presented young people with a vision of the capable citizen that was not perfectly identified with any particular social group and thus was potentially accessible to many youths, including female students and the nonelite. But while many elite, male scouts dutifully learned to cook, clean, and do other manual work, thereby proving their self-reliance,[88] mastery of etiquette and hygiene also provided them with resources for marking their social status.

At the same time, besides the basic skills learned by all Chinese scouts that are discussed above, scouts at higher levels were also able to choose from a vast array of arts, crafts, and athletics that provided outlets for creative development and individual distinction. The higher-level supplemental Scouting curriculum during the 1920s included, but was not limited to, the following lessons: fine arts (*meishu*), music, architecture, sculpture, meteorology, astronomy, photography, aeronautics, agriculture, translation, math, metalwork, woodworking, masonry, mining, leatherworking, engineering, electronics, telegraph, printing, shooting, cycling, physical training, swimming, riding, boat piloting, and driving.[89] A similarly wide array of training—69 special skills in all—was available to scouts in the party-approved curriculum during the Nanjing decade.[90] Higher-

87. Suzhou Zhenhua nü xuexiao, ed., *Zhenhua nü xuexiao sanshinian jiniankan*, 65–66; Zhejiang shengli Hangzhou nüzi zhongxue, *Zhejiang shengli Hangzhou nüzi zhongxue wuzhou jiniankan*, 196–201.

88. E.g., Ling Xuyou, "Tongjun luying zhi yiyi"; Liu Youxiang, "Di sibai tuan tongzijun guoqu he jianglai," 112–13; Zhejiang shengli Jinhua zhongxue chuban weiyuanhui, 30–34.

89. Cheng Jimei, 15–44.

90. Zhongguo tongzijun zonghui choubeichu, 95–112. Cf. Liu Chengqing, 307–32.

level (*gaoji*) scouts in the second and/or third year(s) of lower middle school could obtain belts of ascending status value that were colored blue, blue and white, and then blue, white, and red—the colors of the national flag—as they acquired more badges with mastery of special skills.

Scout leaders throughout this period viewed such training as a forum for nurturing citizens who would be creative, independent, and capable. In his seminal 1922 introduction to Scouting, *Tongzijun zuzhifa* (Organizational method for the Boy Scouts), Wujin's Cheng Jimei argued that, contrary to public perceptions, Scouting was not a form of militarized training that instilled obedience. Rather, he asserted that "the spirit of the Boy Scouts lies in erasing bad habits and cultivating good morals, forming an independent and capable person" and that its training aimed at encouraging voluntary action (*zidong*).[91] Such goals were echoed during the 1930s. For instance, Scouting instructors from Hangzhou Lower Middle School summarized the basic principles of Scout training in the following terms. "The guiding principles of Scout training are to emphasize the cultivation of spirits of self-reliance (*zili*), self-confidence (*zixin*), autonomy (*zizhu*), self-motivation (*zidong*), and self-regulation (*zizhi*), to make [students] perform what they learn and learn by doing (*you xue er zuo, you zuo er xue*), and to seek their own solutions to problems from study and experimentation, thereby acquiring experience and ability."[92] For educators throughout the 1920s and 1930s, Scout training was meant to build student-citizens' capacities for creative problem solving, independence, and self-reliance.

Jamborees, skills competitions, and exhibitions gave students opportunities to demonstrate the skills and arts they had mastered in order to claim individual distinction and to publicize their troop's or patrol's achievements. When groups, subsets of troops, went camping, for instance, there were sometimes skill competitions among groups and/or patrols that were judged by Scout leaders or visiting guests.[93] When a troop or its representatives participated in a county- or province-wide review, they also performed skills, contributed to

---

91. Cheng Jimei, 1–2. Cf. Zhang Junchou.

92. Zhejiang shengli Hangzhou chuji zhongxue bianji weiyuanhui (1937), 32. Cf. Sun Yixin, "Zhongxue tongzijun wenti," 2.

93. Zhejiang shengli Hangzhou nüzi zhongxue, 196–201; Zhejiang shengli Jinhua zhongxue chuban weiyuanhui, 30–34.

exhibitions, and participated in competitions.[94] At the Third Zhejiang Provincial Scout Meeting, in 1934, for example, scouts competed in first aid, semaphore, cycling, knot tying, fire starting, shooting, and a host of other skills. There were also exhibits of scouts' arts and crafts.[95] Skills training and competitions gave students a way to express their own talents and abilities, distinguish themselves in a public forum, and develop a sense of individuality and independence. At the same time, because many competitions were between troops, groups, or patrols, they also encouraged a collective group identity.[96]

Scouting's ability to provide a forum for young people to develop individual talents and claim self-distinction based on public demonstration of their skills is captured in Wuxi County Lower Middle School student Liu Youxiang's report on his troop's performance at the Jiangsu Provincial Scouting Jamboree in November 1932.[97]

In terms of lesson competitions, we participated as a group in bicycle [riding], signaling, first aid, and so forth. Individuals [participated in] knot tying, bugling, pathfinding, map drawing, and so on. We also participated in competitions for the Boy Scout rules' stories, as well as campfire performances. The results of the competitions were that untalented me [i.e., Liu himself] won second-place in map drawing, Gu Baoqi won second place in the rules' stories competition, and the other results were also very good. In particular, Bao Juesheng at the campfire performed a fire-circle dance that won universal applause from everyone assembled, so in the campfire entertainment performances, we took the leading position for the whole province.[98]

Skills training, performance, and competition allowed students to express their creativity and to develop a sense of individuality and

---

94. Interviews with alumni from Yangzhou Middle. Cf. Liu Youxiang, "Di sibai tuan tongzijun guoqu he jianglai," 112–13; Zhejiang shengli Jinhua zhongxue chuban weiyuanhui, 33. For examples of competitions during Scout gatherings from the 1910s and 1920s, see MGRB, May 27, 1923; June 11, 1923; *Shenbao*, October 10–11, 1920; May 8, 11, 1922; Zheng Haozhang, *Shanghai tongzijun shi*, 10, 12, 14, 16, 18, 19–21, 23.

95. Zhongguo tongzijun zonghui choubeichu, 144–45.

96. Groups' "raids" on one another during camping trips also fostered a group ethos (Ling Xuyou, "Tongjun luying zhi yiyi"; Zhejiang shengli Jinhua zhongxue chuban weiyuanhui, 30, 32).

97. Cf. Lu Ying, "Canjia quansheng tongzijun"; Luo Bin, "Tongzijun yewai huodong ji."

98. Liu Youxiang, 113.

independence, even as they demonstrated the range of capacities that were associated with citizenship in the Scouting curriculum.

Social Service. Scouting also taught scouts to serve their communities. Exhortations to social service were repeated throughout Scouting handbooks, and an orientation toward service was embedded within lessons in practical skills. Participating in social service reinforced through practice the dedication to national society that students were taught abstractly in many of their classes, demonstrating to scouts how their individual capacities and talents could benefit society as a whole. Just as importantly, service provided a way to perform citizenship in a public context. Through many quite visible service activities, Scouting contributed to making public service, as a demonstration of commitment to the social welfare, an important aspect of cultural citizenship during the Republican period. Young scouts simultaneously marked themselves as citizens who benefited society.[99]

Starting in the 1910s and 1920s, the message of social service was included within the Scouting curriculum in many different ways. For one, the oft-repeated oath constantly reminded students that their goals were to "fulfill [their] responsibilities as citizens" and to "help others at any time or place." In addition, service (*fuwu*) was one of the main subdivisions of the elective curriculum and walked students through making contributions to public welfare.[100] Patrols, troops, or scouts from multiple troops were often also mobilized for public service. Shanghai's scouts, for instance, engaged in rescue and first-aid work during the Second Revolution, served as sentries and ran first-aid stations at the lantern festivals celebrating the end of World War I, maintained order and managed traffic during the May Fourth protests in 1919, kept order at a mass meeting for disaster relief in 1920 that was organized by Shanghai's business community, and served along with other Jiangsu Scout troops as sentries at the Fifth Far Eastern Athletic Games in 1920.[101]

---

99. In missionary schools, service was also marked as an integral part of modern cultural citizenship. But there, service was also intended to distract students from nationalist protest and may have worked as a kind of displaced religiosity. See Dunch, "Mission Schools"; Graham, 120–39.

100. Cheng Jimei, 44–52.

101. Zheng Haozhang, *Shanghai tongzijun shi*, 4–5, 11, 12, 16–17, 18. See Wasserstrom, *Student Protests*, 82–83, for an account of the role of the Boy Scouts in

During the Nanjing decade, individual and collective service became an increasingly prominent part of the Scouting curriculum and school-based Scouting activities, even as service was actively promoted in other educational programs such as the New Life Movement's Summer Labor Service Movement.[102] Besides calling for student service in their homes, schools, and troops,[103] lessons in individual service encouraged a generalized attitude of selfless help to others rather than the kind of relationship-based ethical demands that were characteristic of Confucian society. Handbooks directed scouts to serve other social members or the general public just as they would serve their family members or friends.[104] One supplemental book, *A Daily Record of a Scout's Good Deeds* by Zheng Haozhang, consisted of a calendar with one "good deed" listed for each day of the year, hammering home the demand for selfless and self-motivated action on the behalf of strangers and the public as a whole.[105] Significantly, all the pictures in Zheng's book were of boy scouts, marking them rather than girl scouts as privileged public actors.

As during the 1910s and 1920s, scouts also mobilized for larger-scale, highly visible public service in their patrols or troops. The advanced Scouting curriculum outlined in detail modes of service such as fire prevention, first aid, relief work, firefighting, keeping order at public gatherings, emergency mobilization and fund collecting, as well as wartime service such as helping refugees and battlefield sup-

---

maintaining order during the May Fourth protests in 1919. Scouts also kept order at many school-based events. E.g., Zhu Junda, 5–6.

102. For details on organization of the Summer Labor Service Movement, see Dirlik, "Ideological Foundations," 951; Jiang Zhongzheng, "Xuesheng ying liyong shujia fuwu shehui"; SMA Q235-1-225, pp. 85–116; Zhejiang shengli Jinhua zhongxue chuban weiyuanhui, 23–24.

103. Hu Liren, 12–13; Zhao Bangheng et al., *Tongzijun zhongji keben*, chap. 2, 12–13, 15–17; idem, *Tongzijun gaoji keben*, 1: chap. 2, 20–25.

104. Anecdotes about Baden-Powell or scouts helping strangers reinforced this idea of universal service. See Fan Xiaoliu, *Xinbian tongzijun chuji kecheng*, 73–74; Zheng Haozhang, *Tongzijun xing shan ri lu*, 26–27.

105. Zheng Haozhang, *Tongzijun xing shan ri lu*, 29–95. Cf. Zhao Bangheng et al., *Tongzijun zhongji keben*, chap. 2, 13–15. Also see elementary-level Scouting handbooks' discussions of "everyday politeness" (Er er wu tongzijun shubao bianyi she, 48–49; Fan Xiaoliu, *Xinbian tongzijun chuji kecheng*, 100–102; *Xinbian nü tongzijun chuji kecheng*, 100–102).

port.[106] In practice, throughout the lower Yangzi region, school-based Scout troops kept order at National Day gatherings, New Life Movement meetings, and provincial and school-based athletic events. They participated actively in promoting the New Life Movement, fought fires near their schools, helped in the organization of antiaircraft drills, and made collections for disaster relief.[107] In all these instances, youths collectively participated in civic action but with a focus on the kinds of constructive activities that were sanctioned and promoted by the Nationalist Party. In doing so, they conformed to a vision of civic action outlined in Nanjing-decade civics textbooks: citizens organized in state-sanctioned groups contributing selflessly to the public welfare under party guidance.[108] Significantly, many Scouting handbooks justified calls for social service by reference to the same organic conception of society that was so prominent in Nanjing-decade civics and language textbooks.[109]

By making service central to scouts' lessons and activities, which were highly visible throughout the 1920s and 1930s, Scouting helped to make the public performance of social commitment a central dimension of China's Republican public culture. Moreover, just as with skills training, competitions, and exhibitions, social service gave students an opportunity to distinguish themselves as important contributors to society, especially when the service was in public arenas, in which uniformed scouts were clearly marked as having official status. Scouts' pride in their contributions to social welfare was reflected in the care with which they recorded their contributions to public undertakings, right down to details such as the number of scouts participating in particular events and the amount of money, food, or clothing collected for relief.[110]

---

106. Zhao Bangheng et al., *Tongzijun gaoji keben*, 1: chap. 2, 27–48. Cf. Hu Liren, 13–17.

107. "Benxiao duan xun"; Hangzhou chuji zhongxue bianji weiyuanhui (1935), 84; Zhejiang shengli Hangzhou nüzi zhongxue, 200–201, 207–8; Zhejiang shengli Jinhua zhongxue chuban weiyuanhui, 30–34; Zhongguo tongzijun zonghui choubeichu, 23–24, 140–41.

108. E.g., Sun Bojian, 1 (1935): 62–64; Wei and Xu, 1: 9–12.

109. Fan Xiaoliu, *Xinbian tongzijun chuji kecheng*, Jiaoxue yaoling: 1–2; Hu Liren, 9; Zhao Bangheng et al., *Tongzijun gaoji keben*, 1: chap. 2, 15; Zheng Haozhang, *Tongzijun xing shan ri lu*, 8–10.

110. See the accounts of school-based Scout troops' service activities cited in note 107.

5.5 "Commemorative Postcard of the Third Group of Scouts from Zhejiang Provincial Hangzhou Girls' Middle School." The Girl Scouts posed facing the camera appear as archetypal young citizens. Their short hair, uniforms, and "at ease" martial posture all reflect the masculinization involved in Scout training and young women's preparation for adopting public roles of civic action. The contrast between the Girl Scouts and the inset picture of the demure, feminine young woman that the Third Group saved from drowning highlights the gender transformations encouraged by some forms of civic training. Zhejiang shengli Hangzhou nüzi zhongxue, *Zhejiang shengli Hangzhou nüzi zhongxue wuzhou jiniankan*, 207.

Public service as a sign of citizenship was captured vividly in a postcard picture taken of members of the Third Group of Scouts from Zhejiang Provincial Hangzhou Girls' Middle next to West Lake, Hangzhou's famous tourist site. The local Scouting council made the postcards to commemorate the girl scouts' rescue of a young woman who had fallen into the lake and nearly drowned. (See Fig. 5.5.) By employing the skills in rescue and first-aid they had learned as scouts, the young women in the Third Group were able to save a life. The postcard commemorating their public service represented them as capable young citizens of the nation.[111] It also captured how the roles of public action for which Scout training pre-

---

111. Zhejiang shengli Hangzhou nüzi zhongxue, 207–8.

pared young women were to some degree masculinized. The uniformed rescuers stand in martial poses, with legs spread shoulder width and arms held loosely at the ready, following the new Republican pattern of masculine physical readiness. By contrast, the young woman who was rescued is shown in the inset picture demurely holding an umbrella, in a pose reminiscent of the highly feminized calendar pictures and advertising posters of the Republican period. This contrast in images suggests how Scouting's gender-crossing training curriculum opened up to female secondary students new activities and forms of social service. These allowed girl scouts to perform an active mode of citizenship but one that was marked by new conventions that either were distinctly masculine or at the very least disrupted conventional ideas of femininity.

III. MILITARIZING CITIZENSHIP

In step with efforts at managed popular mobilization, the Nationalist Party, as several authors have noted, actively promoted "militarization" (*junshihua*) during the Nanjing decade.[112] Students were the prime targets of the party's military-training programs. Yet for party leaders the primary aim of student military training seems to have been less military preparedness through the production of young soldiers than inculcation of a martial form of cultural citizenship. Like modernizing nationalist elites elsewhere, Nationalist Party members sought to use military drill to instill in individual students a new cultural repertoire of regimentation, cleanliness, order, and simplicity characteristic of the soldier. Collective drill in turn organized disciplined individuals into coordinated units that could be used as the basis for, or models of, social order more generally, operating in their drill units as what Timothy Mitchell has evocatively described as an "artificial machine."[113] In these various dimensions, martial training closely mirrored patterns of social mobilization that typified "fascism."[114] However, review of youth military-training programs

112. Dirlik, "Ideological Foundations," 972–74; Eastman, *The Abortive Revolution*, 49, 64–65, 68–69; Kirby, *Germany and Republican China*, 182–84.

113. Mitchell, *Colonising Egypt*, 36–39.

114. Lloyd Eastman first identified fascist tendencies in the Nationalist Party in a chapter on the so-called Blue Shirts in his seminal book, *The Abortive Revolution*. His provocative thesis there has served as the touchstone for an ongoing

suggests that militarization in Nationalist China had different goals than "fascist" militarization in Europe or Japan. Moreover, the externalized discipline, order, regimentation, and absolute obedience to authority that were central parts of Nanjing-decade military training were consistently contradicted or undermined by other dimensions of the very plural forms of training and cultivation promoted by the Nationalist state.

Negotiating Military Training

During the 1910s and 1920s, military training was a relatively marginal feature of many secondary schools' physical education.[115] In the 1912–13 curriculum, "military-style exercise" (*bingshi ticao*) was included as part of the physical education curriculum.[116] In 1923, the New School System curriculum did not explicitly mandate military training as a class in the secondary curriculum.[117] In practice, at those schools that held military training during the 1910s and 1920s, it generally occupied no more than an hour or two of students' time each week, since it split time with regular physical training or physical education classes.[118] Some prominent private secondary schools, such as the Minli and Nanyang middle schools in Shanghai, did not include any military training at all.[119] In terms of the amount of time spent on the classes, as well as the level of intensity and sophistication involved, military training during the Nanjing decade represented a distinct departure from that of the period between 1912 and 1927.

---

debate about the nature of the Nationalist regime. See also Chang, *The Chinese Blue Shirt Society*; Eastman, "The Rise and Fall of the Blueshirts"; Wakeman, "A Revisionist View."

115. I concentrate here on military-training classes (*junshi xunlian*) in high schools. Lower middle school students also participated in military-style drill in their Scout troops.

116. Zhu Youhuan et al., part 3, 1: 354, 361. Instructions regarding exercise classes called for "special emphasis to be placed on military exercise," but for girls' schools not to make it part of their lessons.

117. Quanguo jiaoyuhui lianhehui xinxuezhi kecheng biaozhun qicao weiyuanhui, "Gaoji zhongxue kecheng zonggang," 4–7.

118. E.g., *Jiangsu shengli di'er shifan xuexiao yilan*, 2–3; "Jiangsu shengli diwu zhongxuexiao xueye jiaoshou chengxu biao"; Jiangsu shengli diyi shifan xuexiao, *Xuesheng ruxue xuzhi*, 20–21; Zhu Youhuan et al., part 3, 1: 400.

119. *Nanyang zhongxue*, 1915–16; *Shanghai Minli zhongxuexiao yichou nian zhangcheng*, 45–47.

After 1927, government policy and school practice in the arena of military training developed through the intersection of student demands, state directives, and school administrators' accommodation of both. More extensive school-based military training was first called for by nationalistic students and social groups in the wake of the Ji'nan Incident in May 1928, when Japanese troops interfered with the advance of the Nationalist Party's Northern Expedition in the northern province of Shandong. For instance, a group of citizens from Shanghai petitioned the University Council to start a program of military training in all the nation's schools, citing the fact that Japanese students had received military training since the end of World War I.[120] More importantly, students themselves demanded military training, and the Shanghai Student Union organized a student army that was trained by school athletic instructors.[121] This trend was echoed throughout the lower Yangzi region as schools implemented their own programs of student military training in the context of the nationalist crisis, often under pressure from their students.[122]

Reacting to these popular pressures, the University Council and the First National Educational Congress enacted a program of military training in the summer of 1928.[123] Revisions were made to the training program in January 1929. That year designated military trainers from the Nationalist government's Ministry of Training and Supervision (Xunlian zongjianbu) arrived at most regional schools to start state-run military training.[124]

---

120. Zhongguo di'er lishi dang'anguan, 2: 1239–40.

121. Israel, *Student Nationalism*, 22. Students again militated for increased military training after the Mukden Incident in 1931. See Wasserstrom, *Student Protests*, 173–74. Cf. Israel, *Student Nationalism*, 54–57.

122. For a partial list of the "Student Armies" (*xueshengjun*) at Shanghai area schools, see "Shanghai gexiao xueshengjun diaocha biao (jia)." For implementation or intensification of military training at other schools, see "Benxiao duan xun," 21; Xiaokan bianji weiyuanhui, *Songjiang gaoji zhiye zhongxue*, 103; *Zhejiang shengli gaoji zhongxue xiaokan*, 113; Zhongyang daxuequli Shanghai zhongxuexiao mishuchu chuban weiyuanhui, 134.

123. Israel, *Student Nationalism*, 22; Zhongguo di'er lishi dang'anguan, 2: 1239; Zhongguo guomindang zhongyang weiyuanhui dangshi shiliao bianzuan weiyuanhui, ed., *Geming wenxian*, 54: 5.

124. See, for example, Jiangsu shengli Shanghai zhongxue chuban weiyuanhui (1933), 198; Xiaokan bianji weiyuanhui, *Songjiang gaoji zhiye zhongxue*, 103; *Zhe-*

Though instituted in response to military threats and with the supposed goal of improving national defense, the Nationalist Party's objectives for military training focused as much on augmenting physical and disciplinary training as on cultivating young soldiers.[125] These priorities were expressed in the writings and speeches of the Nationalist Party leadership, at several different levels. Chiang Kai-shek himself, in an address to the participants in the first collective military-training session in Zhejiang in 1935, asserted:

In sum, the significance of military training does not just reside in teaching and learning military regulations and skills. Its main point is to temper the bodies and minds of the trainees and to mold their moral characters, show the proper path of modern life, teach common knowledge for modern citizens, improve the thought and habits of the trainees at the root, and establish complete and flawless characters so that they are able to adhere to discipline, take responsibility, be clear about etiquette and righteousness, know integrity and a sense of shame, and become Chinese citizens (*guomin*) who are suited to modern existence.[126]

Chiang and other party leaders expected that military training would strengthen the nation by making citizens well disciplined, organized, unified, and physically strong, able to live a "military lifestyle." They did not primarily aim at transforming students into soldiers.[127] These same concerns were expressed by some educators and military trainers in local schools.[128]

Yet the evidence we have from interviews and students' writings suggests that the immediacy of the threat from Japan, and the seemingly obvious need for military preparation in a world headed toward war, drove their enthusiasm for military training.[129] For instance,

---

jiang shengli disi zhongxue yilan, 22; Zhejiang shengli Jinhua zhongxue chuban wei-yuanhui, 26–27.

125. Zhongguo di'er lishi dang'anguan, 2: 1241.

126. Jiang Zhongzheng, "Xuesheng jizhong xunlian kaixue xunci." Cf. Dai Ji-tao, *Dai Jitao xiansheng wencun*, 2: 492–95.

127. E.g., "Liuzhou jinian jianyue xueshengjun ji tongzijun xunhua," 9–10.

128. Cao and Liu, "Zhongxue zhi junshi jiaoyu," 1–2.

129. Sheng, "Song junxun tongxue"; Sun Jingkun, "Shenghuo junshihua"; Zhu Ji, "Junxun duiyu guojia minzu de qiantu." Several alumni from Hangzhou High School commented in interviews that the clear Japanese threat after the Mukden Incident and the 1932 invasion of Shanghai drove home to them the purpose of and need for military training.

Jiangsu Provincial Shanghai Middle School student leader Zhou Jian-wen argued that China was facing a coming war and that students, as "core elements" (*zhongjian*) of national society, had the responsibility for leading the nation into the future.[130] To do so, they needed to be well prepared, and the best way to prepare, asserted Zhou, was for students to receive military training to take part in the struggle for national self-defense. A student from Hangzhou High School, writing in 1935, when the Nationalist Party was still pursuing a policy of appeasement toward Japan, proclaimed, "I am willing to take a gun and arm myself to stand on the front line of the national struggle and go to be a brave who will guard the state and protect the nation."[131] These students' orientation toward national defense echoed the calls to protect the nation's borders that were fundamental to history and geography lessons throughout the Republican period. As part of their military preparation, students like those cited here were receptive to some forms of military-style individual discipline and collective drill, which they viewed as necessary to prepare them for war with Japan. But they did not necessarily embrace military-style discipline, regimentation, and absolute loyalty as fundamental attributes of citizenship more generally.

Practicing Military Training

Government policies mandated military training for all male students for the first two years of high school. They were to have three hours of military-training class a week and a three-week period of intensive training at their schools during the summer. Each school was to have a military trainer sent by the Ministry of Training and Supervision and to adhere to a standardized curriculum of classroom lessons and military drill.[132] The three hours of weekly military-

---

130. Zhou Jianwen, "Feichang shiqi zhong qingnian xuesheng."
131. Tan Shaozeng, "Junxun qijian shenghuo huiyi," 147.
132. Zhongguo di'er lishi dang'anguan, 2: 1241. In most instances, military trainers were transferred to high schools by the Ministry of Training and Supervision and often had experience in the Nationalist military forces and/or were graduates of the Central Military Academy. See Jiangsu shengli Shanghai zhong-xue chuban weiyuanhui (1933), 198; *Zhejiang shengli disi zhongxue yilan*, 22, 217; *Zhejiang shengli gaoji zhongxue xiaokan*, 113; Zhejiang shengli Jinhua zhongxue chu-ban weiyuanhui, 27; Zhongyang daxuequli Shanghai zhongxuexiao mishuchu chuban weiyuanhui, Jiaozhiyuan minglu: 8; Zhongyang daxuequli Taicang zhong-

training instruction were divided into two hours of drill and one hour of classroom instruction. Classroom instruction gave students an overview of military topics relating to combat, the production and capabilities of weapons, developments in weapons and material, fortifications, and communications. Students were also instructed as to the aims of military training, the organization of China's national defense, and world military trends. Drill classes focused on individual and group drill that was designed to refine the movements of individual student-soldiers and then to integrate them into units that could be combined modularly to produce larger and larger formations under the command of a single leader.[133] (See Fig. 5.6.) They also included practice in surveying, marksmanship, calculating distance, and using hand flags for signaling, as well as training to set up and run a military camp with sentries and a formal chain of command.

Significantly, besides basic drill, which was also part of the Scouting curriculum, female students were to receive training in nursing or first-aid rather than military training per se.[134] This gender distinction in Nanjing-decade youth military training followed a pattern that has been adopted in many other military systems throughout the world, with men on the front lines and women providing assistance on the "home front."[135] Such a gender dichotomy was particularly sharp in Fascist Italy, Nazi Germany, and early Showa Japan, where girls were prepared for motherhood and care of the home and

<hr />

xuexiao, 7, 12. Schools' military-training units were also subject to regular inspections by representatives of the Ministry of Training and Supervision or the Citizens' Military Training Commission. See "Benxiao duan xun"; Jiangsu shengli Shanghai zhongxue chuban weiyuanhui (1933), 198; "Junshi xunlian xiaoxi"; Xiaokan bianji weiyuanhui, 111–12; *Zhejiang shengli gaoji zhongxue xiaokan*, Gaikuang: 115.

133. Zhongguo di'er lishi dang'anguan, 2: 1242–65. For examples of local implementation, see Jiangsu shengli Shanghai zhongxue chuban weiyuanhui (1933), 199–200; Jiangsu shengli Suzhou zhongxue jiaowu chu, *Jiangsu shengli Suzhou zhongxue geke jiaoxue jindu biao*, "Junshi xunlian jiaoxue jindu biao"; *Shanghai Minli zhongxue sanshizhou jiniankan*, 97; Zhongyang daxuequli Shanghai zhongxuexiao mishuchu chuban weiyuanhui, 134.

134. SMA Q235-1-548; Zhongguo di'er lishi dang'anguan, 2: 1034, 1241, 1268, 1285–87; Zhongguo guomindang zhongyang weiyuanhui dangshi shiliao bianzuan weiyuanhui, *Geming wenxian* 54: 5.

135. Enloe, *Does Khaki Become You?*, esp. 3–7, 15, chap. 4. Cf. Goldstein, *War and Gender*.

學 生 軍 及 童 子 軍

5.6 "Student Army and Boy Scouts (Shanghai Middle School)." Student military trainees and Scouts assemble in their units, forming the collective body of the school community in ordered ranks. They face out of the picture to the right an unseen reviewer—perhaps the Jiangsu Department of Education director Zhou Fohai, who attended a school-wide review of Scouts and student military trainees in the fall of 1933. Jiangsu shengli Shanghai zhongxue chuban weiyuanhui, *Jiangsu shengli Shanghai zhongxue yilan* (1933).

the home front while boys were prepared for battle and political leadership.[136] Only wartime necessity forced these regimes to recruit women into forms of work and military activity that had been reserved for young men and were associated with masculinity in youth-training regimes.[137] This modern link between males in the armed forces and females on the home front was also reinforced by late imperial Chinese tendencies to associate in practice martial prowess (*wu*) primarily with men, despite a rich folk literature depicting women martial heroines.[138]

Girls' middle schools in the Lower Yangzi region dutifully integrated nursing training into their curricula during the Nanjing

---

136. Knopp, *Hitler's Children,* chap. 2; Koon, *Believe,* 97–98; Smethurst, *A Social Basis,* 158.

137. E.g., Knopp, *Hitler's Children,* 108.

138. Louie and Edwards, "Chinese Masculinity."

decade.[139] And, some female students accepted nursing as an appropriate form of civic action for young women in a time of national crisis.[140] But, given Nationalist Party efforts to use military training to cultivate a new generation of citizens, exposing male and female students to distinct forms of training demarcated gender distinctions in kinds of citizenship. Moreover, associating males' battlefield training with preparation for manning the "front," which was clearly the focus of immediate concern, and females' nursing training with "rear" support implicitly privileged the former, even though this relative evaluation was never explicitly stated. As in Nazi Germany, Fascist Italy, and 1930s Japan, in Nationalist China military training seemed to be preparing young women for a separate and second-class kind of citizenship. But the extreme gendering of military training always stood in tension with forms of civic training such as Scouting and self-government, which encouraged female students to be civically active in the same modes as their male counterparts.

Youth military training intensified starting in 1935 and 1936 in reaction to growing imperialist pressure from Japan in North China and in concert with other Nationalist Party programs for civic training such as the New Life Movement. The military management method (*junshi guanli banfa*) organized each high school as a regiment (*tuan*) and aimed at militarizing everyday school routines, such as dress, etiquette, organization, relations of authority, and daily ritual.[141] Students wore military-training uniforms, and teachers and staff wore standard Sun Yat-sen suits (*Zhongshan zhuang*). Students lined up in platoons to enter and leave class and the dining hall, sat in order according to their units, and began and ended the day with flag raising and lowering ceremonies. In addition, starting in the summer of 1935, Jiangsu, Zhejiang, Nanjing, and Shanghai also each conducted a three-month collective military-training camp (*jizhong xunlian*) for the secondary schools in their respective jurisdictions.

---

139. "Benxiao kangri jiuguo yundong zhi shishi banfa"; "Junshi guanli banfa dagang," 10; *Shanghai shili Wuben nüzi zhongxuexiao gaikuang*, 3–4; Zhejiang shengli Jinhua zhongxue chuban weiyuanhui, 21; *Zhenhua nü xuexiao sanshinian jiniankan*, 47.
140. Jin Wanxiang, "Guonan zhong nüzi yingyou de zeren."
141. Cao and Shen, "Gaozhong xuesheng shenghuo junduihua"; Jia [Shaoyi], "Ben xiao gaozhong geji xuesheng zhunbei shixing junshi guanli"; "Junshi guanli banfa dagang," 9–10; Zhejiang shengli Jinhua zhongxue chuban weiyuanhui, 20–21; Zhongguo di'er lishi dang'anguan, 2: 1313–22.

The camp lasted from early April until early July, and all male high school students who had completed the first year were required to attend. Students received noncommissioned officer training that dramatically intensified the kinds of discipline and physical training they received in regular military-training classes.[142] Some groups within the party seized upon collective training as an opportunity for so-called spiritual education, mostly through lectures, and political education with the goal of inspiring loyalty to the party and Chiang Kai-shek.[143] Through both forms of intensified military training, Nationalist Party leaders sought to instill in students militarized habits of daily life and the absolute loyalty of the soldier toward his commander.

IV. CONCLUSION

Through increasing militarization, Nanjing-decade civic training came to resemble the youth training of Japan's and Europe's fascist states. Most similar were the extreme levels of externalized discipline and regimentation, the emphasis on physical fitness, cleanliness, and order, the intense group ethos focused on sacrifice for the nation, and military-style obedience to a political authority that was seen to be embodied in a single party and leader. These qualities, which were prominently featured in military-training programs under the Nationalists, were also central to contemporary youth training in Mussolini's Italy, early Showa Japan, and Hitler's Germany.[144] But in significant ways, Nationalist Party youth training diverged from that of contemporary fascist states.

For one, in Nazi Germany, Fascist Italy, and early Showa Japan, military-style training was initiated and orchestrated by the respective parties and states and yoked to aggressive, expansionist foreign policies. In Nanjing-decade China, by contrast, students and citizens pushed for implementation of military training for a project of national defense about which the Nationalist Party was initially quite ambivalent. Party leaders responded to popular demands for military

---

142. *Jiangsu sheng xuesheng jizhong xunlian gongzuo baogao*; SMA Q235-1-548; Zhongguo di'er lishi dang'anguan, 2: 1287.

143. Huang Yong, "Huangpu xuesheng de zhengzhi zuzhi," 14–15; *Jiangsu sheng xuesheng jizhong xunlian gongzuo baogao*, 1–6; SMA Q251-1-547.

144. Knopp, *Hitler's Children*; Koch, *The Hitler Youth*; Koon, *Believe*; Smethurst, *A Social Basis*, 25–43, chap. 5.

preparation to introduce a program of militarized training, but that training, as we have seen, focused mostly on the domestic goal of creating a new public culture. Nationalist Party leaders did not seek to use youth military training to create the kind of war-hungry ethos that was fundamental to European fascism and Japanese state authoritarianism. Further, because students, educators, and other social groups helped to promote and organize the project of military training, their interests and goals partly shaped its meaning and practice. We see this most clearly in students' calls for training in practical military skills that would prepare them to take part immediately in national defense, a goal that was in tension with the Nationalist Party's appeasement policy toward Japan. Students' enthusiasm for forming "student armies" resonated more with the ethos of a civic republican–style citizens' militia than the state-centered militarization of fascism. In their view, each citizen should personally take responsibility for defending the state.

In addition, because the Nationalist Party built on the moral cultivation and youth training of the 1910s and 1920s, it incorporated into schooling the diverse goals and practices of those educational programs. The eclectic nature of the resulting state-sanctioned civic education meant that military-style training, with its emphasis on regimentation, self-discipline, homogenization, and loyalty to the party-state, always stood in tension with forms of training and cultivation that sent quite different messages about cultural citizenship. Scouting and the extracurricular activities managed by student self-government groups, which had been introduced by Chinese educators inspired by the progressive education movement, encouraged students to participate in arts, academic, and athletic activities that were intended to promote individual creativity and independence.[145] These kinds of athletic and artistic activities constituted a larger proportion of youth training in Nationalist China than in Japan and Europe's fascist states, where military drill tended to predominate, and they provided an alternative to the martial ethos fostered by

---

145. Similar kinds of activities, such as team sports, hiking, gliding, and sailing, were included in youth groups in Nazi Germany and Fascist Italy. But in these countries such activities were used instrumentally to stimulate youths' interest in youth group participation, and they were frequently intended primarily to build a group ethos and/or prepare young people physically and mentally for wartime activities (Knopp, 19–20; Koch, 229–31; Koon, 99–100).

military training.[146] The stress here on pluralism in Nanjing-decade training regimes is reinforced by Andrew Morris's recent work on physical culture in Republican China, which demonstrates that a "liberal democratic" form of athletics persisted in Nationalist China, despite some sporting enthusiasts' and Nationalist leaders' promotion of fascist-style athletics.[147]

At the same time, Nanjing-decade character-development education incorporated many aspects of late imperial moral cultivation. Such moral cultivation built on close relations of guidance and reciprocity that encouraged interpersonal loyalties distinct from the absolute loyalty to party-state and leader that characterized interwar fascism. The ability of the inward self-reflection of late imperial forms of moral cultivation to foster personal moral autonomy also cut against the obedience and regimentation that was central to military training.[148] In short, the undeniable diversity of the training and cultivation in lower Yangzi region schools during the Nanjing decade problematizes efforts to cast the Nationalist Party's youth training policies as exclusively fascistic or militaristic.

Such pluralism characterized regional secondary school instruction about the daily practice of citizenship throughout the Republican period. Much of the training and cultivation in schools was directed at both male and female students and tended to cross or confuse gender distinctions, but other aspects, especially military training, were highly gendered. Scout training in practical tasks such as cooking and cleaning challenged class and status hierarchies, whereas training in etiquette and hygiene seemed to reinforce and mark them. Patterns of moral education and daily life in schools cultivated hierarchical interpersonal relations of respect and responsibility between students and their teachers, and opportunities for service and achievement could mark differences in ability and accomplishment. At the same time, schools' organization of space and time seemed to define students and their classmates as homogeneous and equal elements in the uniform social microcosm of the

---

146. Guido Knopp (15–16) asserts that in Hitler Youth organizations "military-style drill dominated the daily routine." In Japan, one-third to one-half of the instruction in youth training centers was dedicated to military drill (Smethurst, 39, 43).

147. Morris, *Marrow*, chaps. 5–6.

148. de Bary, *Learning for One's Self*; Munro, *Images*, esp. 123–24.

school. The multiple, contrasting patterns of cultural citizenship incorporated in schools' civic training and moral cultivation, even during the Nanjing decade, provided students with diverse idioms of citizenship, which were all sanctioned by the party, state, and educators, and from which students could draw to construct their own modes of civic action.

6

From Mobilization to Routinization: Transformations of Civic Ritual and Performative Citizenship

During the 1932–33 academic year, students at Wuxing County Girls' Lower Middle School participated in 29 major assemblies and ceremonies, nearly one event for every week in the school year.[1] Besides large-scale annual or periodic gatherings such as historical commemorations, Scouting reviews, Confucius' birth date celebration, and various kinds of mass rallies, students and teachers also took part in routine assemblies at their schools. These could include the weekly memorial meeting for Sun Yat-sen, Scout troop swearing-in ceremonies, beginning- and end-of-semester gatherings, and daily flag-raising and -lowering ceremonies. Wuxing Girls' Lower Middle was not exceptional in the density of its ceremonial events. Throughout the Republican period, lower Yangzi region schools were focal points for the performance of civic ritual, and students were primary actors in public ceremonies.

In public ceremonies, students and other social groups enacted their citizenship in a highly visible way through the cultural forms of symbols, performative language, and ritualized action, making civic ritual a fundamental part of Republican cultural citizenship. By "civic ritual" I mean symbolic collective performance that organizes

---

1. Wuxing xianli nüzi chuji zhongxue chuban weiyuanhui, ed., *Wuxing xianli nüzi chuji zhongxue zuijin gaikuang*, 1–5.

social and political relationships, produces cultural patterns, and serves as a context for negotiating social power.[2] At the most basic level, civic ritual in Republican China, as in other modern nations, allowed citizens to perform and reflect upon their membership in the national community. In addition, by adjusting the symbols, forms, and content of civic rituals, social groups and political authorities used public ceremonies to negotiate social place and political power. Changes in the dynamics of civic ritual marked shifts in the meaning and practice of citizenship.

Students, as we will see, were consistently central participants in Republican-period civic rituals. In part, so many students joined in these events because they were a relatively easy group for social elites and political leaders to mobilize through school authorities. More importantly, as a social sector (*jie*), students represented the youth of the nation, its future and hope, so their presence in nation-centered public ceremonies was symbolically essential. For their part, students, both individually and as a group, used ceremonies to claim in symbolic terms a prominent role as active citizens of the national community.

Lower Yangzi region students' participation in civic rituals continued throughout China's Republican period, but the form and significance of those ceremonies changed dramatically over time. Early Republican civic rituals made visible the national community and celebrated the Republic as a polity in which the people played a prominent role. But mass nationalist protest, the National Revolution, and the beginning of Nationalist Party rule each generated new permutations of public ceremonial as different groups, including students, sought to use civic ritual to claim and enact social and political power.

---

2. My definition of "civic ritual" builds on the well-known definition of political ritual formulated by Steven Lukes ("Political Ritual"). I substitute "civic" for "political" in order to emphasize citizens' participation in these public events. Further, whereas Lukes focuses on how ritual represents social patterns, practice theory encourages us to consider how ritual practice is constitutive of cultural structures and relations of power and at the same time is the site where power is challenged and contested. Practice theory also helps us to see how ritual involves active engagement that often entails both consent and resistance on the part of participants. See Bell, *Ritual Theory*, chap. 9.

I. ENACTING
NATIONAL CITIZENSHIP

During the late Qing and early Republican periods, China's social and political elites were exposed to a wide spectrum of the ceremonial forms that were fundamental to modern nationalism. Chinese students in Japan during the last two decades of the Qing dynasty would have witnessed firsthand the pageants of the late Meiji period.[3] Social elites in Shanghai and other coastal cities observed and sought to participate in, reinterpret, and compete with the pageants of the community of foreign residents.[4] Moreover, during the first two decades of the twentieth century, Chinese Christians, under missionary influence, introduced into China key symbols and ceremonies of the modern nation-state, especially the civic rites of Anglo-American republicanism.[5] As Henrietta Harrison demonstrates, by the first years of the Republic, China's citizens had created symbols such as the national flag and anthem and mastered their use, developing ceremonies centered on the nation and the new Republican polity.[6]

After 1911 lower Yangzi region secondary schools quickly integrated the holidays and memorial days of the Republic into their school calendars.[7] October 10, the date of the Wuchang Uprising that had started the 1911 Revolution, was celebrated as National Day in both school rituals and public ceremonies, and the anniversary of the establishment of the Republic on January 1 was marked as a school holiday. Some schools commemorated local events associated with the 1911 Revolution as well. After Yuan Shikai's death in 1916, December 25, the date of the Yunnan Uprising against him, also became a regular holiday at many schools. On May 7 or May 9, the respective dates in 1915 of Japan's ultimatum regarding the Twenty-One Demands for special privileges and Yuan's acceptance of them,

---

3. Fujitani, *Splendid Monarchy*, part 2.
4. Goodman, "Improvisations on a Semicolonial Theme."
5. Dunch, *Fuzhou Protestants*, esp. chap. 4.
6. Harrison, chaps. 1–3. Cf. Goodman, "Democratic Calisthenics"; and Strand, "Citizens in the Audience."
7. For the Ministry of Education's requirements, see "Xuexiao xuenian xueqi ji xiuxue riqi guicheng," 94–95.

students and teachers took part in solemn memorial ceremonies at their schools and in more public arenas.[8]

October 10 was the main focus of nationalist ceremonies during the first decade of the Republic. As in other modern nations,[9] the ceremonies that marked National Day in Republican China, or "Double Ten" (*shuangshijie*), as it was popularly called, enhanced a sense of national community through the simultaneous performance of parallel civic rites throughout the nation. As participatory mass ceremonies, National Day assemblies also demonstrated that the people were a collective political subject in the Republic in a way they had not been in the late imperial period, as they played an active role in the symbolic performance of civic ritual. Students claimed their place as an important social sector in the emergent Republican social order through their visible inclusion in National Day ceremonies and other public political performances.

National Day was established as a national holiday by the National Assembly and the State Council immediately after the founding of the Republic in 1912, suggesting that both the Yuan Shikai government and social elites recognized the power and importance of national founding rituals.[10] It quickly became a popular public event. Cities and towns across the lower Yangzi region, and throughout the country, held commemorative ceremonies on October 10, 1912, to mark the first anniversary of the Wuchang Uprising and the founding of the Republic.[11] Peter Zarrow suggests that Yuan's efforts to use the National Day ceremonies for his own political purposes and his attempt to introduce state ceremonies that drew on late imperial political rituals dampened popular enthusiasm for Double Ten holidays between 1913 and 1915.[12] Yet public ceremonies were revived with great vigor in lower Yangzi towns after Yuan's death early in 1916.

---

8. *Di'er nüzi shifan xuexiao xiaoyouhui huikan*, no. 7 (November 1918), 4–6; no. 8 (May 1919), 1–2; no. 14 (July 1922), 2; no. 18 (June 1925), 1–4; Liao Shicheng et al., 385–402; *Zhejiang shengli diwu zhongxuexiao xianxing guicheng yilan*, 33–34; *Zhejiang shengli diyi zhongxuexiao ershiwu nian jiniance*, 9. Confucius' birth date was also marked as a holiday at most schools, as were the traditional festival days of the lunar calendar.

9. Mosse, *Nationalization of the Masses*; Ozouf, *Festivals and the French Revolution*.

10. Harrison, 93–96; Zarrow, "Political Ritual," 158–61.

11. Harrison, 94–95; *Shenbao*, October 10, 12, 13, 1912.

12. Zarrow, "Political Ritual," 161–72.

Schools and students were active participants in National Day festivities. Students and teachers decorated their schools with flags, lanterns, and pine archways, as did many businesses and government offices. Though nominally National Day was a vacation day when classes were suspended, students were busy then with civic rituals. Some schools held their own ceremonies, during which students honored the national flag, sang the national anthem, and heard speeches that celebrated the Republic.[13] Schools also participated en masse in public gatherings, where they helped to represent "academic circles" (*xuejie*) alongside "commercial associations" (*shangtuan*) and government organizations.[14] These ceremonies included many common elements that paralleled nation-centered civic rituals elsewhere: singing the national anthem; ritualized salutes to the five-color national flag; speeches and addresses by a wide range of government officials and social elites; and cheers for the nation and the Republic.[15] Together these elements provided a basic ritual structure for National Day ceremonies.[16] But many of the other components of these national celebrations were fluid and lively, creating an enthusiastic atmosphere through performances in which students often played a central role. For instance, Shanghai's National Day celebrations in 1916 were enlivened by Nanyang Middle School students' performance of new-style dramas (*xinju*).[17] At Jiaxing's three-day National Day celebration during the same year, Zhejiang Provincial Second Middle School students performed music at the start of the main public gathering, and on the following day female students from the provincial girls' normal school danced during a day of performances by students and others.[18] Schools also held sports competitions or Scout gather-

---

13. *Di'er nüzi shifan xuexiao xiaoyouhui huikan*, no. 7 (November 1918), 6; *Shenbao*, October 12–13, 1916; October 11–12, 1918.

14. For discussion of the division of the Republican citizenry into distinct "sections" or "circles" (*jie*), see Harrison, 117–22.

15. MGRB, October 10, 12, 1925; *Shenbao*, October 10, 12, 13, 1912; October 11–12, 1913; October 12–14, 1916; October 11–12, 1918. See Harrison, 98–111, for analysis of the symbolic elements of early Republican civic rituals.

16. The Ministry of Education mandated a uniform structure for celebratory ceremonies at schools. See "Xuexiao yishi guicheng."

17. "Zaizhi guoqing jinian zhi shengkuang," *Shenbao*, October 12, 1916.

18. "Jiaxing sanri zhong zhi guoqing guan," *Shenbao*, October 14, 1916.

ings.[19] Finally, lantern parades, in which students usually played a central role, were important features of many National Day celebrations, helping to create a colorful and exciting atmosphere.[20]

During these public ceremonies, flags, songs, and cheers symbolically connected the gathered crowds of people to the broader community of the nation. As the same civic rites were simultaneously replicated in many localities, and dutifully reported in newspapers such as *Shenbao* and *Minguo ribao* (Republican daily), the nation as a community was constituted by its citizens and made visible to them. This national community constituted through civic action correlated strongly with the idea of the nation as a functionally integrated social organism and the nation defined by common civic participation rather than through ethnic or racial identification.

The large-scale participation of many different social groups—men and women, old and young, numbering in the thousands or tens of thousands—also demonstrated that the people were members of a political community in ways they had not been during the late imperial period. Late imperial state and community rituals had encouraged a sense of shared culture through the use of common symbols and ritual sequences,[21] but they had not granted the people any political agency, even in symbolic terms. In late imperial rituals, local communities followed set standards of ritual propriety to connect themselves with the legitimizing imperial center, which monopolized sovereignty and authority; in Republican ceremonies, the people *made* the nation as they celebrated it. Still, even after 1911, political authority and power was not distributed equally within the national community, as early Republican civic rituals also made clear. Harrison establishes that "members of certain modern, government-sponsored institutions and associations" dominated National Day ceremonies and other public rituals.[22] Schools were prominent among these privileged modern institutions, and civic rituals such as National Day celebrations gave teachers and students a forum for claiming a central role in public life.

---

19. See, for instance, Liao Shicheng et al., 398; "Quanyi tongzijun huicao ji," *Shenbao*, October 11, 1920; *Shenbao*, October 11–12, 1918.

20. Liao Shicheng et al., 386; *Shenbao*, October 10, 12, 13, 1912; October 11–12, 1913; October 12–14, 1916; October 12, 1918.

21. E.g., Pomeranz, "Ritual Imitation and Political Identity."

22. Harrison, 118.

Smaller-scale and routinized ceremonies within schools echoed and reinforced the messages of mass rituals on October 10, as the same kinds of symbols and ritual forms linked schools to the broader community of the nation and teachers and students acted as citizens. One such school ceremony was the "weekly meeting" (*zhouhui*), which several regional schools instituted during the early 1920s.[23] Weekly meetings at different schools ranged in time between forty minutes and an hour, and they were attended by all students. At schools such as Shanghai's Pudong Middle, symbols, recitation, and song symbolically constituted students as national citizens (*gongmin*).

The content [of the weekly meeting] is divided into the national song, the public reciting of the "Oath of the Young Citizen" (*qingnian gongmin shici*), the school song, reports and lectures, and music. The "Oath of the Young Citizen" is read aloud, sentence by sentence, by the master of ceremonies (*zhuxi*) and then recited in unison by those attending. When reciting one should raise one's arm in a salute as if in a Boy Scout ceremony. This is to express that one will follow and not forget [the oath]. A teacher, administrator, or famous person gives the lecture. Each time, one person reports and the principal serves as the weekly meeting's master of ceremonies. The matters to be reported on are not set: there are things that relate primarily to carrying out school matters; there are explanations of the purpose of a new system or method; and there are explanations of student misunderstandings about the school's administration. But mostly there are admonitory lectures (*xunhua*) to rectify students' conduct.[24]

For teachers and administrators, these assemblies provided an opportunity to lecture students about personal conduct, serving as a technique of training and cultivation. But at the same time, the weekly meetings connected the school community symbolically to the larger national community by singing the national song, reciting "The Oath of the Young Citizen," and saluting the national flag. Consequently, weekly meeting gatherings were in keeping with the National Day assemblies of the early Republican period, where symbolic performance identified local communities as part of the nation and marked participants as political agents.

---

23. Liao Shicheng et al., 33, 385–402; *Pudong zhongxuexiao nian zhou jiniankan*, Xunyu gaikuang: 1–2; Wang Yankang, 15–16. An alumna interviewee also indicated that Zhenhua Girls' Middle School had held weekly meetings before 1927.

24. *Pudong zhongxuexiao nian zhou jiniankan*, Xunyu gaikuang: 1–2.

II. RITUALIZING PROTEST AND
PERFORMING ACTIVIST CITIZENSHIP

In the wake of nationalist protests over Japan's imposition in 1915 of economic and political concessions (the so-called Twenty-One Demands) and its acquisition of the rights to German concession areas in Shandong during the Paris Peace Conference in 1919, students played a central role in introducing patterns of mass activism into civic ritual. Public ceremonies held on key memorial days of protest against foreign imperialism, especially May 4, May 9, and later May 30, increasingly incorporated the dynamics of mass mobilization and protest as symbolic acts within civic rituals. In these rituals, citizenship came to be equated with actions such as marching in demonstrations, shouting slogans, producing and distributing pamphlets, and lecturing to raise popular consciousness. By performing a ritualized form of nationalist protest in the commemoration of mass movements, students enacted an activist form of cultural citizenship that was distinct from both the elite-led early Republican citizenship and the party-led variety that came to predominate from the late 1920s onward.

Starting in the early 1920s, many lower Yangzi region students performed rituals of activist citizenship in commemorations of the May Fourth Movement. The protests that had started on May 4, 1919, to challenge Japan's acquisition of former German concession areas in Shandong came to include a wide range of social groups, but they were closely associated with students, who had in most cases initiated and led them. As a result, by commemorating May Fourth, students celebrated their growing role in political life during the Republican period.[25] In their commemorations of May Fourth, which began immediately in 1920, students throughout the lower Yangzi region ritually replayed the dynamics of mass mobilization that had been central to the initial protest movement.[26] Students in Hangzhou spread throughout the city to give lectures as they had a year earlier. In Yangzhou, students at Jiangsu Provincial Eighth Middle

---

25. Wagner, "The Canonization of May Fourth," esp. 107–8. In 1923, the National Student Union made an explicit appeal for students to memorialize the event. See "Xueshenghui wei wusi jinian zhi tongdian," *Shenbao*, May 4, 1923.

26. *Shenbao*, May 6, 1920.

and Meiman Middle schools paraded through the city in the morning. In the afternoon, students from these two schools, Jiangsu Provincial Fifth Normal School, and the Commercial School spread out to lecture "in order to arouse the common people's ideas." On May 4, 1920, Suzhou's students "[at] 11:00 took separate paths and marched through the streets and distributed leaflets. They also chose active places to lecture about the general outline of last year's student movement (*shiwei yundong*) in Beijing and also informed each social sector about how boycotting Japanese products was patriotic and so forth."[27] Throughout the early 1920s, students in the lower Yangzi region continued to commemorate May Fourth in ways that replayed in ritualized form the dynamics of the original movement.[28]

Commemoration of the May Ninth Memorial Day of National Humiliation (*guochi jinian*) also came to be characterized by symbolic performance of mass protest activities. Before 1919, National Humiliation Day rites on May 9 focused primarily on acts such as abstaining from eating meat as a personal sacrifice to preserve the memory of national shame, listening to lectures about the incident, bowing to the national flag, and singing the national song or the national humiliation song.[29] After 1919, by contrast, protest marches, street-side lectures, distributing propaganda pamphlets, shouting slogans, and issuing telegrams announcing resolutions became fundamental parts of National Humiliation Day commemorations. In Nanjing in 1924, for instance:

At 11:00 they [over six thousand of the city's students] formed teams and marched in the order in which they had arrived at the meeting place. They set off from the meeting place and passed out national humiliation memorial leaflets along the way. Each school had bicycles leading it. . . . When they reached the Confucius Temple complex [Fuzimiao] they dispersed. In the afternoon, each school divided into groups and lectured in various places in order to promote the awakening of the national people (*cu guoren juewu*).[30]

---

27. Ibid., May 7, 1920.
28. Ibid., May 6–7, 1921; May 6, 1922; May 5–6, 1923.
29. E.g., "Guochi jinianhui huizhi," *Shenbao*, May 9–10, 1917; *Shenbao*, May 10–11, 1918. Paul Cohen deftly analyzes how remembering and forgetting became a central issue in National Humiliation Day commemorations ("Remembering and Forgetting National Humiliation").
30. "Nanjing zhi wujiu jinianri," *Shenbao*, May 10, 1924.

In Yangzhou in 1920, during the national humiliation parade, "each student carried a small pennant with an oath to eliminate national humiliation and distributed patriotic publications as they marched to each busy market area, dispersing at the Small East Gate. In the afternoon, they lectured on the streets."[31] In these performances of nationalist commitment, students and other social groups claimed to be citizens by symbolically associating themselves and their community with the nation as a whole, as they had during the National Day celebrations of the early Republican period. But they also acted as citizens through symbolic efforts to protect and sustain the nation that directly paralleled the actions of protest movements.[32]

Civic rituals that borrowed the forms of nationalist protest were often seen to express and evoke strong emotions. In describing the more than 60 lectures delivered at several schools' national humiliation assembly in the northern Jiangsu town of Xuzhou in 1921, *Shenbao*'s reporter stated that "people were crying while speaking, and the whole meeting place was distraught" (*sheng lei ju xia, quan chang qiran*). The reporter also noted that bystanders were moved by students' resolve at marching and handing out pamphlets despite a heavy rain.[33] Emotions at these commemorative gatherings could run the gamut from intensity of commitment to sorrow at the state of national affairs to elation at being part of a mobilized community. In recounting Jiaxing's National Humiliation Day commemoration in 1923, *Republican Daily*'s reporter described the speeches by saying, "All of [the speakers] were emotionally aroused, and it made people feel indignant." He then described the students' feelings of elation after the conclusion of the meeting: "Along the way national pennants swayed in the breeze, and the students carried small flags in their hands and had high spirits."[34] The citizenship performed through activist forms of civic ritual involved not just replaying the

---

31. *Shenbao*, May 11, 1920.

32. For other examples of repertoires from nationalist protest being integrated into civic rituals during National Humiliation Day, see ibid., May 10–12, 1921; May 10–11, 1922; and "Wujiu jinianri zhi youxing (Changzhou)."

33. *Shenbao*, May 12, 1921.

34. "Wujiu jinian zhi qingxing (Jiaxing)."

activities of protest but experiencing intense emotional engagement in the life of the nation.[35]

Each of the acts described above, including marching, shouting and carrying banners with protest slogans, distributing leaflets, and giving street-side lectures, was an element of the political theater of mass protest that transformed Chinese public life during the 1910s and 1920s. During actual protests, these acts were expected to have specific political effects, such as forcing the government to reject agreements deemed damaging to national sovereignty or enforcing a boycott against foreign products.[36] As they became integrated into civic rituals that did not necessarily have an immediate political goal, protest activities eroded and obscured the line, discerned by Joseph Esherick and Jeffrey Wasserstrom in their seminal studies of student protest, between more scripted patterns of political ritual and the fluid forms of political theater.[37] In some instances, the line between symbolic reenactment of protest in a civic ritual and a protest action itself collapsed entirely. For example, in May 1923 students and commercial groups used activities commemorating May Ninth National Humiliation Day to promote an anti-Japanese boycott and advocate for severing economic relations with Japan.[38] Still, we can perceive how the repetition of protest actions in the context of commemorative ceremonies ritualized them by making them formulaic and referential expressions in a primarily symbolic performance. Protest actions in the context of civic ritual were not the May Fourth or May Ninth protests themselves but symbolic references to

---

35. For other descriptions of this emotional intensity, see "Erwanyu ren zhi guochi jinian (Zhenjiang)"; "Guochi jinian zhi youxing (Suzhou)"; "Guochi jinianri zhi guomin dahui (Jinhua)"; "Guochi jinianri zhi Shanghai"; "Hang xuesheng jinian wusi"; "Hang gejie jinian wujiu"; *Shenbao*, May 7, 1921 (Nanjing); *Shenbao*, May 11, 1920 (Suzhou); "Xuexiao zhi wu qi jinianhui" (Schools' May Seventh memorial assembly), *Shenbao*, May 8, 1920

36. See Wasserstrom, *Student Protests*, esp. chaps. 3, 5.

37. They see political ritual as more participatory, structured, and scripted, and political theater as more improvised and aimed at swaying an audience. But they also recognize the potential for political theater and political ritual to borrow scripts and repertoires from one another. See Esherick and Wasserstrom, "Acting out Democracy," 844–46; Wasserstrom, *Student Protests*, 285–86.

38. For a perceptive account of how the May Ninth commemorations in 1923 served as a way to promote a boycott against Japanese products, see Gerth, *China Made*, 158–68. Cf. *Shenbao*, May 10, 1923.

those protests that were meant to evoke their "spirit" (*jingshen*) and to express a particular approach to popular participation in public life.[39] In this way, the activist civic rituals of the early and mid-1920s resembled the dramatizations and mass pageants of the first years of the Soviet Union, when thousands of people "reenacted" revolutionary events such as the "Storming of the Winter Palace."[40]

As a form of symbolic performance, civic rituals of nationalist commemoration during the 1920s represented Republican citizenship in activist terms. Students enacted a form of citizenship defined by going to the people to organize them for action on behalf of the nation. Students engaged the people as leaders to teach them how to be citizens who could claim full membership in national society and take mass political action. In each instance, students worked to include nonelite social groups in civic ritual as they had been integrated into mass nationalist protests since the end of the Qing.[41] Such ritualized mass activism took as its point of departure a conception of the nation as a political community that needed to be created through the active participation of all its members, an idea that students repeatedly encountered in history, geography, and civics classes and in their self-government organizations. At the same time, though, students instantiated hierarchies among citizens by claiming a position of privileged knowledge and leadership as they went to mobilize the people.

Seizing Ceremonies

Popular ritual performances of activist cultural citizenship were created independently by students and other social groups, but the Nationalist and Communist Parties quickly sought to direct these performances of activist citizenship for revolutionary ends. Starting in 1924 and 1925, after the reorganization of the Nationalist Party as a Leninist party and its formation of a United Front with the Chinese Communist Party, members of the two parties began to participate

---

39. Commemorations of May Fourth, in particular, refer to the need to maintain the May Fourth spirit (e.g., "Hang xuesheng jinian wusi"; "Shaoxing minzhong jinian wusi").

40. Binns, "The Changing Face of Power," 591.

41. Wang, *In Search of Justice*; Wright, Introduction to *China in Revolution*.

in and tried to guide commemorations of popular social and political activism. Rather than initiating activist civic ritual, the parties sought to harness dynamics already in motion.[42]

Between 1924 and 1926 party organizations and members assumed a growing role in commemorative ceremonies, contributing to ritual events whose messages were increasingly aligned with the platform of the United Front. In Shaoxing in May 1925, for instance, the May Fourth commemoration ceremony was led by the countywide student union (*xueshenghui*), in which students aligned with the CCP were involved, with the participation of students from local secondary schools. All the members of the local Nationalist Party branch also participated, along with members of the Young Workers' Mutual-Aid Association (Qingnian gongren huzhu hui), suggesting substantial party involvement.[43] A former student of Shaoxing Girls' Normal School from this period observed in her memoir that the student union mobilized students for activist commemorations on May 1, May 4, May 8 (International Women's Day), May 9, and (in 1926) May 30, so that local people called it "Red May" (*hongwuyue*), reflecting popular awareness of CCP influence.[44] At Hangzhou's May Fourth commemoration in 1925, regional CCP leader Xuan Zhonghua was one of the main speakers. He used the opportunity to emphasize that "young students should rise up and fight against this kind of mistaken idea [that students should stay out of politics] and lead the masses to overthrow imperialism and the warlords within the country."[45] Xuan's twofold message echoed exactly the platform of the First United Front. In instances like these, the parties and their members did not invent nor did they necessarily orchestrate commemorative ceremonies, which educated youths had been carry-

---

42. My interpretation differs somewhat from that of John Fitzgerald, who asserts that "mass rituals did not erupt spontaneously" and stresses the party's role in leading civic ritual (*Awakening China*, 274–76). The many popularly generated civic rituals that echoed mass activism in early 1920s Jiangnan suggest that the parties' attempts at mobilization might instead be seen as an effort to take advantage of ongoing popular activism.

43. "Shaoxing minzhong jinian wusi." For Shaoxing secondary school students' membership in the CCP, see Wu Sihong, "Dageming shiqi."

44. Wu Sihong, 228.

45. "Hangzhou zhi 'Wusi.'"

ing out for several years, but they clearly sought to influence and direct them.[46]

Besides May Fourth, International Labor Day on May 1 and May Thirtieth, the anniversary of protests against killings of workers and students by members of the Shanghai Municipal Police, became iconic moments for the temporarily unified CCP and Nationalist Party to commemorate, since both events involved workers.[47] Consequently, by the mid-1920s both memorial days often revealed signs of direct party influence as well as increased involvement of workers. In Hangzhou's May Day memorial meeting in 1924, for instance, CCP-affiliated labor unions were the main participants, with youth participation limited primarily to the members of the Socialist Youth League. Nationalist and Communist party members Shen Dingyi, Xuan Zhonghua, Zhang Qiuren, Yu Xiusong, and others were important participants in what was generally a party-mobilized event.[48] Party organizations within schools themselves also organized students for commemorations on iconic memorial days. At Ningbo's Zhejiang Provincial Fourth Middle School, for instance, the school's Socialist Youth League branch mobilized students for propaganda activities on May Day in 1926 and then organized the school's student association to publish a special commemorative journal and hold a memorial ceremony on the first anniversary of the May Thirtieth protests.[49]

During the late 1910s and early 1920s, then, students and other social groups had developed a growing series of commemorative ceremonies that drew heavily on the scripts of political theater developed during the mass protests of 1915, 1919, and 1925. These ceremonies performed symbolically an activist mode of cultural citizenship that stressed mass mobilization and popular involvement. By joining in, and sometimes leading, activist civic rituals commemorating May 1, May 4, May 9, and May 30, members of both the left wing

---

46. Cf. "Hangyuan zhi guochi jinian hui"; "Liang xueshenghui zhi 'Wusi' jinianhui"; "Shanghai xueshenghui zhi wusi jinian."

47. May Day and International Women's Day (March 8) became prominent festivals in the base area of the United Front in Guangdong. See Gilmartin, *Engendering the Chinese Revolution*, 152–54; Vishnyakova-Akimova, *Two Years in Revolutionary China*, 208–9.

48. "Hangzhou wuyi jinianhui zhi shengkuang"; Schoppa, *Blood Road*, 138–42.

49. Dong Qijun, 92.

of the Nationalist Party and the CCP sought to manage this popular activism, harnessing it for their revolutionary projects. In fact, anniversary celebrations stressing antiforeign nationalism and mass mobilization, which built in frequency and energy over the course of the 1920s, fed into the popular enthusiasm that helped to drive the National Revolution in 1926 and 1927.

III. CAPTURING COMMEMORATION

After the success of the National Revolution, the Nationalist Party followed the precedent of other revolutionary states that had reorganized national calendars by prescribing a particular set of commemorative anniversaries to establish "a new pattern of social time" that helped to legitimize the party and its political projects.[50] But as the Nationalist Party transitioned into a revolutionary party focused on "construction," the party leadership organized the commemorative calendar to reduce, if not eliminate, both memorial days commemorating mass activism and traditional holidays in the lunar calendar. Initially, in 1929, the party mandated 28 national memorial days that would be celebrated at all schools, government offices, and other public organizations.[51] Notably, the first list of commemorative dates included International Labor Day, May Fourth, and May Thirtieth, all dates that the Chinese Communist Party had celebrated as key moments of mass mobilization and revolutionary action. Presumably to preclude possible links to the CCP and leftist appropriation of these commemorative dates, in 1930 the party implemented a more streamlined list, which included only eighteen official revolutionary and party memorial days, with May First, Fourth, and Thirtieth being eliminated.[52]

In local schools in the lower Yangzi region, over the course of the first half of the 1930s, Nationalist Party–centered commemorations gradually displaced both key holidays of the lunar calendar and popular national commemorative days that were associated either with

---

50. Binns, 588–90; Ozouf, *Festivals and the French Revolution*, chap. 7.

51. Harrison, 155–56.

52. "Geming jinianri jianming biao"; Ministry of Education, order number 15522 (December 19, 1934), SMA Q235-1-338. In the 1934 revised version, December 25, the anniversary of the Yunnan Uprising, was characterized as a national, rather than a party, memorial day.

the CCP or nonaligned populist activism. Plotting the holidays celebrated at Jiangsu Provincial Songjiang Girls' Middle School between 1929 and 1934 reveals the nature of this change. In 1929, the school observed International Labor Day on May 1, commemorated both the May Fourth and May Thirtieth movements, and on May 9 sent out publicity teams to lecture and post slogans, following the model of activist civic ritual. Students also had holidays on major anniversaries of the lunar calendar, including Tomb-Sweeping Day (Qingming; usually April 5 or 6), the Dragon-Boat Festival (Duanwujie; the fifth day of the fifth lunar month), and the Mid-Autumn Festival (Zhongqiujie; the fifteenth of the eighth lunar month).[53] In contrast, by 1934 the school did not publicly commemorate May 1, May 4, or May 30 and did not officially observe major lunar-calendar holidays.[54] The erasure of these two kinds of dates from school calendars marked a notable nationalization and partification of time.[55] Elimination of commemorations of mass mobilization was also intended to control the influence of "counterrevolutionaries," which is to say the CCP and other leftist political forces.

Among the eighteen mandated annual memorial days were eight prescribed national holidays that were celebrated by commemorative ceremonies held at each secondary school and/or by mass gatherings orchestrated by the local government or party branch. In addition, there were ten party-related memorial days to commemorate key events in the history of the Nationalist Party, such as Sun Yat-sen's first uprising, and party "martyrs" such as Chen Qimei (Yingshi) and Liao Zhongkai. For party commemorative events, secondary schools generally sent groups of representatives to the local party branch to participate in celebrations.[56] On national memorial days, schools and

---

53. *Songjiang nüzhong xiaokan*, no. 4 (April 10, 1929): 1–2; no. 5 (May 15, 1929): 1–3; no. 6 (July 1, 1929): 1–3; no. 7 (November 15, 1929): 1–4.

54. *Songjiang nüzhong xiaokan*, no. 58 (May 15, 1934): 1; no. 59 (June 1, 1934): 1; no. 60 (June 15, 1934): 1. Cf. Jiangsu shengli Shanghai zhongxue chuban weiyuanhui (1930), 3–4; (1933), 3–5.

55. For restrictions on the May holidays, see Cohen, "Remembering and Forgetting," 11–13; *Shenbao*, May 3–5, 1930; May 3, 1932; May 1, 1934; May 3–4, 1934; May 30, 1934. For Nationalist Party efforts to downplay popular folk holidays, see Poon, "Refashioning Festivals," 202–4.

56. See, for instance, Jiangsu shengli Shanghai zhongxue chuban weiyuanhui (1933), 3–5; *Suzhou Zhenhua nü xuexiao xiaokan* (December 1, 1931), 73–76; Wuxing xianli nüzi chuji zhongxue chuban weiyuanhui, 1–5.

other public and party organizations held hour-long ceremonies that followed the common ritual template (discussed below) of the weekly memorial meeting for Sun Yat-sen.[57] Local governments and party branches also frequently held mass meetings for important national holidays, inviting representatives from local schools, institutions (*jiguan*), and organizations (*tuanti*).[58]

The Nationalist Party sought to impose ideological control over the mandated sequence of civic rituals by dictating the significance and meaning of each commemoration. The Ministry of Education provided schools with summary discussions of each event that was commemorated, including a list of points to be highlighted in the lectures at commemorative meetings. The summary explaining National Day, for example, highlighted party members' activities in the Wuchang Uprising and suggested that Sun Yat-sen had played a pivotal role.[59] In this and other examples, government materials on particular memorial days highlighted the Nationalist Party's role in national events.

The full cycle of national and party commemorations was also designed to enhance the party's legitimacy. Even if we only consider the days of national commemoration, when plotted chronologically, we find that they present a distinctly party-centered narrative of national development.[60] Commemorations of Sun Yat-sen's birth (November 12, 1866), the Yellow Flower Cliff Uprising (March 29, 1911), the Wuchang Uprising (October 10, 1911), and the founding of the Republic (January 1, 1912) traced a historical narrative where Sun's revolutionary parties engineered the revolution that transformed downtrodden post-Taiping Qing China into a modern Republic. Remembering the Yuan Shikai government's acceptance of the Twenty-One Demands (May 9, 1915), the Yunnan Uprising (December 25, 1915), Sun's inauguration as president of the southern government (May 5, 1921), Sun's death (March 12, 1925), and the launching of the

---

57. Cohen, "Remembering and Forgetting," 11; "Guochi jinian banfa."

58. E.g., *Shenbao*, May 4, 1930; October 10, 1930; May 3, 5–6, 10, 1932; May 1, 5, 10, 1934; October 12, 1934; *Zhongyang ribao*, May 5, 1934, 2; May 6, 1934, 3.

59. "Geming jinianri jianming biao," 5788; Ministry of Education, order no. 15522 (December 19, 1934).

60. Ministry of Education, order no. 15522 (December 19, 1934). Cf. Harrison, 155–57. For a more detailed discussion of this narrative, see Culp, " 'China—The Land and Its People.' "

Northern Expedition (July 9, 1926) cast the National Revolution as the culmination of efforts by Sun and the Nationalist Party to overcome the warlords and imperialism to reunite the nation.

By mandating that each school hold commemorative ceremonies for these events, the Nationalist Party sought to include teachers and students in the process of rewriting modern Chinese history so as to make Sun Yat-sen and the Nationalist Party its central agents. This approach to modern history was intended to legitimize Nationalist Party rule and counter alternative narratives of modern Chinese history centered on "the people" as mobilized and led by the Chinese Communist Party.[61] In carefully trying to shape the popular consciousness associated with civic rituals to focus on the governing party and its leaders, the Nationalist Party attempted to implement a form of indoctrination characteristic of contemporary fascist regimes.[62]

However, this Nationalist Party–centered history of modern China was not unchallenged. Few in local schools were bold enough to commemorate holidays in ways that explicitly associated them with the CCP. But during the Nanjing decade, May Fourth and May Thirtieth were evocative nodes for "countermemory," or popular forms of collective remembering that challenge a hegemonic or dominant commemorative narrative.[63] The Nationalist Party was particularly ambivalent about May Fourth as a celebration of a non-party movement. It sometimes mobilized a heightened police presence to guard against "illegal" parades and gatherings, but it also at times, selectively, represented itself as the heir to the May Fourth legacy and it did not ban localized commemoration outright.[64] As a result, May Fourth, and to some extent May Thirtieth, were relatively "safe" days to recall as moments of civic activism and mass mobilization, even as the Nationalist Party tried to downplay them by removing them from the formally sanctioned cycle of annual national holidays.

---

61. For continued CCP efforts to use the May holidays for mass activism, see Stranahan, *Underground*, 73–74. The Shanghai Municipal Social Bureau (Shehuiju) also worried about CCP mobilization on the first anniversary of the Mukden Incident (*Shenbao*, September 18, 1932).

62. E.g., Falasca-Zamponi, *Fascist Spectacle*, esp. chaps. 1, 2.

63. Zerubavel, *Recovered Roots*, 10–11.

64. E.g., Schwarcz, 244–47; *Shenbao*, May 1934; Wagner, 108; Wasserstrom, *Student Protests*, 288–91.

For example, in a May 1933 commemoration issue of the *Jiangsu Provincial Shanghai Middle School's* journal, two teachers wrote articles about May Fourth and May Thirtieth celebrating them as moments of mass action.[65] Lu Renji drew on two other authors to summarize two of the main outcomes of the May Fourth Movement as "the development of mass power" (*minzhong shili de fazhan*) and that "the whole people was in revolution" (*quanmin geming*).[66] Che Mingshen's celebration of the May Thirtieth Movement was even more openly populist, claiming the anti-imperialist protests to have been the autonomous action of China's awakened and politicized "masses" (*minzhong*).[67] According to Che, this high tide of mass activism had provided vital support for the Nationalist Revolution and had stimulated the anti-imperialism of other "weak and small peoples" (*ruo xiao minzu*), most notably in India![68] In these instances, teachers in local schools celebrated the May Thirtieth and May Fourth Movements as examples of popular political action taken at key moments in the nation's history, thereby offering a people-centered alternative to the party-centered narrative cultivated in official histories and commemorations. Such ongoing commemoration of anniversaries that celebrated mass activism suggests the difficulties the Nationalist Party faced in trying to shape the meanings associated with mandated civic rituals and to wholly eliminate or displace popular memorial days.

IV. PARTY POWER AND ORCHESTRATING PUBLIC ORDER

In addition to introducing a new sequence of annual memorial days, the Nationalist Party also established new forms of public ceremony and sought to impose uniform structures on civic ritual. The party's

---

65. Also see the introductory discussion by journal editor Liu Xunmu, "Xie zai wuyue jinian zhuanhao zhi qian," and the essay by Shanghai Middle student Hu Hongzhou, "Jin sui guochi yue zhi ganxiang."

66. Zhanqian [Lu Renji], "Wusi yundong de huigu." Lu was a Central University graduate and teacher of national language, geography, and history.

67. Che Mingshen, "Wusa can'an de yiyi." In an interview, an alumnus of Shanghai Middle related that Che later became a member of the CCP, most likely during the Sino-Japanese War.

68. For a contrasting view of May Thirtieth, see Luo Yiwen, "Wu sa yu xuesheng."

effort to introduce standard ritual practices paralleled Soviet approaches to public ceremony during the late 1920s and early 1930s, where the "imposition of rigid, centralised discipline was accompanied by the 'routinisation' and hyper-organisation of ceremonial."[69] The Nationalist Party developed two basic patterns of ritual practice that could be integrated or used separately. One sequence of action was performed most clearly in the weekly memorial meeting for party leader Sun; the other was the military-style review, which students encountered most frequently at Scouting events but which became a common component of other mass gatherings during the 1930s. Party control over choreographed civic rituals constrained the performance of populist activism. Routinization in public ceremonies also dampened the emotional fervor that had marked students' and the CCP's activist civic rituals during the early and mid-1920s, but it ensured at least the dutiful performance by students of expressions of loyalty, forms of public behavior, and patterns of social order desired by party leaders.

School Ceremony, Ritual Standardization, and Cultivating Modern Citizens

During the Nanjing decade, the most distinctive and well-known civic ritual in secondary schools was the weekly memorial service for Party Leader Sun (*zongli jinian zhou*; hereafter referred to as the "weekly memorial meeting"). The weekly memorial meeting had been prescribed for all government and party offices and military installations under Nationalist government control from as early as 1926.[70] Lower Yangzi region schools began holding the meetings in the late 1920s, soon after the Nationalist government took control of the region.[71] Meetings were held each Monday morning in a large auditorium with a picture of Sun Yat-sen flanked by the party and

---

69. Binns, 596.

70. Harrison, 157–58; Huang, *The Politics of Depoliticization*, 86; *Zhonghua minguo fagui daquan*, 4: 5721.

71. E.g., Yangzhou Middle School began holding weekly memorial meetings in the fall of 1927. Zhongyang daxuequli Yangzhou zhongxue chuban weiyuanhui, 8. Songjiang Girls' Middle established rules for the weekly memorial meetings in the spring of 1929 (*Songjiang nüzhong xiaokan*, no. 4 [April 10, 1929]: 5). Cf. Chan and Dirlik, 103–5; Harrison, 157–58.

6.1 "Auditorium." The Shanghai Middle auditorium was arranged for the weekly memorial meeting for Party Leader Sun, whose picture, hung behind the front table and set between the juxtaposed party and national flags, served as the visual focus of the room. Number plates on the benches marked students' assigned seats. The upright chairs lined up on the podium for administrators and visiting dignitaries represented their authority and status through their vertical elevation and centralized proximity to the iconic image of Sun Yat-sen. Jiangsu shengli Shanghai zhongxue chuban weiyuanhui, *Jiangsu shengli Shanghai zhongxue yilan* (1933).

national flags at the front of the hall.[72] (See Fig. 6.1.) The general order of the ceremony was carefully scripted. All the school's teachers and students filed in and took their proper places in the meeting hall, standing silently. They then sang the party song, saluted the flags and the picture of Sun, and recited his last testament line by line, following the chairman of the meeting.[73] This recitation was followed by a period of silent reflection. Then the principal, a teacher, or a visitor would either lecture or report on matters related to school

---

72. "Shanghai shi geji xuexiao ge minzhong tuanti juxing jinianzhou banfa"; *Zhejiang shengli disi zhongxue yilan*, 20; *Zhejiang shengli gaoji zhongxue xiaokan*, 64. Interviews with alumni from Shanghai's Minli Middle, Songjiang Middle, Shanghai Qingnian Middle, and Yangzhou Middle Schools.

73. For a complete translation of Sun's last will, and efforts by Wang Jingwei and others within the party to use it as a form of symbolic capital in intraparty struggles, see Harrison, 136–37.

life. At the end of the ceremony, students again filed out in order after the chairman of the meeting and any guests had exited.

The many parallels between the weekly memorial meeting and the routine school assemblies of the 1910s and 1920s almost certainly eased implementation of the party's civic rituals in local schools. For teachers and administrators could—and did—use the state-mandated civic rites to organize and exhort the students as they had been doing in their schools' weekly assemblies for a decade or more. However, the symbolism and many other parts of the weekly memorial meeting—including Sun's picture, the national and party flags, the national song, and Sun's final testament—explicitly evoked the Nationalist Party in ways that had not been common before 1927. By drawing together the school community before these symbols of the Nationalist Party, the weekly memorial meeting marked out the party and its government as the main public authorities in China. It served as a forum through which the party could define itself publicly as a legitimate ruling force and prompt at least overt expressions of loyalty on the part of students and teachers. Such opportunities for symbolic expression of the party's authority were vital when its political and economic power were circumscribed in so many ways. At the same time, these assemblies worked to produce the school community as a legitimate social group in the organizational framework of the nation-state through the recognition afforded by rituals where the symbols of party and government authority were present, if not representatives of the party and state themselves.

If teachers and administrators adapted to the weekly memorial meeting and used it as a mechanism for civic training, students expressed little enthusiasm for the ceremony, which is perhaps not surprising given its mechanistic quality. When they recalled the weekly memorial meetings, alumni interviewees living in both the People's Republic and Taiwan uniformly spoke of being compelled to attend. School administrators had to enforce attendance in the meetings with strict penalties on students' conduct grades if they were late or absent, suggesting many students participated only reluctantly.[74]

In prescribing uniform symbols and practices for the weekly memorial meetings, the Nationalist Party sought to establish a new

---

74. *Suzhou zhongxue gaikuang*, 4–5; *Zhejiang shengli gaoji zhongxue xiaokan*, 64.

regime of ritual practice in modern China. By focusing so heavily on ritual action, this effort to standardize civic ritual resembled in certain ways the late imperial state's approach to rites, which James Watson argues was centered on orthopraxy, or correct practice. Watson suggests that through the end of the Qing, emperors, imperial officials, and social elites sought first and foremost to standardize popular symbols, deities, and sequences of ritual action. The resulting consistency in ritual practice generated some degree of cultural unity in imperial China.[75] Nationalist Party cadres and officials may, in fact, have been influenced by Qing-era, practice-centered approaches to public ritual when they fashioned their modern civic rites, for their concern with ritual protocol seems to reflect the late imperial assumption that standardizing ritual practice is an effective mode of governance.

As with late imperial states, the Nationalist Party coupled promotion of orthopraxy with attempts to instill orthodoxy, or correct ideas.[76] As we have seen, the party widely disseminated Sun Yat-sen's Three Principles of the People through textbooks, Scouting handbooks, lectures, and a variety of other methods while banning texts considered "counterrevolutionary." Yet the Three Principles of the People were themselves highly eclectic and open to interpretation.[77] Moreover, the topics of the lectures at the weekly memorial meeting were decided at each local school and ranged from reports on school affairs to instruction on student behavior, to lectures on national affairs, to purposeful inculcation of the Three Principles of the People and other Nationalist Party agendas such as the New Life Movement.[78] Such widely varying content could not have related a uniform and consistent orthodoxy.

---

75. Watson, "Rites or Beliefs?"

76. Besides publication and distribution of canonical texts to examination candidates, dissemination of texts such as the Sacred Edict and promotion of community lectures by local gentry reflect a concern by late imperial states with instilling some degree of orthodoxy (Elman, *Cultural History*, chap. 2; Mair, "Language and Ideology").

77. Even Sun's short final testament was subject to multiple interpretations (Harrison, 158–59).

78. *Minli xunkan*, nos. 2–3, 5–7 (March 20–April 10, April 30–May 20, 1936); *Zhejiang shengli gaoji zhongxue xiaokan*, 64–65. Interviews with alumni from Hangzhou High School, Shanghai Middle, Suzhou Middle School, and Yangzhou Middle.

Consequently, the Nationalist Party's new ritual regime was most reliable for standardizing ritual performance in public contexts. At one level, the parallel performance of the exact ceremonies at the same time each Monday demonstrated the idea of the national community as interconnected through common cultural practices. These uniform ritual performances also ensured that the party received public expressions of loyalty, through recitation of Sun's final testament and salutes to the party and national flags, from captive audiences such as school students. In addition, the methodical steps of the ceremony and detailed regulations governing the dress and behavior of participants in the ceremonies choreographed enactment of the kind of public decorum and etiquette the Nationalist Party sought to instill in its citizens.[79]

The ceremonial order of the weekly memorial meeting became a template for a wide range of civic rituals in Nationalist China.[80] Singing the party/national song, saluting or bowing to the flags and the picture of Sun, reciting his last testament, observing a moment of silent reflection, and listening to lectures (*xunhua*) or a report (*baogao*) were fundamental procedures of commemorative ceremonies and many other public events.[81] Even when party leaders instituted rites to honor Confucius each August 27, near the start of the school year, they followed the basic pattern of the weekly memorial meeting. They altered it only by adding a picture of Confucius to the constellation of Sun's picture and the party and national flags and by prescribing a lecture on the significance of honoring the sage and the singing of a commemorative song.[82] Instead of continuing late imperial state ritual in a traditionalistic way, as Yuan Shikai attempted by reviving imperial rituals between 1914 and 1916,[83] party leaders invented new rites that referred metaphorically to China's imperial and Confucian past. At the same time, by promoting stan-

---

79. See *Songjiang nüzhong xiaokan*, no. 4 (April 10, 1929): 5; Xiaokan bianji wei-yuanhui, 46; Zhejiang shengli Jinhua zhongxue chuban weiyuanhui, 22.

80. Even a ceremony meant to substitute for the popular Ghost Festival in Guangzhou followed the weekly meeting template (Poon, 218).

81. See the sources listed in notes 57 and 58.

82. "Xianshi Kongzi danchen jinian banfa."

83. Levenson, "The Suggestiveness of Vestiges"; Zarrow, "Political Ritual," 167–73.

dard ritual practice, the Nationalist Party followed a structural pattern common to late imperial states, which had stressed ritual orthopraxy.

The Military Model of Civic Ritual

The military-style review (*jianyue*) provided another template of standard ritual action, one that could order large groups of people in the context of mass assemblies. Lower Yangzi region secondary students participated in reviews most frequently at collective Scout gatherings or joint military-training sessions. Our discussion here will focus on Scouting reviews, since they were much more frequent and involved large numbers of young people. But over the course of the Nanjing decade, party leaders sought to extend the patterns of the military-style review to other kinds of public gatherings, including some commemorative ceremonies and "citizens' assemblies" (*shimin dahui*). As with the weekly memorial meeting, military-style reviews demonstrated the Nationalist Party's authority publicly, mapped out an ideal sociopolitical order, and orchestrated public expressions of loyalty from participants.

Large-scale gatherings of scouts for reviews, collective drill (*huicao*), and "jamborees" (*da luying*) became a regular feature of Chinese Scouting during the 1910s and 1920s. Scout gatherings during these two decades were extremely eclectic collections of group and individual performances and competitions that might or might not include a period of group drill. Scouts built fellowship through group camping, exhibitions, or skills competitions, performed demonstrations of drill movements, and assembled in group formations for reviews.[84] Significantly, when reviews of scouts in military formation were included in early Republican collective training assemblies, the reviewing authority was often local government officials or their representatives but might also be composed of Scouting leaders, local notables, or foreign dignitaries. In general, it seems that before 1927 scouts in drill formation performed as much for their communities as for government officials, a dynamic that

---

84. See MGRB, May 27, 28, 1923; June 11, 1923; *Shenbao*, October 10–11, 1920; May 8, 11, 1922; Zheng Haozhang, *Shanghai tongzijun shi*, 10, 12, 14, 16, 18, 19–21, 23.

reflects the absence of a stable state authority during the 1910s and 1920s.[85]

During the Nanjing decade, the Nationalist Party integrated Scouting reviews into its repertoire of mandated civic rituals. Under party direction, collective Scout gatherings were frequent events and usually included a drill review of all the attending scouts by party leaders and government officials. Reviews were held during Scout meetings at the school, county or municipal, provincial, and national levels, and they were sometimes integrated into the ceremonies during commemorative assemblies.[86] With many different levels of Scout gatherings happening annually or semiannually, Scout troops might participate in or send delegations to reviews every several months.[87]

Two national Scouting reviews were held in the capital during the Nanjing decade, the first from April 16 to 21, 1930, and the second on National Day in 1936.[88] At the second national review, all the symbolic resources of the new Nationalist government were mobilized.[89] The review was held at the Central Stadium facing the Sun Yat-sen Memorial, which was the epicenter of symbolic legitimacy for the Nationalist Party and Chiang Kai-shek.[90] Attending the review were 11,000 scouts, from every province and municipality, and an estimated crowd of over 100,000 spectators. At 8:00 on the morning of October 10, the scouts assembled in their troops and provincial groupings for the review, which was the central event in the gathering. The scouts saluted the party and national flags, a pic-

---

85. Reviews of student military-training groups operated in similar ways. In Nantong during the 1910s and 1920s, students performed in drill formation in front of provincial military officials but also before audiences of local people and guests at sports meets. See Shao, *Culturing Modernity*, 164–66.

86. During Shanghai's National Day celebrations in 1930, for instance, party and government leaders conducted a full-scale review of the city's Scout troops (*Shenbao*, October 10, 1930).

87. E.g., Liu Youxiang, 112–13; Zhejiang shengli Hangzhou chuji zhongxue bianji weiyuanhui, 84; Zhejiang shengli Hangzhou nüzi zhongxue, 201–6; Zhejiang shengli Jinhua zhongxue chuban weiyuanhui, 30–34.

88. For an account of the first national review, see Liu Chengqing, 47–48.

89. "Quanguo tongzijun di er ci da jianyue da luying." For a student's perspective, see Ru Gengbo, "Nanjing luying shenghuo."

90. On the symbolic value of this site, see Wang, "Creating a National Symbol."

ture of Sun Yat-sen, and Scouting General Association Chairman Chiang Kai-shek as well as the assistant chairmen, party theorist Dai Jitao and General He Yingqin. Then Chiang led a group, which included Dai, He, and many other leading party and government figures, in reviewing each Scouting unit arrayed on the field. Each Scouting unit saluted as Chiang and his entourage passed. After the review there was a march-past ceremony, in which each Scouting unit marched in prescribed order past the reviewing stands, saluting as they passed. When all the groups had returned to their places on the field, they listened to an address by Chiang, sang the Scout song, shouted slogans, and marched in order out of the meeting ground. During the Scouting review, the party and government leadership's oversight of a field of thousands of young people from all provinces, arrayed in regular formations and moving in unison, provided a dramatic representation of party authority over the orderly body of national society.

Reviews followed the same pattern at the county, municipal, and provincial levels. Wuxi County, for instance, held its first county-wide Scouting review from April 24 to 26, 1931.[91] A general drill and review occurred on the morning of April 25, and exhibitions and competitions went on throughout the afternoon. The scouts' drill formations were reviewed by representatives of the party-controlled Chinese Scout Headquarters (Silingbu) and the Provincial Party Bureau, members of the Wuxi County Party Bureau, the county magistrate, and the head of the County Bureau of Education. In the course of the review, these national, provincial, and local party and government figures represented the party-state's authority in a way that paralleled the central party leadership's symbolic supervision during the national Scout review.

Shanghai Middle student Zhang Minghe's account in the student self-government association's journal of the third municipal review in Shanghai in 1931 relates a concrete sense of the experience from the student's perspective.

At about nine o'clock the whole battalion reached the field. After halting at our designated places we paused for a breath, then each group that was participating entered the field in succession. Afterward the whole group of

---

91. Wuxi xian jiaoyuju, *Wuxi jiaoyu zuijin gaikuang*, 95–99. Cf. Zhongguo tongzijun zonghui choubeichu, 139, 144–45.

participants marched one circuit around the field. [We were] led by our school's marching band, which was solemn and powerful, making everyone feel a deep respect! After marching around the field, we saluted and then held the review.

The review was split into two parts: (1) a march-past and (2) a march-forward. The march-past was divided into three aspects: first, reporting the number of people; second, the review; and third, doing the basic drill movements. The march-forward was split up into two aspects: first, giving the order to begin; second, the whole group's passing before the reviewing stand, turning a circuit. When the review was over there were lectures (*xunhua*) [from the official observers].[92]

The dynamic of the review enacted the process by which each school's patrols, companies, and troops fit with others to make an integrated collective body. Their merging to form a single group under the gaze of party and government dignitaries demonstrated how individuals and primary groups formed the organic totality of the national community, when led by the Nationalist Party and its government. Party theorist and Scouting promoter Dai Jitao explicitly captured the intended parallel between the two forms of integration in an address to a provincial Scouting review in Zhejiang in 1933: "In a large country like this with such a large population, the Scouts' large-group movements are an initial foundation for attaining large scale group life (*dadao da jituan shenghuo de chuji*), starting small to train the whole nation's citizens in group life habits."[93] Building up from individual and patrol-level group training, Scouts could be combined into "large-group movements" that could serve as a basis for and model of national order and unity. The organized body of scouts at review ceremonies demonstrated what Nationalist Party leaders saw to be the ideal form of national community, one that was "militarized," orderly, and responsive to Nationalist Party control.[94] The structural relationship these review ceremonies constructed between the party and the people was similar to that described by Fujitani in his analysis of late Meiji-era victory parades and reviews, where the "panoptic gaze" of

---

92. Zhang Minghe, "Canjia Shanghai shi tongzijun disanci da jianyue," 55. Cf. Ru Gengbo.

93. Dai Jitao, *Dai Jitao xiansheng wencun*, 2: 822. Cf. ibid., 2: 823.

94. Dirlik, "Ideological Foundations," 972–74.

the leader was meant to discipline disparate social groups to create a sense of public order.[95]

With the start in February 1934 of the New Life Movement, the Nationalist Party increasingly emphasized "militarization." In the area of civic ritual, party leaders sought to impose the patterns of the military-style review exhibited in Scout gatherings and student military-training sessions on all public gatherings. In the rules for organizing mass meetings that were issued in conjunction with the New Life Movement, citizens' groups representing each social sector were directed to organize like military units that were on parade and review. Assemblies were to share elements of the military-style review, including parading in predetermined order and assembling in neat groups.[96] (See Fig. 6.2.) The promulgation of these rules was premised on the assumption that "a nation-state's prosperity (*guojia zhi sheng*) is closely related to whether or not the thought and actions of the majority of people can operate in an orderly way (*zhengqi huodong*) under unified direction (*tongyi zhihui*)."[97] Even if public assemblies during the Nanjing decade were never carried out in quite this orderly a fashion, these rules reflect party leaders' goal of creating civic rituals that would demonstrate the orderly arrangement of a militarized society under Nationalist Party direction.[98]

How students responded, emotionally and intellectually, to Nanjing-decade public ceremonies is difficult to gauge. Explicit reflections on Scouting reviews or other collective rituals are rare, but those we have express strongly how participation in these mass gatherings reinforced students' sense of being part of a national community. One student, recounting a trip to a provincial review in Zhejiang, commented, "There were three [train] cars that were filled just with khaki Scout uniforms." This observation suggests that traveling together and interacting with so many other uniformed Scouts could bring home to students that they were part of a larger collective.[99]

---

95. Fujitani, 128–45.

96. "Choukai shimin dahui xuzhi."

97. Ibid., 353.

98. In fact, the dynamics of the military-style review seem to have been integrated even into some festive gatherings, such as Hangzhou students' celebration of Chiang Kai-shek's 50th birthday in 1936 ("Relie qingzhu lingxiu wushi shouchen").

99. Dong Qicai, "Canjia quansheng tongzijun da jianyue ji," 15.

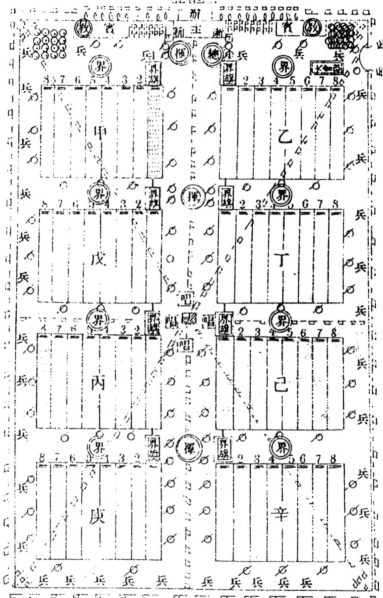

6.2 "Diagram of the Meeting Place for a [Citizens'] Mass Meeting." The labeled boxes indicated the places for different social sectors (*jie*), which in turn were composed of groups of citizens organized in neat, numbered rows. They face a central podium where party and government leaders were to sit to review the gathering. The square flags that were pictured ringing and crossing the meeting place signaled the symbolic saturation that was supposed to characterize such gatherings. "Choukai shimin dahui xuzhi," no. 529: 363.

One Jiangsu Provincial Yangzhou Middle School alumnus recalled to me his awareness of being part of a national group as he participated in the 1936 Second National Scouting Review in Nanjing on National Day and met scouts from sequentially numbered troops from all over the nation. The marching and collective drill of troops from different schools and localities further encouraged students to see themselves as part of a large collective body of Scouts that was identified with the national community and mobilized for action.[100] In the words of Shanghai Middle student Zhang Minghe, reflecting on the Scouting review described above, "I think that if all our country's people could be like this [i.e., unified together as in a review], I would dare to guarantee that in less than ten years we could become strong and flourishing!"[101]

Even if the nationalist consciousness voiced by Zhang and other students was widespread, growing awareness of and identification with the nation did not necessarily correlate with strong feelings of loyalty to the state or party leadership. But then, stirring passionate emotions was not necessarily the goal of party-mandated civic rites, with their emphasis on carefully orchestrated repetitive behavior, their limited aesthetic appeal, and their built-in constraints on ritual actors' ability to improvise and alter the rituals to appropriate them. The Nationalist Party's emphasis on enforcing standardized practice rather than creating emotionally stirring spectacles can be seen as a strategic choice given the political environment of the 1930s. As Elizabeth Perry shows, the CCP from very early on proved quite attentive to and effective at stirring people's emotions and generating passionate commitments to revolutionary struggle, often through ritualized forms of political activism.[102] We have seen here that the activist civic rituals that educated youths initiated and participated in during the 1920s, and that the CCP and left wing of the Nationalist Party later sought to direct, inspired passionate feelings in student-citizens. Further, repeated Japanese imperialist incursions during the Nanjing decade sparked powerful emotional responses in a wide spectrum of China's citizens, especially students, often leading

---

100. Such feelings were common among participants in collective military-training events as well. See Qiu Qiqin, "Jixun shenghuo yizhou ji"; Hu Zunong, "Gao tongxue shu."

101. Zhang Minghe, 55.

102. Perry, "Moving the Masses."

to critical protests against the Nationalist Party's policies of appeasement.[103]

In such an environment, the politics of emotion was a dangerous game for the Nationalist Party to play. Instead, party leaders focused on regulating public behavior to ensure that people obeyed Nationalist Party directives and at least outwardly expressed loyalty to the state. These goals of Nationalist Party policies regarding civic ritual are captured in Chiang Kai-shek's statement to the nation's scouts at the Second National Review in Nanjing in 1936:

Today Scouts from the whole country have come to the capital from each province, not shirking from travel over land or water, to participate in the national Scouting jamboree and review. That everyone together was able to obey orders and adhere to discipline, making everything neat and orderly (*zhengzheng qiqi youtiao youli*), indicates that the whole country's youths are able to be of one mind to support the Nationalist government and obey the state's orders and are able to unite with camaraderie to take up the mission of reviving the nation.[104]

For Chiang, the review itself served as a demonstration of loyalty to the party-state and the enactment of a form of unity and public order that would contribute to reviving the nation. The Nationalist Party's focus on routinizing ritual practice rather than creating engaging and aesthetically appealing ceremonies can be seen as another key point of difference between it and contemporary fascist regimes in Europe, for which rousing mass spectacles were a central technology of power.[105]

V. CONCLUSION

By creating new forms of ritual and imposing standardized templates of ritual practice, the Nationalist Party did, in fact, transform many dimensions of public life during the Nanjing decade. The party's commemorative calendar increasingly governed the rhythms of pub-

---

103. E.g., Coble, *Facing Japan*; Israel, *Student Nationalism*. By contrast, Chiang's release at the end of the Xi'an Incident in December 1936 and his agreement to resist further Japanese aggression led to an enthusiastic public ceremony in Shanghai. See Wakeman, *Policing Shanghai*, 240–43.

104. "Quanguo tongzijun di er ci da jianyue da luying," 127.

105. E.g., Falasca-Zamponi; Mosse, *Nationalization of the Masses*; Sontag, "Fascinating Fascism."

lic institutions, such as schools, and public ceremonies. The symbols and sequences of Nationalist Party ceremonies such as the weekly memorial meeting organized more and more public gatherings during the 1930s. Through ritual performances, the Nationalist Party was able to model the corporatist and militarized sociopolitical order that party leaders envisioned. Moreover, students' frequent repetition of standard ritual action in Nationalist ceremonies drilled them in new forms of public decorum. Carefully prescribed civic ritual behavior enabled the Nationalist Party to enact some degree of order over observable practice in the public domain and orchestrate expressions of loyalty even if it did not "move the masses" emotionally.[106]

The shift from the improvised, emotional, and broadly participatory performances of symbolic civic activism of the early 1920s to the state-orchestrated, routinized civic rites of the Nanjing decade represented a major transformation in the dynamics of citizenship enacted by lower Yangzi region students. The transformation paralleled changes in student self-government, where a voluntarist ethos and student self-determination, which modeled a radical form of direct democracy, gave way to strict party supervision and party-led mobilization during the 1920s and 1930s. In both domains, party tutelage replaced populist politics with students in command.

However, as a mode of cultural citizenship, the periodic, public expressions of loyalty and order that were fundamental to standardized civic ritual under the Nationalist Party required very limited commitments. Although party leaders certainly hoped that students would develop, through repeated exposure, faith in the Three Principles of the People, they demanded minimally that students perform their roles in clearly delineated public ceremonies. Outside of this arena of visible performance, which was easily monitored, cultural citizenship and civic action, as we have seen, were much less clearly defined and uniform. Throughout the Republican period, including during the Nanjing decade, students had flexibility to negotiate the diverse approaches to citizenship introduced through their schooling and to fashion their own ways to act as citizens. We turn now to consider how they did so.

---

106. For a parallel approach that stresses the power enacted through public forms of mandated ritual practice, see Wedeen, *Ambiguities of Domination*.

7

Enacting Citizenship: Youth Activism and National Politics

Many schools in the lower Yangzi region with a history dating back to the early twentieth century have published commemorative volumes that highlight their alumni's contributions to the nation. The schools celebrate revolutionary activists and underground party members, artists and athletes, and engineers and scientists, many of whom received their education during the Republican period. In one commemorative volume, Jiangsu Provincial Yangzhou Middle School proudly records the biography of 1935 graduate He Fang, who died fighting in the CCP's Youth Anti-Japanese Vanguard Battalion in Shandong in 1938. But it also boasts of eight graduates from the 1930s who became members of the Chinese Academy of Science after 1949.[1] Shanghai's Nanyang Middle School celebrates Yu Changzhun (attended from 1922 to 1925), who joined the CCP through Yun Daiying in 1926 and became a party leader in Wuhu during the Northern Expedition. It also recalls 1938 graduate Chen Haiguang, who went to the United States as a representative of the Chinese Scouts in 1935 and led Shanghai's Scouts in wartime service at the start of the Sino-Japanese War.[2] As these school histories suggest, many educated youths in the generations that came of age during the 1920s and 1930s distinguished themselves through their civic action and contributions to the nation. They threw themselves into the

---

1. *Jiangsu sheng Yangzhou zhongxue jianxiao jiushi zhounian jiniance.*
2. *Shanghai shi Nanyang zhongxue jianxiao 100 zhounian*, 57–58. Cf. *Shanghai shi Songjiang xian di'er zhongxue jianxiao jiushi zhounian jiniance*, 27–28; Zhao and Qian, comps., *Hangzhou sizhong 95 zhounian xiaoqing*, 102–6.

National Revolution, committed themselves to public service during the Sino-Japanese War, and played prominent roles in projects of national construction both before and after 1949.

The call to active citizenship, as we have seen, was one of the most consistent and powerful messages of Republican-period schooling. In lower Yangzi region secondary schools, history and geography instruction, self-government activities, and the model of the social organism together delivered the mutually reinforcing message that citizens needed to act to produce a healthy society and a strong nation. But we have also seen that schools provided each cohort of Republican-period students with multiple sanctioned ways to act as citizens. The pluralism of citizenship education allowed students to combine elements from their schools' curricula and activities to fashion their own approaches to civic action; the diverse paths they took are suggested in the life trajectories outlined above.

Students, through multiple approaches to civic action, cast themselves as archetypal citizens, the "core elements" (*zhongjian fenzi*), of national society. From the 1920s into the period of the Sino-Japanese War, the Nationalist and Chinese Communist Parties vied, with varying success, to mobilize educated youths by linking students' chosen forms of civic action to their competing projects of social and political change. But during the decisive opening stages of the War of Resistance (1937–45), the Chinese Communist Party attracted young nationalist activists' support by giving students and recent graduates outlets and an organizational infrastructure for performing the kinds of national salvation work that their Nationalist Party–mandated schooling had encouraged them to undertake.

I. CREATIVE RADICALISM AND THE POLITICS OF SOCIAL TRANSFORMATION

Students facing the maelstrom of debates about social, cultural, and political change during the late 1910s and early 1920s struggled to find their own perspective on China's emerging social order, to determine where they fit within it, and to decide how they should act in this changing context. Working as intellectual bricoleurs, students drew on the disparate intellectual perspectives circulating in their schools and local society to construct their own worldviews, define social roles for themselves, and express their plans for changing society.

Two essays from school journals published in the early 1920s suggest how New Culture Movement ideas fed but did not determine student thought during that period. In a 1923 piece titled "Analyzing Human Life," Jiangsu Provincial First Normal School student Pang Guoliang fashioned a distinctive worldview from a wide array of reform discourses.[3] Pang combined elements from various trends of thought that emerged during the New Culture Movement and were common fare for students in lower Yangzi region schools. Faith in the scientific pursuit of truth and Cai Yuanpei's and Tolstoy's stress on joining art and work in daily life merged in his own unique perspective on human life: "My own proposal for the ultimate aim of human life is to emphasize spiritual life (*jingshen shenghuo*), to be active, happy, and stress science, work, and art as the goals of human life, to respect the dignity of myself and others and place the greatest value on human life."[4] Rather than borrowing whole a philosophy of life from one of the dominant streams in New Culture Movement discourse, Pang drew on many different elements. Integrating ideas such as the dignity of the individual, the scientific worldview, and living a life focused on work and art, he formulated his own eclectic perspective on how to live as a modern person. Lower Yangzi region students like Pang could formulate their own worldviews because of the multiple threads of social and political discourse they were exposed to during the early 1920s that allowed them to escape the hegemonic patterns of the imperial period, the controls of local social elites, or what Arif Dirlik has called the "strategies of containment" later undertaken by the Nationalist and Communist parties.[5]

In this open milieu, students sometimes also fashioned sharp critiques of social institutions and constructed roles for themselves as instigators of radical social transformation. For instance, in a 1921 article called "Inheritance and the New Youth's Own Revolution," Wang Weihua, a graduating senior at the elite, private Nanyang Middle, criticized the system of inherited private property and cast himself and his fellow educated youths in the role of social revolu-

---

3. Pang Guoliang, "Rensheng poujie," 12–23.
4. Ibid., 23.
5. Dirlik, *Anarchism*, 38–40.

tionaries.[6] Wang argued that revolution was the process of rejecting old practices that no longer fit with the current times. He called on youths to recognize the need for revolution and to carry it out by refusing to participate in the system of inheritance, which kept young people from striving for advancement and thus obstructed social evolution. Social revolution, Wang argued, could only occur through individuals self-consciously conducting their own personal revolutions against old social institutions, such as property inheritance, which the modern world had made obsolete. For Wang the modern citizen in a progressive society was necessarily a social activist.

Wang's essay, like Pang's, revealed clear influences from New Culture Movement discourse. Most obviously, he challenged the institution of inheritance and the intergenerational dependence to which it contributed. But he also disavowed any connection with socialist parties or "movements to destroy familism" (*pochu jiazu zhuyi de yundong*). Rather, Wang's discussion resembled Pang's in that he combined ideas from intellectual traditions as diverse as classical Confucianism and evolutionary thought to fashion a distinctive outlook on Chinese society. During the late 1910s and early 1920s, students like Wang and Pang were formulating their own perspectives on society and charting courses of social critique and transformative personal action by drawing on the many visions of social order that circulated in their schools. While Pang and Wang undertook very personal searches to construct a meaningful vision of social order and to define individual paths of social activism students at nearby schools were taking collective approaches that created powerful reverberations throughout the lower Yangzi region.

Zhejiang Provincial First Normal School and Radical Social Politics

Student activists at Zhejiang First Normal played a leading role in defining the patterns of student social activism in the lower Yangzi region. As Wen-hsin Yeh illustrates, the convergence during the late 1910s of social, political, and cultural reformers Jing Ziyuan, Chen Wangdao, Xia Mianzun, and Liu Dabai at Zhejiang First Normal created an environment that encouraged experimentation

---

6. Wang Weihua, "Yichan."

and activism.[7] The exposure to social reformist texts at Zhejiang First Normal through the book-buying societies, library acquisitions, and extracurricular reading described in Chapter 4 also contributed to that atmosphere of intellectual experimentation and helped to sustain it after those figures were forced to leave the school at the beginning of 1920. Zhejiang First Normal students responded to the visions of social change intersecting in their school with their own attempts to enact social and cultural reform and chart new trajectories of political action.

The most politically active students' views on society and politics were captured in the November 1919 inaugural preface (*fakanci*) of the journal *Zhejiang New Tide* (*Zhejiang xinchao*), a collaborative effort with students from Zhejiang First Middle School that students from Zhejiang First Normal dominated. This essay offered a critique of existing social inequalities in order to arouse the working classes' self-awareness so that they would rise up in league with the students and transform society.[8] Writing in an anarchist mode, the journal's editors argued that human freedom and welfare could only be secured by destroying restrictive social institutions such as religion, family, and the state.[9] Only the laboring classes, who made up the majority of people within society and who experienced its bitterness most deeply, had the power to change society, once they became "self-conscious" (*zijue*) and "unified" (*lianhe*). The students' journal sought to effect this awakening through "investigation" (*diaocha*), "criticism" (*piping*), and "guidance" (*zhidao*), viewing education, in good anarchist fashion, as a key technique for social change, and to realize unification by building up larger groups from smaller groups. Using the common image of a nested hierarchy of communities, students related the particular local struggle for social progress in Zhejiang to broader processes of world development. Significantly, radical political commitments for social equality were framed not in terms of individual or sectional interests, competition, or class conflict but through a language of unification, mutual aid, and universal

---

7. See Yeh, *Provincial Passages*, chaps. 6, 7. Cf. chap. 3 above.

8. "*Zhejiang xinchao* fakanci," 72–77.

9. Yeh, *Provincial Passages*, 167–71. For Chinese anarchist approaches to these social institutions and human liberation, see Chan and Dirlik, *Schools*, chap. 2; Dirlik, *Anarchism*, 28–29, 90–91, 96–100, 129–30, 146, 164–66, 180–81, 184–89; Zarrow, *Anarchism*.

social harmony.[10] Despite avoiding the rhetoric of class conflict, *Zhejiang New Tide* was shut down after its second issue because of the publication of an iconoclastic article by Zhejiang First Normal student leader Shi Cuntong called "Decry Filial Piety!"[11]

In the radical student journals published at Zhejiang First Normal and elsewhere, students consistently assigned themselves the responsibility to "awaken" other social classes in order to create a society with more justice, freedom, and unity. So, even though the journals promoted greater social equality, their student editors cast themselves as the leaders of social reform, claiming in print a privileged position, as they did in the activist civic rituals of the same period. Students' repeated claims of sociopolitical leadership suggest that social and cultural self-production was a fundamental part of their activism.[12]

Starting in 1920, Zhejiang First Normal students took more active steps to "awaken" and "unify" the region's laboring classes. Through the publication of their journals, they had come in frequent contact with Hangzhou's print industry workers. In 1920, the Zhejiang Printing Company Work Mutual-Aid Association (Zhejiang yinshua gongsi gongzuo huzhu hui) began publishing a workers' journal, *The Qu River Worker's Tide* (*Qujiang gongchao*), with Zhejiang First Normal's Qian Gengxin, Wei Jinzhi, and Chen Lewo serving as editors. With Zhejiang First Normal students writing at least some of the articles in the journal, it extended the earlier journals' attempts to "awaken" Hangzhou's workers by offering a critique of capitalism and calling for workers' unification.[13] In 1921, Zhejiang First Normal

---

10. For similar themes and rhetoric from other Zhejiangese students, see "*Yuesheng* fakanci"; and "*Qianjiang pinglun* fakan zhiqu."

11. For a full account of the controversy surrounding Shi Cuntong's piece, see Yeh, *Provincial Passages*, 174–85.

12. Geisert similarly argues that "the 'would-be' elites of the GMD [during the 1920s] were using the party as a weapon in their 'fight for the rice bowl'" (*Radicalism and Its Demise*, 36). This argument applies most directly to young local party leaders whose activism promised to catapult them directly into positions of influence. Activist youths such as Zhejiang First Normal's students seem to have been more concerned about carving out a civic role for themselves as social leaders and exemplary citizens. The symbolic social status and cultural capital associated with civic leadership were prime motivators for many student activists.

13. Dong Shulin, "'Zhe yishi xuechao' de yingxiang," 7–8; Xu Xingzhi, "Dang chengli shiqi Zhejiang de gong nong yundong," 38–39. Xu, who was a print work-

students and alumni assisted the print workers in establishing a workers' school. Qian Gengxin, Yu Datong, and others taught at the school, and former student leader Xuan Zhonghua went there to lecture.[14] Zhejiang First Normal students also helped Hangzhou's barbers hold their first strike against shop owners, leading the barbers in marching on the Provincial Assembly to demand better wages and conditions.[15] Zhejiang First Normal graduates continued their engagement in social and political reform by assuming a central role in running the Yaqian Village School and Peasant Association that Shen Dingyi (Xuanlu) spearheaded in 1921 and publishing the radical journal *Responsibility* (*Zeren*) starting in 1922.[16] Through their participation in these rural organizations and this journal, Zhejiang First Normal graduates contributed to one of modern China's first efforts to mobilize peasants to claim political agency and challenge rural class relations.

Students at Zhejiang First Normal began engaging in grassroots social politics in 1919 and 1920, preceding by several years the activism of students at many other regional schools. Moreover, they invented their own form of radical social politics that stressed using revolutionary education and mass mobilization as means to create greater social equality in China. Zhejiang First Normal students' activism had profound impacts, at both the national and regional levels, which have been masterfully traced, by Wen-hsin Yeh and Keith Schoppa.[17] When we broaden our perspective from this one excep-

---

er at the time, refers to the group as the Zhejiang Print Workers' Club (Zhejiang yinshua gongren julebu).

14. Xuan Zhonghua had graduated from Zhejiang First Normal in the summer of 1920 (Zhao Zijie et al., "Xuan Zhonghua," 87). It is unclear whether Qian Gengxin and Yu Datong graduated with Xuan, or whether they were still enrolled in Zhejiang First Normal in 1921. Dong Shulin refers to them uniformly as "First Normal students" (see 8).

15. Dong Shulin, 8.

16. Chow, *Research Guide*, 109; Dong Shulin, 8; Schoppa, *Blood Road*, 97–123, 130; idem, "Contours of Revolutionary Change"; Shao Weizheng, "Yaqian nongmin xiehui shimo," 462–63; "Yaqian nongcun xiao xuexiao xuanyan," 21–22; "Yaqian nongmin xiehui xuanyan"; Zhao Zijie et al., 88–95; Zhonggong Zhejiang shengwei dangshi ziliao zhengji yanjiu weiyuanhui and Zhonggong Xiaoshan xianwei dangshi ziliao zhengji yanjiu weiyuanhui, eds., *Yaqian nongmin yundong*, 80. Shen Dingyi provided financial support for the journal *Responsibility*.

17. Schoppa, *Blood Road*; Yeh, *Provincial Passages*.

tional school, the activism of Zhejiang First Normal students can be seen as the leading edge of a form of student social radicalism that continued to gain momentum throughout the lower Yangzi region during the 1920s. This growing wave of activism was quickly recognized and exploited by the Nationalist and Communist parties during their first period of cooperation.[18]

Conducting Revolution

The forms of student social consciousness and political activism that grew organically at schools such as Jiangsu First Normal, Nanyang Middle, and Zhejiang First Normal during the early 1920s were consciously nurtured by party organizers starting in the mid-1920s. In the years leading up to the Northern Expedition, the CCP and the Nationalist Party sought to channel educated youths' social reformist fervor and used secondary schools to build their party organizations.[19] The socialist youth groups and party cells that were centered in the region's schools could both funnel student activists into the national party organizations to generate new revolutionary cadres and provide local bases for organizing workers and peasants.

Ningbo's Zhejiang Provincial Fourth Middle School provides one prominent example of such a school-based party organization. As discussed in Chapter 4, at Zhejiang Fourth Middle during the early 1920s, student and teacher reading groups focused on socialist and Marxist readings combined with lectures by prominent Communist and Nationalist Party intellectuals to influence student thought. The

---

18. For other examples, with varying degrees of party influence, of lower Yangzi region secondary students' social activism, see the following: Chen Heting; Fan Chongshan, "Dang lingdao de Yangzhou zaoqi geming douzheng," 52; "Yangzhou zaoqi dang zuzhi de lingdaoren—Cao Qiqian," 47–48; Fan and Su, "'Wusi' yundong zai Yangzhou," 45–46; Jin Yiqun, "Ningbo qingyun de xianfeng," 131–37; Lin Mei, "1924–1927 nian Wenzhou xuesheng yundong"; Mao Liyuan, "Wo suo zhidao de Jiangsu shengli Suzhou Zhongxue"; Shao, *Culturing Modernity*, 228–338; Wen Liangzhi, "Ershi niandai Zhenjiang jiaoyujie de yichang fengbo"; Zhonggong Jiangsu shengwei dangshi gongzuo weiyuanhui and Jiangsu sheng minzhengting, *Jiangsu geming lieshi zhuan xuan bian*, 25–29, 97–100, 116–18, 287–91.

19. Geisert, *Radicalism*, 36–37. For example, of the party members entering the Nationalist Party in Songjiang County early in 1924, 18.6 percent were secondary school students.

Huoyao Society, which students formed at Zhejiang Fourth Middle in the 1924–25 school year, most likely functioned as the school's first Communist Youth League organization.[20] The fact that the school's party cells grew in step with, if not directly out of, student and teacher reading groups suggests that the circulation of radical publications was an important factor encouraging student radicalism, along with the undeniable importance of the human networks traced by Yeh and Schoppa.

Influenced by ideas of worker mobilization and social equality, students at Zhejiang Fourth Middle became key student-movement leaders in Ningbo during the antiforeign May Thirtieth protests in 1925, and during the mid-1920s they engaged in union and political organizing among Ningbo's workers. For instance, student-movement leader Chen Hong, who was a member of the Socialist Youth League and then the CCP, helped to form a workers' night school and led workers in organizing a union at the Hefeng Textile Factory.[21] New principal Chen Shijue tried to stifle activism but was continually foiled by student radicals. For example, student leader Zhang Lingqian ignored Chen's restrictions and on May Day in 1926 rang the school's bell to call out students to march and spread propaganda as they had in previous years. Conflicts between Chen and student radicals led to the school being closed in the fall of 1926, only to reopen with the arrival of the Nationalist Party's Northern Expedition forces in March 1927. At that time students became pivotal leaders and organizers in the labor unions, peasant association, and student union that blossomed in the wake of the Northern Expedition.[22]

This revolutionary upsurge was short lived, however, as the Nationalist Party purge of CCP members within revolutionary organizations hit Ningbo in the spring and summer of 1927. Students Chen Liangyi and Wu Deyuan, who along with others had organized a common people's school, were both killed, and Zhejiang Fourth Middle was again closed. After it reopened in July 1927, police conducted a night raid in November that netted seven members of the Communist Youth League, with eight others escaping.[23] The inten-

---

20. Dong Qijun, 81.
21. Ibid., 82–83.
22. Ibid., 90–94.
23. Ibid., 94–95.

sity of the purge and the string of very politicized school closings and openings suggest the extent to which schools such as Zhejiang Fourth Middle had become the focus of local activist politics and how seriously that activism was viewed by local authorities, who treated students as seasoned revolutionaries.

Social and political activism was not limited to male students at boys' schools, as is illustrated by the example of Shaoxing Girls' Normal School.[24] Located on the site of Mingdao Girls' School, which revolutionaries had established during the late Qing, Shaoxing Girls' Normal had a long revolutionary tradition. During the mid-1920s, the principal, Zhu Shaoqing, was a member of the left wing of the Nationalist Party, and many of the teachers were Nationalist or Communist party members. The school provided a supportive context for female students to challenge gender norms, a process enacted symbolically by each student cutting her long braid and wearing her hair short or in a short braid.[25] These students' new hairstyles signaled their adoption of a new role of public social action.

As teachers, lectures, and politicized public ceremonies extended messages of radical politics into Shaoxing Girls' Normal, the Nationalist and Communist parties successfully recruited female activists who contributed to the Northern Expedition.[26] One of the school's CCP members, Huang Chaochang, was a leader in the student movement. Wang Ruozhen, who was also a party member, worked in the county Nationalist Party Women's Department, leading the women workers' movement in Shaoxing. Another student, known for her fine calligraphy, wrote slogans for the County Party Bureau. In the heady revolutionary atmosphere created by the May Thirtieth Movement and the Northern Expedition, many female students responded by taking on the role of activist citizens, working to change the social order within the revolutionary framework provided by the Nationalist and Communist parties.[27]

---

24. Wu Sihong, "Dageming shiqi," 225–36.

25. For the symbolism of women cutting their long braid, see Gilmartin, *Engendering the Chinese Revolution*, 152, 180.

26. Cf. Gilmartin, 123–27, 135–39; Jin Yiqun, "Ningbo qingyun de xianfeng," 131–37; Zhonggong Jiangsu shengwei dangshi gongzuo weiyuanhui et al., 304–9.

27. Gilmartin (part 2) characterizes the period between the May Thirtieth Movement and the Northern Expedition as the high tide of women's political activism during the 1920s.

Thus, at Zhejiang Fourth Middle and Shaoxing Girls' Normal, as at Zhejiang First Normal, study societies, political groups, and contact with inspiring teachers or visitors motivated students to plunge into society to organize unions and associations to seek equality for workers, peasants, and women. The networks of people and texts mapped in these examples reached throughout the region, such that schools both inside and outside the regional core became sites for the development of student radicalism and party organizations.[28] The greater density of those networks in core areas meant that student radicalism probably developed somewhat earlier and was somewhat more common there than in the periphery, as is suggested by these examples from closely connected areas in northern Zhejiang.[29] In joining and leading social and political movements, students enacted a form of activist citizenship that used revolutionary propaganda and mass organizations to mobilize nonelite social groups to level social inequalities.

Despite similarities in their modes of activist citizenship, the developmental trajectories of early student social activism at Zhejiang First Normal and the later experiences of students at schools such as Zhejiang Fourth Middle, Shaoxing Girls' Normal, and many others were quite distinct. Zhejiang First Normal's mature student radicalism, and the shoots of social reform cultivated by Pang Guoliang and Wang Weihua, grew organically out of the fertile soil of intellectual experimentation enriched by people and texts that were carried through dense regional social networks. By the mid-1920s, student social radicalism became formalized into a pattern of Nationalist and Communist recruitment managed through school-based party or youth group organizations. Rather than "awaken" or spark students' activism, then, the parties channeled and managed students' emer-

---

28. For examples from outside the lower Yangzi core, see Barkan, "Patterns of Power," 206–7; Shao, *Culturing*, 235–37; Zhejiang sheng Lishui diqu geming wenhua shiliao zhengbian bangongshi, *Lishui diqu geming (jinbu) wenhua shiliao huibian*, 77–81; Zhonggong Jiangsu shengwei dangshi gongzuo weiyuanhui et al., 25–29, 97–101, 304–9. For parallel examples from Jiangxi, see Averill, "The Cultural Politics of Local Education."

29. Cf. Geisert, *Radicalism*, 107–8; Yeh, *Provincial Passages*. But Wen-hsin Yeh stresses that students going to core areas from more culturally conservative "middle counties" had the greatest propensity to become radicalized.

gent social and political radicalism, as they did with students' activist civic ritual during the same period.[30]

Student activists recruited by the Nationalist or Communist parties during the mid-1920s often became lifelong party members and valuable cadres. Chen Hong, the student-movement leader at Ningbo's Zhejiang Fourth Middle, transferred to the Wusong General Labor Union and took part in the Third Shanghai Workers' Uprising in March 1927. After an imprisonment by the Nationalist Party that lasted for most of the Nanjing decade, Chen worked as a labor organizer and underground activist during the Sino-Japanese War.[31] Shaoxing Girls' Normal's Huang Chaochang served as an underground party operative for the CCP in Chongqing during the War of Resistance.[32] Student leaders inspired to activism by their school experiences were vital contributors to China's revolutionary parties.[33]

II. SERVICE AS CIVIC ACTIVISM AND YOUTHS AS SOCIETY'S CORE ELEMENTS

Not all students joined the radical social movements and political revolution of the 1920s, but many of them still acted as engaged citizens through participation in moderate social reform activities. In keeping with the civic republican imperative to make concrete contributions to society, student-citizens of the 1910s and 1920s joined in food and clothing drives for victims of flood, famine, or war, traveled to the countryside to give lectures or classes on national affairs or hygiene, and engaged in firefighting and rescue activities in their own communities.[34] The discussions of Scouting and student self-government activities in previous chapters have included examples of some of these varied forms of social service. But the idea of enacting

---

30. Elizabeth Perry makes a similar point when she notes that CCP labor organizers in Shanghai during the 1920s were confronted with workers who already possessed sophisticated mechanisms for economic and political action (*Shanghai on Strike*, chap. 4).

31. *Zhejiang renwu jianzhi*, 321–22. Cf. ibid., 318–19, 336.

32. Wu Sihong, "Dageming shiqi," 236.

33. For many other examples of student activists during the 1920s who became party cadres, see the references in note 18.

34. For examples, see "Jiangsu shengli diba zhongxue renxu ji dashiji"; Liao Shicheng et al., 328; and Zheng Haozhang, *Shanghai tongzijun shi*, 5, 12, 16–18.

citizenship through concrete social action was perhaps encapsulated best in students' organization and management of literacy schools.

Mass literacy education was introduced in China during the late Qing and promoted with increasing intensity during the New Culture Movement and various other social reform programs as a way to create educated new citizens, reform social customs, and, in some cases, raise workers' and peasants' social status.[35] By the early 1920s, students at many schools in the lower Yangzi region organized some kind of common people's school.[36] As an institution for spreading ideas and organizing people, the common people's school could focus on literacy education itself as a mode of social reform or be used as a tool of political indoctrination and mobilization, a tension that Glen Peterson shows persisted in Chinese literacy movements after 1949.[37] These contrasting approaches marked student-run literacy schools as well. For example, activists at Zhejiang First Normal and Zhejiang Fourth Middle both established literacy schools for workers or peasants as a first step in mobilizing them for other forms of social or political activism.[38] But in most instances, student organizers viewed literacy schools as a less politicized form of "service to society" and a concrete method of social reform.

This practical focus on social reform was exemplified by the common school at Wuben Girls' Middle School in Shanghai.[39] Established in October 1919, in the wake of the May Fourth Movement, the common school had a staff of 25 student-teachers and a student body of 78 that was divided into two classes. Classes were held for two hours each day and covered the standard curriculum of the first two years of primary school.[40] The school undoubtedly pro-

---

35. Bailey, *Reform the People*; Chow, *The May 4th Movement*, 193–94; Hayford, *To the People*; McElroy, "Transforming China," 178–81, 213–23.

36. Benji quanti, "Wuben yiwu xuexiao de gaikuang"; Jia Fengzhen, "Nian nian lai benxiao changchu duanchu," 2; Liao Shicheng et al., 327–28; *Pudong zhongxuexiao ershi zhou jiniankan*, 21; "Yiwu xiaoxue xiaoshi"; Zhejiang shengli diyi shifan xuexiao xuesheng zizhi hui, 15. Students in missionary schools also ran charity schools and night schools. See Dunch, "Mission Schools," 126; Graham, *Gender*, 135–36.

37. Peterson, *The Power of Words*, esp. chap. 3.

38. Chen Heting, 150; Dong Qijun, 83; Dong Shulin, 8.

39. Benji quanti, "Wuben yiwu xuexiao."

40. The curriculum included conversation, mathematics, local geography and history, national language, singing, and physical exercise.

vided a valuable resource for uneducated poor people in the school's neighborhood. But it was also a way for students to claim publicly the role of active citizen through concrete social action. Wuben's students supported their claim by reprinting supportive testimonials from observers, quoting at length a *Shishi xinbao* reporter who praised the Wuben common school as an effective example of literacy education, which he regarded as "the vanguard of the culture movement and an effective tool for reforming China."[41]

Many secondary schools initiated or reestablished literacy schools soon after the completion of the National Revolution, continuing this popular form of elite reformist activity from the early Republican period. Student-run common schools flourished during the Nanjing decade, in part because literacy education was a form of civic action promoted in multiple arenas under Nationalist Party calls for constructive service.[42] Nanjing-decade literacy schools varied widely in size, mode of organization, and method. Some, such as those at Zhejiang's Hangzhou High School, Jiangsu Provincial Shanghai Middle School, and Yangzhou Middle, were run by the student self-government association and taught 50 to 100 students.[43] On the other end of the spectrum was the common school at Shanghai's Fudan University Affiliated Middle School.[44] That school grew from having 30 to 40 students at its start in 1919 to having several hundred students attending both lower primary and upper primary divisions by the early part of the Nanjing decade. In all of these schools, secondary school students filled key positions as administrators and teachers on a voluntary basis, demonstrating Nanjing-decade students' commitment to constructive civic action. At Fudan Affiliated

---

41. Ibid., 7.

42. E.g., Jiang Zhongzheng, "Xuesheng ying liyong shujia fuwu shehui"; SMA Q235-1-225; Zhongguo Guomindang zhongyang zhixing weiyuanhui xuanchuanbu, *Xuesheng funü wenhua de tuanti*, 7. Literacy education was also a sanctioned form of social outreach at schools with activist traditions, such as Labor University. See Chan and Dirlik, 190–98.

43. Chen Zhaokui, "Minzhong xuexiao baogao"; "Pingmin yexiao gaikuang"; Zhou Jianwen, "Dijiujie de minzhong yexiao." Cf. *Songjiang nüzhong xiaokan*, no. 5 (May 15, 1929), 10; Suzhou Zhenhua nü xuexiao, ed., *Zhenhua nü xuexiao sanshinian jiniankan*, 86–87.

44. Li Chongbi, "Fudan yiwu xiaoxue de guoqu ji jianglai"; "Yiwu xiaoxue xiaoshi."

Middle's common school, for instance, as many as several dozen secondary school students served as teachers.[45] The seven committee members who organized and ran the much smaller common people's school at Yangzhou Middle were responsible for teaching or working in the school office on a daily basis, whenever they were not in class or studying.[46] By participating in mass literacy education, lower Yangzi region secondary students performed an active mode of citizenship within the parameters permitted by the Nationalist Party.

Because of Nationalist Party oversight, common people's literacy schools during the Nanjing decade were generally apolitical. Instead, student organizers identified literacy education itself as a key means of strengthening the nation both by improving the knowledge and capacities of the common people and by raising their consciousness of the nation.[47] When ideology was injected into the curriculum, it was in the form of the Three Principles of the People.[48] Rather than focusing on politicization, student teachers of the common people's schools often expressed an interest in instilling discipline and civility in the illiterate common people and poor children who attended the schools. These points of emphasis reflected the Chinese scholarly elite's goal of reforming the people and elite secondary students' concern with crafting a position of authority and distinction for themselves as the people's teachers.[49] In the words of the students involved in managing the common people's school at Yangzhou Middle, "Most uneducated adults and children have had no household education, and it can be said that their actions and habits of conduct usually have no restraint. The greatest objective in this school's training is to bring their lawless behavior (*wu dinglu zhi xing-*

45. Li Chongbi, 181; "Yiwu xiaoxue xiaoshi."

46. "Pingmin yexiao gaikuang," 56–57.

47. See, for instance, Chen Zhaokui, "Minzhong xuexiao baogao"; Luo Jiangyun, "Shizi yu guonan"; Wei Zhen, "Xin shenghuo yundong he qiangpo jiaoyu de xiaoguo," 12; Wu, "Fight for the Children."

48. E.g., Suzhou Zhenhua nü xuexiao, ed., *Zhenhua nü xuexiao sanshinian jiniankan*, 87; Zhou Jianwen, "Dijiujie de minzhong yexiao," 61. Cf. Yu Zixiang, "Zhongxuesheng yu xiangcun (luntan)," 92.

49. Chen Zhaokui, "Minzhong xuexiao baogao," 1586; Zhou Jianwen, "Dijiujie de minzhong yexiao," 61–64.

dong) onto the proper path and to gradually improve their habits and promote good character."[50] The belief that the customs and habits of the common people were superstitious, backward, and in need of reform through mass education was expressed by many lower Yangzi region secondary students during this period, reinforcing the disciplinary aspect of their mass education schools.[51] By using literacy education to transform the habits of the common people according to their own standards, student educators reinscribed the elite-common class distinction, based on differences in knowledge and refinement, that had characterized Confucian society and that had been reproduced in the dynamics of mass mobilization during the early twentieth century.

The commitment to social service expressed by student volunteers in activities such as literacy schools was also voiced in many students' writings. For instance, in an open forum on how to use summer vacation in the independent journal *Middle School Student* (*Zhongxuesheng*), several students from lower Yangzi region schools independently argued that, rather than sleep in and waste time during summer vacation, students should serve society because students all "have the responsibility to reform and develop society."[52] In these articles, each of which contained several ideas for how to use the summer break, the authors suggested that their fellow students manage or teach in literacy or summer schools, "organize fly-extermination teams, hygiene teams, or popular lecture societies . . . [or] organize groups to visit factories or institutions in order to investigate them and understand their [systems of] organization." This imperative for social service and constructive civic action was repeated in many lower Yangzi region students' writings in their school journals as well.[53]

---

50. "Pingmin yexiao gaikuang," 60.

51. For examples of secondary students' negative reactions to the customs and habits of the common people, see Feng Heyi, "Pochu mixin"; Jiang Xianduan, "Shimin dahui"; Liu Dejue, "Xin shenghuo yu xin jingshen"; Wang Jie, "Ruhe dapo xisu"; Yu Zixiang.

52. "Zenyang liyong women de shujiaqi (Wenti taolunhui, diwuci taolun)."

53. E.g., Gu Ruwen, "Xuesheng fuwu zizhi tuanti"; Xu Yangben, "Wo duiyu qingniantuan de xiwang"; Yu Zhenjian, "Zhongxuesheng de zeren"; Yu Zixiang; Zhou Jianwen, "Feichang shiqi zhong qingnian xuesheng yingshou de xunlian."

Significantly, many female students during the Nanjing decade adopted a rhetoric of active national service that mirrored that of their male counterparts. Students at Zhenhua Girls' Middle School were told "do not study my example" by their principal Wang Jiyu, who remained single throughout her life and did not establish her own family. But a number of Zhenhua graduates did in fact emulate Wang in aspiring to future lives of public service, in careers in fields such as law, medicine, and education.[54] These students' approach to citizenship is captured in an essay titled "Women's Method of Saving the Nation" by Zhenhua student Shen Zhilin.[55] Shen argued that as one-half of the nation, women had a great responsibility to contribute to national survival through a variety of activities. She suggested that young women boycott foreign products (especially foreign cosmetics), "do things for the public" in the same ways as men, act diligently and withstand hard labor, "abandon all unnecessary embellishments and eliminate all luxurious and idle habits" that were characteristic of "young misses and wives" (*xiaojie* and *taitai*), and strengthen their bodies through athletics to be able to fulfill their social tasks. Shen acknowledged that personally doing the household work could be a basic first step in contributing to national strength, but she also called for women to discipline and train themselves in preparation to fulfill equal duties with men in working for national salvation. Students like Shen called for women's entry into civic action, but they also suggested that it entailed their masculinization or de-gendering by purging feminized habits of consumption and leisure, as well as overcoming forms of physical and moral weakness that were considered harmful to the nation.[56]

Educated youths who expressed a responsibility to participate in social service claimed that their ability to do so, based on their education, intellect, and youth, marked them as privileged citizens. As expressed by one student from Zhejiang Ninth Middle School in Jiande:

---

54. Three separate alumnae interviewees; "Wo lixiang zhong de geren shenghuo," 56–59.

55. Shen Zhilin, "Nüzi jiuguo de fangfa."

56. Cf. Lu Miaojin, "Dadao qingnian de tuifei bing"; Peng Kunyuan, "Xiandai nüzi yingyou de juewu"; Shi Yide, "Dadao qingnian de tuifei bing"; Tao An, "Wo ren zai xiao ruhe xiuyang"; Zhang Bihua, "Nü qingniantuan huodong jilüe."

It goes without saying that all Chinese people have a responsibility [to save the nation]. . . . [But] the responsibility of people in the intellectual class (*zhishi jieji*) is especially great, because they must invent various methods to construct China and then make common people act according to their method. Students are part of the intellectual class, so students are people who carry great responsibility, and our responsibility as middle school students is even greater.[57]

Students could empower themselves by integrating into their writing school rhetoric that granted them both ability and responsibility. In the words of one student, "Youths are society's core elements. Their power is great, and they are able to influence society's thought and action."[58] As Nanjing-decade students participated in government-sanctioned arenas of civic action, they assumed the role of "society's core elements."

Critical Voices

Lower Yangzi region secondary students' frequent use of the rhetoric of service and participation in party-sanctioned modes of civic action does not necessarily indicate that they accepted, without question, party rule and its citizenship ideal. We know, for instance, that many students during the Nanjing decade rejected the party's policies of appeasement toward Japan.[59] But beyond challenging Nationalist Party foreign policy, many lower Yangzi region students offered wide ranging critiques of other government initiatives and party rule during the Nanjing decade. Students' criticisms often centered on the party's failure to implement effectively policies that it had promoted or to live up to a standard that it had set for itself and China's citizens.

For instance, some students criticized how the New Life Movement was organized and executed, even as they sympathized with

---

57. Yu Zhenjian, "Zhongxuesheng de zeren." Cf. Dineng, "Jinri de zhongxuesheng"; Wang Yongquan, "Gaozhong biyesheng de zeren," 12; Weilutsu, "Foreword." Thøgerson suggests that students at Shandong's Changshan Middle School were also "imbued with an ardent sense of their own sociopolitical role and with strong progressive nationalist sentiments" (p. 68).

58. Zhang Xiujun, "Feichang shiqi zhong qingnian xuesheng yingshou de xunlian," 61. Cf. Zhou Jianwen, "Feichang shiqi zhong," 50. Other student authors identified youths or students as the "masters of tomorrow." See Yang, "The Students' Part in the National Crisis."

59. E.g., Israel, *Student Nationalism*; Wasserstrom, *Student Protests*.

the goal of generating national reconstruction by reforming popular mores and behaviors, goals that students themselves pursued in their literacy schools and other outreach activities. One student deplored the campaign-oriented nature of the movement and the way local organizations responded only to prompts from superiors. "It should be," he asserted, "that every person first examines himself (*hecha ziji*), then goes to exhort his subordinates or others."[60] Some students pointed out more fundamental failings in the Nationalist Party's program. For instance, Chen Erchun, a student at Zhenhua, followed the pre-Qin philosopher Guanzi (d. 645 BCE) in asserting that people needed to be economically secure before they could concentrate on their morals and demeanor. Although the relatively wealthy and educated elite should be moral exemplars in order to lead the people, the people could be expected to learn these virtues only after their needs had been addressed.[61] Another student similarly criticized the New Life Movement by questioning whether Chiang Kai-shek himself could act out the four virtues he promoted and provide a proper example for the people to follow.[62] These students used the party's own rhetoric and standards of economic development and moral responsibility to criticize the party's failings. Further, some students based their criticisms on the Confucian assumption that leaders should be moral exemplars, a pattern of authority and moral leadership that was inscribed in schools' systems of character-development education, in which teachers and administrators were expected to serve as mentors and teach by example.

Students could be equally critical of the party's approach to fundamental social problems of poverty and social inequality. For instance, a student at Pudong Middle School, Yu Fu, dismissed as inadequate the government's efforts to use foreign experts to solve the problems of China's rural economy and suggested that the government itself may have contributed to rural poverty through excessive taxes and unstable fiscal policies.[63] A student at Minli Middle School identified capitalism as the root of China's economic troubles and argued that changing society depended on the power of the masses, especially the

---

60. Wei Zhen, "Xin shenghuo yundong," 11.
61. Chen Erchun, "Xin shenghuo yundong zhi wojian."
62. Yang Hongsheng, "Wode riji," 23.
63. Yu Fu, "Tan fuxing nongcun," 11–12.

peasantry, who would be awakened by the students and led out from under the control of the exploiting class.[64] Other students concerned with current social inequalities, such as Zheng Ruochuan at Jiangsu Provincial Songjiang Girls' Middle School, called for equalization of land prices, reduction of rents, formation of cooperatives, and the implementation of self-government to redress the rural crisis, all of which were elements of Nationalist Party policies.[65]

Thus, despite party censorship, students at a range of schools across the region challenged the party using its own ideals and policy statements. Their criticisms suggest that they questioned the Nationalist Party's viability as a leading agent of change and, by extension, its legitimacy as a ruling authority, despite their repeated exposure to history textbooks, civics lessons, lectures, civic training, and civic rituals that represented the party as China's sole legitimate ruling authority. By drawing on elements of party-sanctioned citizenship training and moral cultivation, as well as from the Nationalist Party's own ideology, these students crafted a discourse critical of party leaders and national policies that could be expressed in school journals that were reviewed by school authorities. At the same time, students like those who contributed to literacy schools and other social reform projects drew selectively from their civic education to practice a mode of citizenship focused on constructive civic action in arenas promoted by the party.

III. COMPETING ARTICULATIONS

During the mid-1920s the united Nationalist and Communist parties had sought to channel nascent forms of radical activism into the common project of the National Revolution. As war with Japan loomed ever closer a decade later, the competing Nationalist and Communist parties fashioned divergent models of "national salvation" (*jiuguo*) work through which each sought to align students' widespread commitment to active citizenship with their party's platform and projects. Starting in the early 1930s, Nationalist Party

---

64. Dajun, "Women de chulu."

65. Zheng Ruochuan, "Woguo nongcun jingji pochan." For similar criticisms regarding the Nationalist Party's handling of antisuperstition campaigns, see Feng Heyi, "Pochu mixin," 37–41.

leaders repeatedly told students that "to save the nation one must study; studying is saving the nation" (*jiuguo bixu dushu; dushu bian shi jiuguo*). By contrast, the Chinese Communist Party encouraged anti-Japanese nationalism and organized students and recent graduates for forms of direct activism that included grassroots propaganda work, mass organization, and, eventually, guerrilla warfare. Each approach provided an outlet for the active citizenship geared toward national service that many students discussed and practiced during their time in school.

Studying for National Salvation

The Nationalist Party's vision of national salvation through study and self-cultivation had its intellectual roots in Dai Jitao's 1928 interpretation of Sun Yat-sen's ideology for China's students, *The Road for Youth (Qingnian zhi lu).*[66] Dai exhorted students to focus on acquisition of technical knowledge that they could use to contribute to national construction. Such calls intensified in the wake of student protests critical of the Nationalist Party's appeasement policy after the Mukden Incident in 1931 and the Japanese invasion of Shanghai in 1932. In this context, party leaders such as Chen Lifu, then a member of the Central Party Organization Committee, and Zhou Fohai, the head of the Jiangsu Department of Education, directed students to study diligently, rather than strike and protest, as a way to save the nation.

Zhou's essays and Chen's lecture on the theme of study stressed that foreign powers were able to influence and control China because of their material strength, which was rooted in technical superiority.[67] To combat these foreign powers' military and industrial strength, China needed more people with technical learning to contribute to its economic reconstruction. Chen and Zhou urged Zhejiang's and Jiangsu's students to dedicate themselves to acquiring such advanced knowledge. At the same time, they encouraged educated youths to continue their efforts at moral cultivation in order to ensure that their technical knowledge would be used in morally up-

---

66. Yeh, *Alienated Academy*, 262–64.
67. Chen Lifu, "Dushu yundong zhi zhenyi"; Zhou Fohai, "Ruhe zhengdun xuefeng?"

right ways to contribute to the welfare of the social whole.[68] Chen's and Zhou's comments certainly aimed, along with a number of other educational policies adopted at the time, to divert students' attention away from nationalistic protests that were highly critical of the Nationalist Party.[69] But they also sought to channel students' sense of mission to the nation and spirit of social activism into "study" (*dushu*) that would contribute to national economic construction, which was a central focus of Nationalist Party policy during the Nanjing decade.[70]

In practical terms, students could interpret these exhortations to study as calls to pursue higher learning (*shengxue*). Such calls would have resonated with many lower Yangzi region youths, the vast majority of whom identified further study as their main aspiration.[71] By linking study to China's struggle for survival, party leaders such as Chen and Zhou made further study even more attractive to students by equating the personal pursuit of higher learning with patriotism. Impulses to further study as the best course for young students may have been particularly strong in the lower Yangzi region, with its long tradition of scholarship and academic success during the late imperial period.[72]

In fact, many lower Yangzi region secondary students during the 1930s did pursue further study. Statistics on Jiangsu students' life courses after high school during the Nanjing decade suggest that the majority of students in the highly competitive regular course of study (*putong ke*) in high schools went on to further study after graduation.[73]

---

68. Chen Lifu, "Dushu yundong zhi zhenyi"; Zhou Fohai, "Women zenyang qu zuoren?"

69. See Israel, *Student Nationalism*, chaps. 3, 4.

70. See Kirby, "Engineering China"; Zanasi, "Chen Gongbo."

71. E.g., Xu Minzhong, "Jiaxing zhongdeng xuesheng shenghuo zhuang-kuang," 92–94.

72. Brook, "Family Continuity and Cultural Hegemony"; Elman, *Classicism, Politics, and Kinship*; idem, *Cultural History*, 256–60.

73. See Jiangsu sheng jiaoyuting, *Jiangsu jiaoyu gailan*, 3: 63–64, 74–75, 96, 105, 113, 120, 124, 135, 147. For female students, the picture was mixed. Statistics from Jiangsu Provincial Songjiang Girls' Middle School and Zhenhua Girls' Middle School in Suzhou show the majority of the young women entering college or university (Jiangsu sheng jiaoyuting, 3: 105; Suzhou Zhenhua nü xuexiao, ed., *Zhenhua nü xuexiao sanshinian jiniankan*, 126). By contrast, over half of the high school class of 1930 at Provincial Nanjing Girls' Middle School were engaged in "ser-

Students in the top tier of provincial and private high schools, such as Yangzhou Middle and Suzhou Middle School in Jiangsu, were funneled directly into the nation's best universities, such as Communications, Central, Qinghua, Fudan, and Beijing universities.[74]

Some of the lower Yangzi region students who went on to college during the Nanjing decade were there drawn to the CCP. For example, Li Yueshan attended Hangzhou Lower Middle School and then Hangzhou High School during the mid-1930s. In 1936, he began studying at Daxia University in Shanghai, participated in anti-Japanese activities, and joined the CCP in 1938. He later took part in mass mobilization work for the CCP's New Fourth Army during the Sino-Japanese War.[75] Similarly, some of the student activists of the December Ninth Movement, who were radicalized in university as a result of Nationalist Party suppression of their national salvation activities, came from the lower Yangzi region.[76] But the university students from the lower Yangzi region who found their way to the CCP capital of Yan'an during the mid-1930s were clearly a minority. Most of their fellow students who had gone on to university focused their efforts on their studies and contributing to the war effort through "constructive work" in their fields of study.[77]

As alumni from lower Yangzi region secondary schools studied further and acquired specialized skills, many of them in fact contrib-

---

vice," most likely teaching in primary schools or kindergartens (Jiangsu sheng jiaoyuting, 3: 60).

74. Jiangsu sheng jiaoyuting, 3: 113, 120; *Jiangsu shengli Suzhou zhongxuexiao gaikuang*, 14–15; Shen and Xiao, 29.

75. Zhao and Qian, 102–3. Cf. Shaoxing shi wenhuaju and Zhonggong Shaoxing shiwei dangshi bangongshi, *Shaoxing geming wenhua shiliao huibian*, 152–53; Zhonggong Jiangsu shengwei dangshi gongzuo weiyuanhui et al., 344–47.

76. Most notable of these educated youths from the lower Yangzi region were Lu Cui and Huang Jing. See Israel and Klein, *Rebels and Bureaucrats*, 55–59, 72–75. Also see the example of Wang Wenbin (Zhonggong Jiangsu shengwei dangshi gongzuo weiyuanhui et al., 46–50).

77. Jiangsu shengli Yangzhou zhongxue liushi zhounian jinian tekan bianji weiyuanhui, comp., *Jiangsu shengli Yangzhou zhongxue liushi zhounian xiaoqing jinian*, 398–482, records life stories for dozens of Yangzhou Middle graduates from the 1920s and 1930s who went on in school and dedicated their lives to social service in professional careers. By contrast, *Jiangsu sheng Yangzhou zhongxue jianxiao jiushi zhounian jiniance*, published in the PRC in 1992, records only ten revolutionaries who studied at Yangzhou Middle during the Nanjing decade. Cf. Zhao and Qian, 102–6.

uted in concrete, and sometimes dramatic, ways to China's defense and reconstruction through their work during the late 1930s and 1940s. Biographies of students who graduated from Yangzhou Middle or Zhejiang First Middle (along with one of its successor schools, Hangzhou Lower Middle) and went on to study for higher degrees either before or during the war confirm this assertion.[78] These graduates, many of whom studied for technical degrees in engineering or medicine, worked on railroads or in public health, served as doctors or agricultural researchers, managed defense factories, or worked in urban planning after 1937. In the context of total war, lower Yangzi region middle schools' alumni working in these technical fields had ample opportunities for heroic constructive service along the lines they had imagined and written about during their time in school during the 1930s.

For instance, class of 1931 Yangzhou Middle graduate Chen Zhongfu went on to earn a degree in civil engineering at Communications University (Jiaotong daxue). After the start of the war, he was assigned to help run the railroad at the strategic town of Tongguan in Shanxi. There, he rallied his fellow engineers and workers to labor, sometimes through the night and under attack, to keep the line open when it was under almost constant Japanese bombardment.[79] Chen's younger schoolmate, Dai Zhijun, graduated from Yangzhou Middle in 1933 and went on to complete a civil engineering degree at Qinghua University in 1937. When the Nanjing Public Health Office where he was training came under attack that fall, he returned to teach middle school in his home of Tianchang in Anhui. When it, too, became threatened by Japanese attack, Dai fled through Hong Kong and Vietnam to Chongqing, where he worked on the Yangzi River Water Conservancy Committee and taught at Central Political University.[80] For Chen, Dai, and many others in their generation of lower Yangzi region secondary graduates, "education for national salvation" that focused on advanced learning and technical skills fed into forms of social service for national welfare that expressed the ethos of civic activism they had created in their schools.

---

78. Jiangsu shengli Yangzhou zhongxue liushi zhounian jinian tekan bianji weiyuanhui, 398–463; Zhao and Qian, 102–6.

79. Jiangsu shengli Yangzhou zhongxue liushi zhounian jinian tekan bianji weiyuanhui, 408–9.

80. Ibid., 426.

Mobilization for National Salvation

In contrast to the Nationalist Party's program of study for national salvation, during the two years before the Second Sino-Japanese War began in July 1937 and in the war's early stages, the Chinese Communist Party developed a program of youth mobilization for national resistance.[81] Through its wartime mobilization, the CCP offered students who were primed for national service outlets for civic action. Although the Nationalist Party flirted with a mobilization-oriented youth policy at times, its officials and sympathizers frequently suppressed or disrupted extracurricular student activism. Moreover, after the start of the war, the party sought to funnel educated youths into the Three Principles of the People Youth Corps, where their activities could be safely controlled by party leaders.[82] In response to these contrasting strategies, many lower Yangzi region secondary students and recent graduates who were active in prewar and wartime anti-Japanese activities came to operate within the CCP or its front organizations, as educated youths' independent nationalist resistance work merged with CCP mass mobilization.

In the years between 1931 and 1937, many secondary students in the lower Yangzi region expressed concern about China's tenuous international situation and anticipated their country's coming war with Japan.[83] Some lower Yangzi region students, as we have seen, dedicated themselves to study, self-cultivation, and military training, but others also took part in various forms of propaganda and anti-Japanese resistance activities. These student activists formed propaganda troupes that toured the countryside to perform patriotic dramas, sing nationalistic songs, and distribute wall posters and pamphlets promoting resistance to Japan.[84] Such protest and mobilization activities often ran counter to the Nationalist Party policy of

---

81. Israel, *Student Nationalism*, chaps. 5, 6; Israel and Klein, 109–22, chap. 5; Van Slyke, *Enemies and Friends*, 66–71.

82. Huang, *The Politics of Depoliticization*, 100–164; Israel and Klein, chap. 4.

83. See, for instance, Yang Lieyu, "Xie zai qianmian"; Yu Hetong, "Huabei teshuhua yu zhong ri gongtong fanggong liang da wenti"; Yu Zhongyin, "Fendou ba! Pengyou!" 7; Zhou Jianwen, "Feichang shiqi zhong."

84. E.g., Ji Le, "Zhanshi Lishui wenyi xuanchuan huodong diandi," 86–87; Zhonggong Jiangsu shengwei dangshi gongzuo weiyuanhui et al., 46, 136, 213, 222, 277.

appeasement and national construction, but they nonetheless built on the modes of civic action students had learned at school.[85]

An account of one student propaganda team from the elite Hangzhou High School gives some sense of how students drew eclectically from the various forms of citizenship training that they received in school—in this case Scouting or Youth Group activities such as cycling and first aid, as well as basic knowledge about defense from military training—to fashion independent forms of civic action.[86] During spring vacation in April 1937, twelve students took an extended, round-trip bicycle tour of Hangzhou, Zhapu, and Jiaxing.[87] They carried with them propaganda materials, as well as medicine for treating simple medical problems. At each town and village they visited, they tried to arouse the interest and trust of the local people to promote national consciousness and resistance to Japan.

When we reached [the village of] Sanbao, everyone stopped and began to work: some treated the common people medically while others gestured and lectured to them about patriotic pictorials (*huabao*). For a while we were surrounded by many people. I pointed to a pictorial pasted to a wall and said, "This is our national flag, and we must respect it!" "This is our leader, and we must protect him!" "We must help our own people to fight the foreigners!" "When the enemy spreads poison gas, everyone must run to the mountains! Poison gas is . . ." Our words were serious and our hearts thoughtful, and we ultimately gained the common people's sincere welcome. Although we exhausted our intelligence to explain to them, their attitude spurred us on so that we had to keep going.[88]

As this passage suggests, propaganda activities like these made students feel they were fulfilling their responsibilities for national service and acting as engaged citizens. Students' participation in mobilization efforts fueled an enthusiasm for further activism. Ultimately, this inspiration was perhaps as important an outcome as any consciousness-raising among the common people, for it

---

85. For the ways in which student self-government associations could provide a framework for nationalist protest, see Culp, "Rethinking Governmentality."

86. The focus of provincial youth group activities at Hangzhou High School was training in skills such as horsemanship, driving, boating, photography, cycling, and climbing, each of which recorded large numbers of participating students ("Qingniantuan teji").

87. "Zhexi nongcun xunli," 1681–88.

88. Ibid., 1682.

consolidated students' commitment to practicing an active mode of citizenship.

After war began in July 1937, students continued their efforts at mobilization. Student nationalist activities at central Zhejiang's Xinchang County Lower Middle School exemplify the vigor and the breadth of many students' activism during the first year of the Sino-Japanese War.

Xinchang Middle School's ardent youths who were not willing to be subjugated people felt passionately that "everyone has responsibility for the rise and fall of the state" (*guojia xingwang, pifu youze*), so they took the path of anti-Japanese national salvation. . . . They issued a wall-poster newspaper, pasted up slogans, painted cartoons, collected scrap copper and iron [for weapons production], made collections of winter clothes [for soldiers and refugees], supported the front lines, and comforted the wounded. Their thought was progressive, and their work was active.[89]

Students in local schools throughout Zhejiang and Jiangsu engaged in similar activities in the first year after the start of the war.[90] In their activism, they drew on many repertoires of social service and messages of nationalism that had been central to their schooling.

During the first year of the war, Japanese forces progressively occupied most of the core areas of the lower Yangzi region between the Yangzi and Qiantang rivers, taking Shanghai by November and Hangzhou and Nanjing in December. Even as the disruptions of war forced educated youths, students, and whole schools to relocate to more peripheral inland areas, they continued their efforts at nationalist mobilization.[91] For instance, in the fall of 1937, Jin Haiguan, principal of Xiaoshan's Xianghu Normal School in north-central Zhejiang, led four hundred students on a march to mountainous

---

89. Yu Yueyin, "Kangri zhanzheng shiqi Xinchang zhongxue," 255–56.

90. E.g., Ji Le, "Zhanshi Lishui"; Pan Pi et al., comps., *Subei kangri genjudi jishi*, 2–3, 5, 6, 26–27; Taizhou diqu wenhua ju geming wenhua shiliao zhengji bangongshi and Taizhou diqu wenwu guanli weiyuanhui, comps., *Taizhou geming wenhua shiliao xuanbian*, 19–22; Zhejiang sheng Lishui diqu geming wenhua shiliao zhengbian bangongshi, *Lishui diqu geming (jinbu) wenhua shiliao huibian*, 88, 90, 95, 183–84.

91. Chen Shiying, "Cong Daji minjiaoguan dao Lianji minzu xiaoxue"; Zhejiang sheng Lishui diqu geming wenhua shiliao zhengbian bangongshi, 95; Zhou and Zhou, "Kangzhan chuqi zhonggong Sheng xian dixiadang de jianli yu fazhan," 69.

southwestern Zhejiang, reestablishing the school in Songyang County.[92] Along the way and while in Songyang, students published a mimeographed daily newspaper and organized the "Awaken the People Theater Troupe" (Xingmin jutuan), "Harmonious Sounds Chorus" (Jiejie geyongtuan), "Sound of the Hoe Literature and Art Troupe" (Chusheng wenyituan), and "Wartime Art Troupe" (Zhanshi huihua tuan). Each group from Xianghu Normal sought to raise popular consciousness about the war and to mobilize people for wartime activism.[93] A similar relocation of educated youths from core areas occurred in northern Jiangsu. According to the CCP's *New China Daily* (*Xinhua ribao*), "Because there has been a loss of many central cities during the resistance war—especially with Shanghai and Nanjing falling one after the other—many youths enthusiastic about national salvation returned in succession to northern Jiangsu's inland areas from the cities. In widespread villages, they have spread the seeds of resistance to Japan and national salvation."[94]

In the first months of the war, efforts by students and recent secondary graduates in Jiangsu and Zhejiang to produce anti-Japanese propaganda and mobilize for resistance were largely independent. In most instances, neither political party directed educated youths, and, as we have seen, their activities generally followed prewar precedents. In fact, Nationalist Party authorities and government-aligned school administrators often obstructed students' activism, conforming to the pattern of Nationalist Party treatment of student protesters during the prewar period. At Xinchang Middle in 1937 and 1938, for example, principal Zhang Mengdan still called upon students to "study to save the nation" and threatened those undertaking social activities with expulsion.[95] At Funing Middle School in northern Jiangsu in

---

92. Xiaoshan xianzhi bianzuan weiyuanhui, *Xiaoshan xianzhi*, 799; Zhejiang sheng Lishui diqu geming wenhua shiliao zhengbian bangongshi, 189. High schools from the occupied coastal core areas of Hangzhou, Jiaxing, and Huzhou also relocated inland to Lishui at the start of the Sino-Japanese War to set up unified secondary schools (Gao and Lou, "Kangzhan shiqi de shengli liangao"). Cf. Pan Pi et al., 20.

93. Zhejiang sheng Lishui diqu geming wenhua shiliao zhengbian bangongshi, 175, 188–90.

94. *Xinhua ribao* (April 13, 1938), as summarized in Pan Pi et al., 16.

95. Yu Yueyin, 256. Cf. Zhejiang sheng Lishui diqu geming wenhua shiliao zhengbian bangongshi, 271.

November 1937, anti-Japanese student protests were disrupted by local Nationalist government authorities. Police attacked student protesters, detained a number of students, and locked the campus so students could not go in or out. Though the students were released, a number were expelled, and the school authorities declared a "temporary vacation" (*linshi fangjia*) after students vandalized the campus in anger over the government's actions. The practical result was that students dispersed to their home areas to undertake anti-Japanese activities there.[96]

In contrast to the Nationalist Party, the CCP encouraged student activism, developed youth organizations linked to the party, and systematically sought to recruit students and educated youths as cadres. The party's goal was clearly expressed in a July 1938 letter to local party branches from the CCP's Taizhou Special Committee in coastal southern Zhejiang: "The party must take advantage of all possibilities to mobilize the broad youth masses (*guangda de qingnian qunzhong*) and organize youth national salvation groups, youth wartime service groups, student resist-Japan associations (*xuesheng kangri hui*), children's army service teams (*tong jun fuwu tuan*), and so forth, and in our work recruit many outstanding youths into our party."[97] As during the 1920s, the party had to "organize" and "recruit" more than "inspire" educated youths, for both Jiangsu and Zhejiang witnessed widespread nationalist activism among students and recent graduates during the mid-1930s and the first years of the war.

The CCP systematically directed youth organizers to court politically active educated youths and to infiltrate nationalist organizations in order to form party-affiliated groups. At Lishui's Chuzhou Middle School, where students had engaged in anti-Japanese nationalist activities since the start of the war, young party member and former Chuzhou Middle student Dong Xinchang was directed by party leaders to initiate a branch of the National Liberation Vanguards of China (Zhonghua minzu jiefang xianfengdui; hereafter NLVC). Dong first set up a Chuzhou Middle Fellow Students' Association (Tongxuehui) to develop NLVC group members. Through

---

96. For the incident in Funing, see Pan Pi et al., 6. For examples from other areas, see ibid., 3, 25, 30.

97. As quoted in Taizhou diqu wenhua ju geming wenhua shiliao zhengji bangongshi et al., 21.

discussions about current events in this front organization, Dong eventually built up the membership of the NLVC group, which subsequently took the lead in anti-Japanese activities at the school. During school breaks NLVC members carried out anti-Japanese propaganda activities in their home areas outside of Lishui and became ready recruits for CCP membership.[98] In this way, the CCP built on existing patterns of youth wartime service and nationalist activism to draw students and other educated youths into the party. Significantly, the primary focus in these efforts at mobilization was anti-Japanese resistance rather than the program of social revolution that had captured many youths' imagination and attracted them to the CCP during the 1920s.

The CCP also set up mass organizations and training schools to coordinate young people's nationalist activism and to connect them with the party. For instance, the Fujian-Zhejiang Border Area Resist Japan National Salvation Cadre School (Min-Zhe bian kangri jiuwang ganbu xuexiao) was formed by the CCP's provisional border area provincial committee in January 1938.[99] The school posted a recruiting advertisement in the *Wenzhou Daily News* (*Zhe'ou ribao*) and admitted a class of 150 youths, most of whom were middle school students or graduates from southern Zhejiang counties and the prefectures of Taizhou, Shaoxing, and Ningbo.[100] The focus of the school was preparing students for guerrilla resistance warfare and propaganda activities, and students conducted anti-Japanese propaganda activities in surrounding villages. But the school's curriculum included many familiar activities such as physical education, military drill, and surveys of conditions in local society, and students were organized into ten-person squadrons that formed the basic units of school life. Thus, the CCP adapted fundamental elements of the civic training in Nationalist-controlled schools for its own project

---

98. Zhejiang sheng Lishui diqu geming wenhua shiliao zhengbian bangongshi, 271–74. For Chuzhou Middle School students' prior anti-Japanese activities, see ibid., 88; Ji Le, "Zhanshi Lishui." Cf. Yu Yueyin, 256–57.

99. Huang Xianhe, "Yi Min-Zhe bian kangri jiuwang ganbu xuexiao," 1–5. Huang Xianhe (previously named He Wei) had been assistant principal of the school. For a similar school in northern Jiangsu, see Pan Pi et al., 17.

100. For a biography of one student, Yi Bing from Sheng County in Shaoxing Prefecture, see Shaoxing shi wenhuaju et al., 157–58.

of mobilization for national defense, suggesting the congruence between those modes of training and mass nationalist politics. After the cadre school concluded on March 15, 1938, "most of the students split up to go to various places in southern Zhejiang and throughout the province to participate in the anti-Japanese National Salvation Movement. Of these, about ten people were led by [regional CCP leaders] Lian Zhen and Lin Fu to form a New Fourth Army Mobile Propaganda Team (Liudong xuanchuan dui), deeply penetrating the townships and towns of counties such as Pingyang, Rui'an, and Taishun."[101] Such institutions drew students into the CCP by providing them with outlets for active citizenship centered on national service that had been promoted in the schools throughout the Nanjing decade.[102]

The CCP also courted local nationalist activists who were recent graduates of middle or normal schools and were working in local communities. Because normal-course graduates were expected to serve as local schoolteachers in order to repay the tuition and fees from which they were exempt during their schooling, many served as teachers or administrators in rural primary schools.[103] Graduates of second- or third-tier middle schools who could not afford or test into the better universities frequently also taught in rural schools or took other administrative jobs in their home areas in the regional periphery. At the start of the war, educated youths who were working in rural towns and villages and took the lead in mobilizing their local communities for the Sino-Japanese War were supported and actively recruited by the CCP.[104] For example, Tang Shuhong (1915–39), a native of Jiangsu's Shuyang County, attended Huai'an Middle School and subsequently graduated from the normal course at Jiang-

---

101. Huang Xianhe, 5.

102. CCP-sponsored women's movement and national service organizations similarly provided secondary students and other "progressive youths" (*jinbu qingnian*) with outlets for wartime civic action while drawing them into the CCP's infrastructure. E.g., Wang Xiangqin, "Xuesheng shuqi fuwutuan gaikuang"; Wang Xuancheng, comp., "Kangri shiqi Jinhua de fuyun"; Zhou and Zhou.

103. Jiangsu sheng jiaoyuting, *Jiangsu jiaoyu gailan*, 3: 74–75, 105, 113, 120, 147, 174–75, 180–81, 208–9, 220, 230; *Jiangsu shengli Suzhou zhongxuexiao gaikuang*, 14–15.

104. For examples, see Shaoxing shi wenhuaju et al., 157–58, 163–64; Zhonggong Jiangsu shengwei dangshi gongzuo weiyuanhui et al., 222–25, 277–80.

su's Donghai Middle School in 1933, after which he returned to his home of Tanggou to teach.[105] In 1937, at the start of the war, Tang led a reading group based in his school that engaged in anti-Japanese mobilization by singing patriotic songs, lecturing, and performing dramas. Tang was also instrumental in forming a local self-defense militia of peasants in the area surrounding Tanggou in 1938. For a time Tang carried out his activities independently, but in January 1939 he joined the CCP and led his militia to serve as a guerrilla regiment of the Eighth Route Army.[106] The CCP provided an institutional framework for Tang's active service in the national struggle. In these and many other cases, educated youths whose formative experiences were in lower Yangzi region secondary schools threw themselves into resistance activities in local communities in northern Jiangsu or southern Zhejiang and found an outlet for their nationalist efforts within the CCP.

Like unoccupied areas of the regional periphery in northern Jiangsu and southern Zhejiang, "solitary island" Shanghai became a key center of mobilization for educated youths. Shanghai's concession areas remained governed by foreign authorities and thus insulated from direct Japanese control from the summer of 1937 to the beginning of the Pacific War in December 1941. Consequently, many of the city's premier schools continued to operate, and secondary schools from occupied Shanghai and other areas of Jiangsu, including Suzhou Middle, Yangzhou Middle, and Songjiang Girls' Middle, relocated to the foreign concessions.[107] At schools such as Provincial Shanghai and private Minli and Nanyang middle schools, students' patriotic activities during the early years of the war focused on preserving their schools' autonomy by resisting registration with the collaborationist Wang Jingwei government and waging protests against pro-Wang and pro-Japanese sympathizers in their

---

105. Zhonggong Jiangsu shengwei dangshi gongzuo weiyuanhui et al., 34–38.

106. Tang Shuhong's recruitment was part of a conscious effort by two CCP operatives to cultivate young nationalist activists as party members. See Pan Pi et al., 35.

107. Hang Wei, "Shanghai zhongdeng xuexiao de fazhan," 14–15; *Jiangsu sheng Yangzhou zhongxue jianxiao jiushi zhounian jiniance*, 3; Zhonggong Jing'an quwei dangshi ziliao zhengji weiyuanhui bangongshi and Zhonggong Jing'an quwei jiaoyu gongzuo weiyuanhui, *Shanghai shi Jing'an qu jiaoyu xitong geming douzheng shiliao*, 5.

midst.[108] These schools also became hosts to party organizations, which successfully developed student nationalists into party members or activists for work in Shanghai or outside base areas. In September 1942, for example, in anticipation of intensified arrests of student activists by Japanese and collaborationist forces, 24 party members and student activists from Minli, Shanghai, and Nanfang middle schools, as well as from St. John's University, transferred to CCP base areas.[109]

The pattern of post-1927 CCP recruitment of educated youths in Jiangsu and Zhejiang seems to have been distinct from that of other regions. For instance, Xiaoping Cong's insightful analysis of CCP recruitment of Shandong and Hebei normal school students charts a somewhat different trajectory, reflecting divergent regional conditions.[110] In those areas of limited Nationalist Party control, CCP cells and members operated with relative freedom in rural normal schools. As students were exposed to Marxist thought and party propaganda, many joined the party before the war and were then assigned to CCP military forces or mobilized for activism in local communities during its early phases.

By contrast, in much of Jiangsu and Zhejiang, the Nationalist Party seems to have effectively uprooted or at least severely constrained CCP organizers during most of the Nanjing decade.[111] As a result, during the early stages of the Sino-Japanese War, students attending school in contested or threatened rear areas, students driven

---

108. Zhonggong Jing'an quwei dangshi ziliao zhengji weiyuanhui bangongshi et al., 6, 32–35; Zhonggong Shanghai shiwei dangshi ziliao zhengji weiyuanhui, *Huo hong de qing chun*, 56–65.

109. Zhonggong Jing'an quwei dangshi ziliao zhengji weiyuanhui bangongshi et al., 41. Cf. Zhonggong Shanghai shiwei dangshi ziliao zhengji weiyuanhui, 137–38. The CCP also set up Jiancheng Middle School in Shanghai in July 1939 to facilitate movement of party members between eastern Zhejiang and northern Jiangsu and to recruit nationalistic students for anti-Japanese resistance (Zhonggong Shanghai shiwei dangshi ziliao zhengji weiyuanhui, 145–49).

110. Cong, "Localizing the Global," 372–76, 400–417.

111. For accounts of the decline in CCP organizational activity in Jiangsu and Zhejiang from the early 1930s through 1937, see Geisert, *Radicalism*, 198–203; Schoppa, "Contours of Revolutionary Change," 783–85. In Shanghai, CCP activities were scaled back dramatically after 1928 and eventually came to focus mostly on national salvation work, which gathered the greatest momentum between 1936 and 1938 (Stranahan, *Underground*).

to the rural periphery from their schools in core area cities, and recent graduates working in unoccupied local areas all undertook independent nationalist propaganda and self-defense work. Once the CCP was able to operate more freely under wartime conditions, it effectively appealed to students and recent graduates by aligning its program with the forms of nationalistic civic action that youths themselves had initiated.[112] Viewed from educated youths' perspective, participation in the CCP or party-aligned organizations enabled them to live out the active form of national citizenship that they had crafted for themselves while they studied in secondary school during the Nanjing decade.

IV. CONCLUSION

Throughout the Republican period, different groups and cohorts of lower Yangzi region secondary students drew creatively from their schools' eclectic civic education and citizenship training to fashion roles for themselves as active citizens. China's Leninist parties persistently sought to harness students' civic action for their own political projects, which included the National Revolution, national construction, and anti-Japanese resistance. In each instance, specific groups of students responded and became basic cadres of revolutionary projects. During the 1920s the dense infrastructure and interpersonal networks in the lower Yangzi region exposed students to texts and people that encouraged their social activism and later drew them into the Nationalist and Communist parties. In the 1930s, students at elite, core-area secondary schools pursued "study for national salvation" and came to use their academic and technical skills to contribute to China's economic development and then its war effort. Students and graduates in threatened local communities who responded to Japanese invasion with nationalist activism found within the framework of the CCP and its mass organizations an outlet for wartime contributions that incorporated many dynamics of civic action promoted in Nationalist government educational policies and party-supervised schools.

---

112. In Shandong's Zuoping County, as well, wartime conditions allowed for successful CCP recruitment of nationalist intellectuals and students at Changshan Middle School. Thøgerson, 122–24.

Both the Nationalist and the Communist parties were heir to what James Scott has called the "high modernist" goal of Leninism to map out and direct all stages of the revolutionary project from the center. But party mobilization was most effective when it accommodated popular energies for revolutionary action—just as Lenin had been able to do in 1917, despite his penchant for centralized planning—by adjusting to different social groups' motivations for and dynamics of civic action.[113]

Educated youths became main actors in several of the major political dramas of the Republican period, but students' civic activism has also had long-term and more diffuse, though no less profound, effects that are harder to trace. Socially and culturally, lower Yangzi region students' construction of themselves as "core elements," by virtue of their academic learning, cultural literacy, and political leadership, re-inscribed within the national community a hierarchical distinction between learned elites and the illiterate masses. At the same time, by participating in nationalist protest, enacting community self-government, engaging in social service, performing civic rituals, or embodying new forms of civility, students and educated youths demonstrated the practices of citizenship for a wider public. Students' performance of some of the lessons of civic education they encountered in school extended specific models of citizenship out into their communities. Young students acting as modern citizens expressed in public the hopeful possibility that China would one day be a modern nation-state, with a working polity, egalitarian society, and civilized public. Students' embodiment of modern citizenship may explain why disciplined Boy Scouts raising the national flag, nationalistic young protesters, and educated youths instructing the people became some of the most powerful images of the Republican period.

---

113. Scott, *Seeing Like a State*, chap. 5.

The Trajectory of Modern
Chinese Citizenship

In reconstructing the project of educating lower Yangzi region youths to be citizens, this book explains how citizenship was configured in Republican China and how two generations of educated youths acted as citizens. Reinforcing messages in history and geography instruction, student self-government associations, and instruction about the structure of the social order taught that citizenship entailed concrete action taken to benefit society. But schools also presented students with diverse avenues of civic action that were taught through Scouting, school routines, civic ritual, and military training. Students drew creatively from these school-sanctioned approaches to fashion their own modes of citizenship that enabled them to constitute themselves as "core elements" in Republican society and politics. As prototypical active citizens, students were persistently targeted for mobilization by both of China's Leninist parties, which sought to align their political projects with students' patterns of civic action.

Republican Chinese approaches to engaged citizenship drew eclectically from foreign models. Continental European approaches to civic republicanism, Anglo-American Scouting, Stalinist modes of civic ritual, and fascist-style military training were variously incorporated into Republican-period civic education and citizenship training. But these diverse and often contradictory approaches to citizenship were juxtaposed with each other in distinctive ways. They were also integrated with practices and perspectives that persisted from China's late imperial period, such as moral cultivation through mentoring relationships and viewing the sociopolitical order as grounded in the ethical choices and actions of individuals.

These juxtapositions and processes of merging generated forms of citizenship that do not neatly parallel any other national pattern. For example, in contrast to the liberal model of citizenship that is most prominent in the minds of many Americans, Chinese Republican elites consistently downplayed protection of individual rights and freedoms, calling instead for more direct forms of political participation and selfless contributions to an assumed common good. This penchant for engaged civic action was rooted in evolving patterns of local-elite management that themselves grew out of a long statecraft tradition of Confucian discourse. The resulting civic republicanism later became coupled with Leninist forms of party tutelage after 1927, leading to party-state guidance in self-government, civic ritual, and social service. Although Nanjing-decade civic training also incorporated elements such as an organic conception of society and intensive military training, it cannot be easily equated with a fascist or state-socialist model of citizenship either. For schools still included modes of training based on progressive education models and pragmatist approaches to social change, which promoted students' actions as creative and independent individuals, and forms of cultivation influenced by late imperial Confucianism that encouraged moral autonomy and more proximate personal loyalties. Further, despite clear expressions of Han cultural chauvinism in some history and geography textbooks, the territorially based nationalism taught widely in Republican secondary schools was quite distinct from, for example, Nazi forms of racialist nationalism. Republican-period civic education borrowed liberally from global discourses of citizenship but, because of persistent historical legacies and China's specific social and political dynamics during the first half of the twentieth century, it incorporated disparate elements in distinctive ways.

The Republican project of training educated young citizens bridged elite intellectuals' early efforts to define citizenship during the last decade of the Qing dynasty and the dramatic growth of popular engagement in national affairs and public life that took place during the Sino-Japanese War and the first decades of the PRC. By stressing this long-term trajectory of change that reached from the end of the Qing through the Mao period (1949–76), I build on a conception of the Chinese revolution first formulated by Mary Wright in her classic introduction to *China in Revolution*. There, she described the period between 1900 and 1913 as the "first phase" of the Chinese revolution, which she saw stretching into the Mao period.[1] Much of what Wright saw to be revolutionary in that period and later was the rapid growth of nationalist consciousness and active popular participation in civic and political life.

> It seems to me that there has been one single revolution, the salient features of which were rooted in this early twentieth century experience: the persistent prominence of educated youth, the unusually prominent role of women, the insistence on strong leadership and revolutionary discipline (for the initial absence of these had been tragic), suspicion of all foreign powers, and finally a kind of "great leap" psychology, a disinterest in classic Marxism and other theories that assumed a slow and to us rational development of society, and a conviction that by superhuman effort of an indoctrinated elite, China could bypass the usual stages and achieve its own kind of good society through sheer application of human energy and willpower.[2]

Like Wright, I see this historical period as a long revolutionary process that pivoted on creating modern citizens as publicly active people (*gongmin*) who were committed to serving the nation. However, where Wright emphasized the origin, or "roots," of this revolutionary process in the final decade of the Qing, I have explored how the content and trajectory of citizens' "superhuman effort" were formulated and contested multiple times during the Republican period, leading to complex social and political outcomes. This book has focused on one period and dimension of this overall process, analyzing how Republican civic education consolidated and spread new ideals of public action and how many lower Yangzi region students became dedicated to acting as citizens. Sketching out this longer revolutionary trajectory will clarify the pivotal nature of the story about citizenship training and student civic activism told here by relating that education and activism to broader social and political processes. It will also suggest how the dynamics of active citizenship reconstructed in this book both continued and changed up through the end of the Maoist period.

Extending Wright's suggestive thesis, then, I propose a new periodization of the Chinese revolution based on stages in the process of

---

1. Wright, introduction to *China in Revolution*, 62–63.
2. Ibid., 62.

forming China's people into citizens who were engaged with national issues and participated actively in civic life. Reconstructing this trajectory of growing popular engagement in national life and civic action transforms our understanding of twentieth-century Chinese history. It suggests that disparate, and sometimes apparently contradictory, social and political projects—from the progressive education movement of the 1910s and 1920s to the Nationalist Party's New Life Movement to the CCP's rural revolution—contributed to a common process of making a nation of citizens who could and would participate in public life. The activist approach to citizenship developed over the first half of the twentieth century both culminated and reached a point of crisis in Maoist radicalism. The process of formulating a Chinese version of citizenship and creating Chinese citizens transformed many features of social and political life, ranging from table manners to political dynamics. Further, the ideas and practices associated with citizenship created new opportunities (and demands) for diverse social groups to craft public identities for themselves and chart new courses of social and political action.

Extending Republican Citizenship

The Introduction to this book demonstrated that late Qing intellectuals, political leaders, and social elites introduced ideas of modern citizenship to China because they felt that conventional late imperial patterns of ethics and politics gave them limited tools for confronting China's internal and external crises. Their primary goal was securing popular engagement in national affairs to strengthen the polity. But intellectuals and elites also debated and experimented with many different approaches to political participation and association, opening up a discussion of the nature of citizenship that would continue for several decades. During the late Qing, social and political elites were the primary group to act as citizens by forming their own professional associations, social organizations, and political parties, and by participating in Qing government-sponsored representative assemblies.[3] Elite groups also used the press and schools to spread ideas about the nation and citizenship to other sectors of society,

---

3. Fewsmith, "From Guild to Interest Group"; Rankin, *Elite Activism*; Thompson, *China's Local Councils*.

and they sought to mobilize a broader spectrum of society for mass protests such as the anti-American boycotts of 1905 and the Railway Rights Recovery Movement (1904–7, 1910–11).[4] However, social and political elites maintained exclusive control over nascent democratic political institutions and treated other social groups, including students, as a force to be led rather than as fully entitled citizens who could act on their own.[5]

The early Republican period witnessed continuing debate over the nature of citizenship and redoubled efforts to draw a broader spectrum of people into the realm of public action. As this book has demonstrated, political leaders, intellectuals, and educators sought to educate students to be citizens, and they experimented with various models of political, cultural, social, and national citizenship. Moreover, many lower Yangzi region students came to self-identify as citizens and take part in public life in the ways described in the previous chapters.

But students were by no means the only group to use the language and practices of citizenship to enter public life during the early Republican period. In urban China, many different social groups—including professionals, small businessmen, and industrial workers—contributed to creating a growing arena for civic action and negotiated the parameters of citizenship in practice. Civic rituals, mass protests, native-place organizations, and commercial associations all became sites where disparate social groups could claim a public role and debate the dynamics of citizenship.[6] Students, as we have seen, played a pivotal role in this process by using education and various techniques of mass mobilization to expose other social groups to ideas about national identity, social reform, and political participation, and to draw them into civic action. As part of this ambitious project of creating mass citizenship in China during the 1910s and 1920s, some lower Yangzi region students formulated an activist

---

4. See Judge, "Citizens or Mothers of Citizens?"; idem, *Print and Politics*; Wang, *In Search of Justice*; Wright, ed., *China in Revolution*.

5. E.g., Esherick, *Reform and Revolution*; Ichiko, "The Role of the Gentry"; Judge, *Print and Politics*.

6. Carter, *Creating a Chinese Harbin*; Dunch, *Fuzhou Protestants*; Gerth, *China Made*; Goodman, "Democratic Calisthenics"; Harrison, *Making*; Perry, "From Paris to the Paris of the East"; Strand, *Rickshaw Beijing*; Tsin, *Nation*.

approach to citizenship that combined an open, participatory form of civic republican self-government, social mobilization in pursuit of mass equality, and civic rituals that symbolically enacted a politics of mass protest. We have also seen many examples where female students eagerly joined in this process of expanding mass citizenship by participating in self-government, marching in ritualized protests, and mobilizing the masses for the National Revolution.

Yet even when a wider range of social groups joined in commemorative assemblies or civic rituals during the early Republic, the participants were primarily urban social groups or the rural elite. Moreover, for nonelite groups civic action was often limited to symbolic acts of joining in marches or assemblies or changing their modes of dress and behavior. These expressions of civic participation marked a dramatic change from the late imperial period. But, outside of exceptional moments such as workers' central role in the establishment of the Shanghai Citizens' Representative Congress in 1927, nonelite groups were frequently still excluded from other dimensions of citizenship, such as participation in political institutions, civic associations, and voluntaristic social service.

To whatever extent mass participation in public life expanded in China during the 1910s and 1920s, avenues of civic participation for all social groups narrowed during the Nanjing decade. Nationalist Party state corporatism and party tutelage placed constraints on urban civic association and stressed control over autonomy in local self-government.[7] Even in mass assemblies for national commemoration days, which had served to extend nonelite civic participation during the 1910s and 1920s, the groups involved narrowed considerably. As we saw in Chapter 6, most state-sponsored commemorative assemblies during the 1930s included a limited number of representatives (*daibiao*) of institutions, organizations, schools, and each social circle (*gejie*), a formula that seemed intended to limit true mass participation. The Nationalist Party leadership instead focused its nation-building efforts on exhorting the people to change their daily habits through campaigns such as the New Life Movement and to contribute to building the national economy.[8]

---

7. Fewsmith, *Party, State, and Local Elites*, 164; Kuhn, "Local Self-government."
8. Dirlik, "Ideological Foundations"; Friedman, "Civilizing the Masses"; Zanasi, "Chen Gongbo," 142–43.

One major exception to this narrowing trend seems to have been schools. As we have seen, secondary schools continued to provide students with multiple arenas for positive civic action, albeit under party-state supervision, through self-government activities, military training, state-sponsored civic ritual, and voluntaristic social service activities. Students were being prepared to participate in many different arenas of public life, even as the Nationalist Party increasingly excluded most social groups from taking part in meaningful civic action.

By contrast, the threat to national survival posed by the Sino-Japanese War drove all kinds of sociopolitical elites to commit to a nationwide mobilization of China's people. Chang-tai Hung, for instance, demonstrates that the goal of early wartime propaganda efforts was to organize all social groups to contribute to the resistance war.[9] This entailed using spoken dramas, stories, songs, cartoons, and newspapers to teach people to imagine themselves as members of the national community and citizens who could take positive action in public life. Author Lao She captured the core message of these efforts to extend the message of citizenship to nonelite groups in his patriotic drum song, "Wang Xiao Drives a Donkey": "As he [Wang Xiao] turns around and looks at his home again,/He sees his mother standing stiffly at the doorstep./Choosing between being a loyal citizen and a filial son is hard,/But [at last] he stamps his feet and leaves his hometown."[10] Peasants and other nonelite social groups were called to leave the parochial communities of their families and villages, either metaphorically or in fact, to become citizens dedicated to public action for the nation's welfare.

Students, as we have seen, played an active role in mobilizing a wide cross-section of society to take political action on behalf of the nation, thereby expanding the community of Chinese people who viewed themselves as citizens. In the lower Yangzi region between 1937 and 1939, educated youths who were enrolled in local schools in contested areas or who had fled to rural inland areas spread anti-Japanese propaganda and organized self-defense militias, independently at first and then increasingly under CCP guidance. Insofar as they successfully mobilized local communities, their efforts contributed to expanding the universe of people who considered themselves

---

9. Hung, *War and Popular Culture.*
10. Ibid., 200.

to be citizens of the nation (*guomin*) and people who participated in public life (*gongmin*).

The Nationalist Party proved ambivalent about mass mobilization efforts after the war resolved into a stalemate in 1939.[11] By contrast, the CCP persistently spread the message of national citizenship and anti-Japanese resistance while also creating institutions in rural villages that allowed peasants to participate fully in local politics and engage in social reform.[12] Hung, for instance, suggests that the CCP's wartime propaganda, on one hand, promoted national consciousness and anti-Japanese resistance. On the other, it also educated peasants in new forms of political action that included building a new society, establishing gender equality, founding peasant associations, holding local elections, improving rural hygiene, maintaining harmonious relationships between the Red Army and the people, and trusting the correct leadership of Mao and the CCP.[13] This vision of citizenship included identification with the nation and mass participation in public life under party leadership.

Both before and during the Sino-Japanese War, in rural soviets the CCP coordinated peasants' full involvement in political life, through participation in militias, mass associations, and local elections, and by assuming leadership roles in their communities. Edgar Snow, in his sympathetic portrayal of the central base area in Yan'an in 1936, illustrates that the soviet's interlocking institutions of representative government, mass organizations, and local militias made all members of local communities full participants in civic life, under CCP guidance. In Snow's words, "The aim of soviet organization obviously was to make every man, woman, and child a member of something, with definite work assigned to him [*sic*] to perform."[14] Chen Yung-fa has also argued persuasively:

The CCP's wartime program improved the livelihood of many peasants, but in the process the peasants also acquired new responsibilities in the mass associations, militia, administration, and Party branches. Furthermore, peasants were required to provide manpower and revenues for the war, and mili-

---

11. Ibid., 214–15.
12. The classic formulation of the argument that the CCP drew peasants into national life through wartime resistance is Johnson, *Peasant Nationalism*.
13. Hung, 230–32.
14. Snow, *Red Star over China*, 220–22.

tary service, in spite of its attractiveness for the very poor, required them to leave home and to risk their lives in combat.[15]

Chen describes here a full engagement in public life that echoes in many ways a civic republican approach to citizenship, where citizens are obligated to fulfill wide-ranging commitments to public service. Significantly, it was a civic order in which the party defined the parameters, goals, and institutions of public activity, whether political, military, or social. Suzanne Pepper has described a similar situation for the Civil War period, arguing that the struggles of land reform not only unseated the rural elite; they also taught peasants how to be political actors and introduced peasant associations and other village organizations that served as a new political infrastructure and mobilized local communities to contribute to the war effort.[16]

Wartime mobilization, then, made large segments of Chinese society conceive of themselves and act as citizens for the first time, so that many came to believe "'It is shameful for a person to be a noncitizen.' 'Without citizenship, one will be shunned.'"[17] Through elite- and party-led mobilization, growing numbers of people identified themselves as members of the national community and engaged in civic action ranging from voting in village and township elections to participating in mass organizations to assuming administrative positions to fighting in the militia or the army. As in prewar conceptions and practices of citizenship, social service and contributions to public welfare were emphasized more than open policy debate or protection of individual rights and interests. Indeed, in all of these instances, the forms of civic and political organizations and the nature of political action were defined and dictated by the CCP. The Communist Party's emphasis on party-led civic participation can be seen as continuous with the Nationalist Party's efforts to channel students' civic action into self-government organizations, civic rituals, service programs, and youth organizations that were managed by

---

15. Chen, *Making Revolution*, 16. For a complete discussion of the civic and party institutions the CCP introduced to rural villages, and their relationship to the dual projects of social revolution and the resistance war, see ibid., chaps. 3–6. Cf. Friedman, Pickowicz, and Selden, *Chinese Village*, chap. 2.

16. Pepper, *Civil War in China*, 329–30.

17. Chen, *Making Revolution*, 238. Citizenship, in this instance, was equated with registering to vote in village elections.

the party. But the CCP departed from Nationalist Party precedents by seeking to mobilize for social service a much broader cross-section of Chinese society, and it drew those people into a more sustained and intense engagement in public life than anything attempted by the Nationalist Party.

Socialist Citizenship

Similar patterns of party-led mobilization for civic action continued throughout the first decade and a half of the PRC, embracing almost everyone within the society. The CCP perpetuated or created anew institutions to both inspire and manage civic activity. In rural areas, the party committees and peasant and women's associations of the Sino-Japanese War and Civil War periods continued and were expanded to villages throughout China. Periodic mass labor and rectification campaigns mobilized whole communities either for development projects or for political activism. Mass labor mobilization for collective projects was most common during the years of the Great Leap Forward (1958–60), when huge campaigns for pest control, irrigation, militia organization, and ill-fated backyard steel production were carried out.[18] In these campaigns, all rural residents were to dedicate themselves to public service on behalf of their communities and the nation as a whole.

Rapid collectivization of agriculture during the mid-1950s further broke down barriers between private and public life in rural areas. Under collectivization, most forms of production and consumption became public activities that had direct impact on collective welfare. By the end of the 1950s, rural people worked together on collectively held land and shared the yield of that land. In this context, each individual's productive work was also transparently a form of public service, and the welfare of individuals and families was inseparable from the common good in ways it had never been before.[19]

In urban communities as well, political organizations and state control of the economy led to high degrees of engagement in public life in ways that were always guided by the party. Within residential

---

18. Friedman, Pickowicz, and Selden, chap. 9; MacFarquhar, *The Origins of the Cultural Revolution*, vol. 2.

19. Chan, Madsen, and Unger, *Chen Village*; Parish and Whyte, *Village and Family*.

neighborhoods, small groups and neighborhood committees provided an organizational infrastructure for state-directed local self-government activities.[20] Work units organized workers in both state-owned and collective enterprises into permanent social and political groups that were, to varying degrees, self-regulating and under CCP direction.[21] Both work units and neighborhood organizations mobilized people for mass campaigns, such as enforcement of birth control and sending urban educated youths to the countryside, which demanded urban citizens' commitment to act for the public welfare.[22] Worker pickets or cross-class militias were organized under trade unions, the military, or the public security apparatus to maintain order, ferret out class enemies, or assist in defense.[23] In urban areas, state ownership and the work allocation system (*fenpei zhidu*) broke down the distinction between labor and social service, just as collectivization did in rural areas. Throughout the Mao period and beyond, most urban residents were appointed to positions in state-owned or collective enterprises for work. Because people's career paths were governed by state needs, any individual's work was directly linked to the broader public welfare, which was now seen as far more important than the success of their immediate company/unit or individual fulfillment.[24]

To prepare people for work lives of public service, education during the early Mao period came to focus on instilling in youths an awareness of the broader community and their responsibility for contributing to it. The emphasis on social service in Republican-period civic training and the voluntarism embedded in the ideals of student self-government were echoed in Mao-era education, which presented an image of the ideal citizen as selflessly dedicated to serving the common good.[25] Ridley, Godwin, and Doolin summarized this vision of Chinese socialist citizenship in their study of primary school textbooks during the 1950s and early 1960s in the fol-

---

20. Parish and Whyte, *Urban Life*, 22–24.

21. Lü and Perry, "Introduction: The Changing Chinese Workplace," 8–12.

22. Parish and Whyte, *Urban Life*, 240–45.

23. Perry, *Patrolling the Revolution*, chap. 4. According to Elizabeth Perry, citizenship as constituted through militia activities was variously defined in terms of class or community, depending on the political circumstances.

24. Parish and Whyte, *Urban Life*, 39–42.

25. Munro, *The Concept of Man in Contemporary China*; Whyte, *Small Groups*.

lowing way: "Within the 'new' society, the child-citizen is made aware of his particular social and political responsibility to contribute to the overall good of the community. Any goals of personal achievement must remain secondary to the pre-eminent responsibility of contributing to the common good of the new society."[26] Model soldier Lei Feng, or his creators, captured this all-encompassing commitment to social service with his declaration that he would be happy to be an anonymous screw working in the machine of the nation-state.[27]

Civic training in Taiwan during the 1950s and 1960s similarly emphasized student-citizens' dedication to the social welfare in ways that echoed the calls of the Republican period. In the words of one study of child socialization, "Children in Taiwan, even at a very young age, have a developed sense of their membership in society as a whole. Furthermore, at least ideally, they conceive of themselves as having a primary duty and loyalty to society."[28] Beyond this commonality with Mao-era PRC civic training, though, Nationalist Party citizenship education on Taiwan emphasized two of the dynamics that had characterized it during the Nanjing decade. Specifically, under martial law the Nationalist Party continued to foster a military ethos, using the threat of invasion by the PRC as justification for implementing militarized youth training and developing an elaborate civil defense system.[29] Further, schools placed great emphasis on cultivating loyalty to leaders, culminating in devotion to Chiang Kai-shek and Sun Yat-sen, directly continuing dynamics common to military training during the 1930s.[30]

When viewed in longer historical context, Mao-era civic action in the PRC can be seen as the culmination of early twentieth-century Chinese elites' efforts to construct citizens who would be publicly engaged. We can see many parallels between Mao-era civic action and the ideal of active citizenship that had developed by the 1930s and was voiced and practiced by many lower Yangzi region educated

---

26. Ridley, Godwin, and Doolin, *The Making of a Model Citizen*, 186–87. Cf. Chan, *Children of Mao*.

27. For a discussion of Lei Feng and other revolutionary exemplars from the Maoist period, see Sheridan, "The Emulation of Heroes."

28. Wilson, *Learning to Be Chinese*, 44.

29. Bullard, *The Soldier and the Citizen*.

30. Wilson, *Learning to Be Chinese*, chap. 2.

youths. Citizens during the 1950s and 1960s were expected to view themselves as integral members of the national community and to commit themselves to serving the public good. Institutions ranging from neighborhood committees and work teams to the Communist Youth League and peasant associations organized citizens' public service. All these activities occurred under the guidance and prompting of the Leninist party-state. But the CCP effectively broadened participation in public life so that members of nearly every social group throughout China came to consider themselves and act as citizens in the sense of being publicly engaged people (*gongmin*). The CCP expanded mass participation in public life in ways the pre-1949 Nationalist Party did not, and perhaps could not, because of its successful organization of overlapping networks of mass organizations and its control over the economy. Both features allowed few opportunities for withdrawal into private life and personal networks.[31]

However, Mao-era public life diverged from Republican precedents insofar as the cult of Mao and class struggle served to enforce and shape public activism and social service. The Mao cult's ideal of sociopolitical order sought to constitute all members of society as part of a homogeneous mass that would be dedicated to collective action for the common good and responsive to the direction of the exalted leader, Chairman Mao. The Maoist social model entailed the atomization of individuals and their separation from personal commitments as well as their recombination as a mass public that could respond to party leadership. In a seminal article written on the eve of the Cultural Revolution, Ezra Vogel argued persuasively that in the PRC the impersonal and atomistic participation of people in public life as homogeneous "comrades" had displaced many close personal connections, or "friendship." Fundamental to this process were the struggle and rectification campaigns of the early Mao years, which had cut close personal ties by demanding that people prove their loyalty to the state by providing damning information about friends and relations. Given the dangers that close relationships posed to both oneself and others in an environment of cyclical campaigns, people understandably came to act as "comrades" rather than as friends. According to Vogel, comradeship constituted "a

---

31. On the impact of CCP economic controls and mass organizations on social practice and state-society relations, see Siu, *Agents and Victims*.

universalistic morality in which all citizens are in important respects equal under the state, and gradations on the basis of status or degree of closeness cannot legitimately interfere with this equality."[32] The Maoist CCP's attempt to displace "friendship," or close personal relations, with "comradeship" that Vogel describes took early twentieth-century Chinese intellectuals' hopes of creating a publicly oriented person (*gongmin*) to an extreme degree.

At the same time, the mass spectacle and pageantry of the cult of Mao similarly aimed to draw China's people wholly into public life and detach them from any interior space or personal identity that was separate from their role as committed member of the cult. In her insightful analysis of the Mao cult, Mayfair Yang notes that the uniform fixation on Mao as the paramount leader was intended to displace commitment to horizontal ties, erode individuals' sense of internal space, and form all people into a homogeneous mass, "the People-as-One," constituting them as a collective through their dedication to Mao.[33] The merging of all citizens in the unified mass activism of a movement or the Mao cult drew women into public action that eroded gender distinctions through imposed uniformity. As with the nationalist movements of the first half of the twentieth century, revolutionary mobilization allowed women to escape some restrictive features of femininity in dress and action but often required conforming to male patterns of dress and behavior, such as wearing military uniforms and engaging in masculine-gendered political violence.[34]

Of course, the uniform, selfless, passionately dedicated comrade of the Mao cult was an ideal of the time and a caricature. In practice, many social distinctions persisted. In terms of gender, although women were expected to assume fully the citizen's public duties, expectations regarding work in the private sphere were still highly gendered, male authority was not fundamentally questioned, and femininity was still marked in subtle ways in dress and decorum.[35] Further, attempts at social engineering during the Maoist period simultane-

---

32. Vogel, "From Friendship to Comradeship," 46.

33. Yang, *Gifts, Favors, and Banquets*, chap. 7.

34. Evans, "'Comrade Sisters'"; Honig, "Reassessing the Red Guards"; Young, "Chicken Little in China."

35. Evans, "'Comrade Sisters'"; Young, "Chicken Little," 255–56.

ously generated many new distinctions and conflicts within Chinese society. Class labels based on a family head's source of income at the time of the founding of the PRC created caste-like forms of stratification that privileged some (the "five red categories") politically and economically while seeming to resign others (the "five bad elements") to perpetual second-class citizenship.[36] In addition, stratification by occupation differentiated the population by income, access to services, and political authority. Campaigns to impose labels on disenfranchised categories of people and privileges granted to honored categories of people created groups that had varying degrees of group identity and either a stake in or resentment toward the sociopolitical order.[37] All citizens were to be equal in selflessly serving society and contributing to building socialism, but society was internally differentiated by access to economic privileges and political authority. The tensions among groups with different labels strained the ethos of collective striving that was a central aim of Maoism.

Those tensions exploded powerfully in the Cultural Revolution, which was characterized by violent clashes between privileged and disenfranchised groups. In Shanghai, for instance, "contract workers or ex-rightists, some unionized individuals who disliked their bosses, and transformed or radicalized capitalist youths tended to join together in supporting this call for vaguely defined 'revolution' as a justification for airing their grievances."[38] Marginal, disenfranchised, and underprivileged workers sought to claim political authority and/or economic privilege through the forms of activism encouraged by Mao and others within the party leadership.[39] In many schools, students excluded by their backgrounds or behavior from the political elite of "good–class background" students who monopolized Communist Youth League slots, party membership, and student political leadership viewed the Cultural Revolution as "a genuine 'revolution' in which to manifest their activism."[40] From the perspective of our discussion of citizenship, these conflicts can be seen as attempts by diverse social groups to assume full citizenship as it

---

36. Kraus, *Class Conflict in Chinese Socialism.*
37. White, *Policies of Chaos.*
38. Ibid., 236.
39. Perry and Li, *Proletarian Power.*
40. Chan, *Children of Mao,* 139.

was constituted in the PRC at the time. They did so by joining in political struggle as well as by claiming the positive social labels—as "rebels" and "revolutionaries"—and economic privileges that constituted full political and social citizenship. As millions of people demonstrated and fought during the Cultural Revolution, they performed the kinds of activism associated with Maoist citizenship and assumed public political roles.

But even as the Cultural Revolution involved broad segments of society in mass politics, its repeated upheavals and twists and turns led to disenchantment, alienation, and social disintegration. The generalized conflict justified by the language of revolution and devotion to Mao during the Cultural Revolution ended up damaging nearly everyone in urban society, even the most committed public activists. As youths were attacked and punished with imprisonment or long years in the countryside for public activism that they had believed was in line with Mao's directives, they became further alienated from the CCP and abandoned a commitment to civic action for the public welfare.[41] Young people's disenchantment as a consequence of feeling that they had been "used" and "betrayed" during the Cultural Revolution became a society-wide phenomenon by the early 1970s. In the words of William Parish and Martin Whyte, "the emphasis on class struggle and random political attacks had destroyed a sense of participation in a grand national purpose."[42] The cult of Mao, the hunt for class enemies, and endless cycles of rectification undermined the passionate national commitment and patterns of civic activism that were developed gradually over the first 60 years of the twentieth century.

Reinventing Citizenship in Post-Mao China

With the end of the Mao era came the end of a form of citizenship centered on social service for the common good carried out under direction from the party-state. Starting in the early 1980s, under the leadership of Deng Xiaoping, the Chinese Communist Party introduced market-based economic reforms, dismantled rural collectives, began to reform the state-owned industrial sector, and sought to

41. Ibid., chaps. 4, 5.
42. Parish and Whyte, *Urban Life*, 320.

develop the economy quickly through infusions of foreign investment and technology and by joining the global marketplace.[43] The politics of the campaign and mobilization for mass action were gradually abandoned over the course of the 1980s and 1990s by a CCP leadership that feared returning to the chaos of the Cultural Revolution and hesitated to divert popular energies away from economic development. At the same time, by encouraging private economic activity, the Reform-era PRC state has allowed a private sphere of social activity to develop. Individuals have more personal freedoms than ever before under the CCP as well as limited legal protections as producers, consumers, and property owners.

The CCP leadership's declining enthusiasm for an approach to citizenship centered on mass mobilization for public service under party guidance has created an opportunity to define citizenship in fresh terms.[44] Over the past 25 years, intellectuals, different elements within the CCP, and various social groups have begun an intensive reexamination of and multisided negotiation over each dimension of citizenship. In many cases, the alternative conceptions and practices of citizenship that are now being debated or enacted have Chinese precedents that found expression during the Republican period but were deemphasized during the intervening Maoist period. The tentative nature of current constructions of citizenship, and the tenuous or even contradictory ways in which each dimension of citizenship is related to others, parallels the situation of the Republican period. There are especially close parallels to the Nanjing decade, when another Leninist state concerned with economic development and sociopolitical modernity took a great interest in defining citizenship but struggled with reconciling diverse and often incommensurate models. Because of these parallels, this book's approach to analyzing the dynamics of constructing citizenship in Republican China may hold clues to guide our understanding of the renegotiation of citizenship that is occurring in the PRC today.

---

43. For an overview of the post-1978 reforms, see Lieberthal, *Governing China*.

44. Elizabeth Perry demonstrates that mobilizational strategies were not abandoned entirely during the post-Mao era, as worker pickets were used to quell student protests in Shanghai in 1989 (*Patrolling*, chap. 6). However, the dismantling of state-owned enterprises and growing worker unemployment may have eroded the feasibility of worker-centered mass mobilization since the economic restructuring of the 1990s.

Perhaps the most striking shift since the Mao era is that, in the environment of reform, some intellectuals and a growing number of nonelite social groups have pushed for an expansion of political and civil rights in relation to the party-state. This process began with the Democracy Wall Movement of 1978 and 1979 and continued most visibly with the nationwide student protests in 1989.[45] In recent work, Merle Goldman has shown how critical intellectuals working within state or state-sanctioned institutions and "disestablished" intellectuals, who have challenged the party and been shut out of official positions, have advocated for civil liberties and political rights, using nongovernmental organizations, journals, and the Internet.[46] Disestablished intellectuals have also tried to form an independent political organization, the China Democracy Party, and sought to connect with community-based protest groups that are challenging state policies regarding the economy and the environment and are calling for protection of local rights and interests.[47] At this point, the main goal of these intellectuals seems to be to create institutions that will allow some degree of critical voice and in the process to influence state policy. But Goldman takes this farther, suggesting that such institutions may be the opening wedge of systemic change in Chinese politics.

Village and township elections, too, suggest a shift in the dynamics of representational politics at the local level. Kevin O'Brien, for instance, shows that people in rural communities are increasingly taking a competitive approach to local elections, using them as leverage to unseat unpopular cadres and to make all local cadres accountable.[48] In reflecting on recent local elections, Elizabeth Perry and Mark Selden observe:

Grassroots elections are also a two-edged sword for the state. State leaders hope that such procedures will help curb the corrupt excesses of local

---

45. For Democracy Wall, see Goldman, "The Reassertion of Political Citizenship"; Nathan, *Chinese Democracy*. The literature on the 1989 protests is vast and varied. For essays that outline many of the central interpretive positions, see Perry and Wasserstrom, eds., *Popular Protest and Political Culture*.

46. Goldman, *From Comrade to Citizen*.

47. Local communities' protests based on recognized policies and standards of rights and privileges have become a powerful mode of challenging the state in recent years. See Jing, "Environmental Protests."

48. O'Brien, "Villagers, Elections, and Citizenship."

tyrants and thereby dampen the fires of rebellion in the countryside. Yet giving villagers an enlarged sense of their own political efficacy, and providing a public forum for open discussion of civic affairs, can also act to stimulate local resistance to higher-level dictates.[49]

Graduated levels of indirect election limit the immediate impact of local elections on national policies. But O'Brien argues that villagers are using the language of rights and the contractual logic of the market to defend their local interests and critique state policies. Interestingly, in Taiwan, limited local elections, which were introduced by the Nationalist Party during the 1950s, contributed to building the democratic political culture that underpins the island's current liberal democracy and feeds a conception of political community based on shared civic action.[50]

Both disaffected intellectuals and villagers at the local level, then, are beginning to espouse civil rights and forms of political influence that have been central to a liberal, democratic mode of citizenship.[51] This advocacy for civil rights and representative politics in some ways marks a return to the more democratic approach to citizenship that was introduced by China's late Qing reformers and revolutionaries. Moreover, throughout the Republican period, liberal reformers such as Hu Shi continued to push for "human rights, political pluralism, representative institutions, and constitutionalism."[52] But, as we have seen, this liberal, democratic ideal of political citizenship was marginalized during the Republican and Maoist periods by a civic republican approach that stressed mobilizing citizens for social service that would benefit the common good. As a result, we might better view this emergent liberal, democratic form of politics as the product of social, economic, and political conditions that in some ways parallel those of the Republican period rather than as a direct continuity of the Republican liberal tradition. In both eras, the most important conditions for experimentation with liberal democracy may have been exposure to outside models of democratic politics and a

---

49. Perry and Selden, *Chinese Society*, 11.

50. Rigger, "Nationalism Versus Citizenship," 369–71.

51. Significantly, China's growing affluent middle class has played a marginal role in this push for expanding political rights. See Dickson, "Do Good Businessmen Make Good Citizens?"; and Goldman, *From Comrade to Citizen*, 14–15.

52. Fung, *In Search of Chinese Democracy*, 10.

vibrant, privatized commercial economy that allows both elaboration of a sphere of private life and the formulation of collective economic and political interests.

As with political citizenship, social citizenship in post-Mao China is in significant flux. During the Mao period, relative social status and modes of social membership were determined by state controls over access to publicly controlled entitlements that were determined by urban or rural residence, family class background, and work-unit membership.[53] Market reforms, dismantling of state-owned enterprises, abandonment of class designations, and deterioration of the household registration system, which had restricted rural people from relocating in urban areas, have opened up a complex field for renegotiating societal membership.[54] Within this relatively open sphere of social interaction governed by the consumer market, differential access to economic resources and social networks contribute to demarcating anew levels of citizens within society.[55] The marketplace is also a context where gender difference is being reconstituted through feminized and masculinized patterns of consumption.[56] This growing degree of income-based socioeconomic differentiation marks a return to the forms of social stratification, and the accompanying disenfranchisement of the poor, that characterized the pre-1949 period.[57] Market-driven differentiation could potentially also fuel the reemergence of socioeconomically defined interest groups that could seek to use the political process to protect their rights and interests in ways similar to Anglo-American liberalism.

In step with the resurgence of market-driven social differentiation that accompanied economic reform has come a renewed discourse of cultural citizenship defined in terms of levels of civility. As during the Republican period, international standards of "civilized" behavior and Chinese social elite and state desires to be part of a "civilized" global community serve as a reference for civil behavior

---

53. E.g., Parish and Whyte, *Urban Life*; Solinger, *Contesting Citizenship*.

54. Davis, ed., *The Consumer Revolution*, 2–3. Cf. Yang, *Gifts, Favors, and Banquets*, esp. chap. 8.

55. Davis, ed., *The Consumer Revolution*.

56. Evans, "Marketing Femininity"; Hershatter and Honig, *Personal Voices*.

57. Significantly, recent research suggests that the unemployed and marginalized sometimes challenge their poverty and disenfranchisement in terms of socialist citizenship. Lee, "Pathways of Labor Insurgency," 52; Perry, *Patrolling*, chap. 6.

within China today.[58] Refinement in terms of dress, bearing, and public behavior differentiate a sociocultural elite from the supposedly uncivilized masses that serve as their "inappropriate other." In this way, embodied forms of civility become a way for elite social groups to monopolize positions as modern subjects in a globalized society and thereby exclude most people from full cultural citizenship. Casting most people as "uncivilized" (*buwenming*) or of "low quality" (*suzhi di*) also serves the needs of the CCP party-state, for it can use such designations to present them as "not ready" for modern civic roles and also cast itself as the educating force that will teach the people civility, much in the way the Nationalist Party did during the 1930s and 1940s with the New Life Movement.[59]

Standing in a complex relationship to both the civilizing project and nascent trends of representative politics is a resurgent nationalism. Many commentators have described the explosion of a powerful antiforeign nationalism in China during the 1990s after some intellectuals' brief flirtation with the West as a potential model of openness, reason, and progress during the 1980s.[60] This new nationalism was often expressed by elite and middlebrow intellectuals in the increasingly open mass media of the 1990s, especially in commercial publishing and television. Commentators see the new nationalist discourse as giving voice to a century-old desire for Chinese parity with Euro-American powers and Japan in the international order and expressing an urge to reestablish China's national prestige after a long period of "national humiliation" (*guochi*). Underpinning much of this new nationalism is a hegemonic conception of the Chinese nation as a Han-centered, "Confucian civilization" composed of "descendents of the Yellow Emperor."[61] Recent forms of Chinese nationalism, in their powerful antiforeignism and impulse to ground the national identity in culture and history, echo the antiforeign nationalist movements and some nationalist discourse of the Republican period.

---

58. Anagnost, *National Past-Times*, chap. 3; "The Corporeal Politics of Quality (Suzhi)."

59. See Anagnost, *National Past-Times*, 78–79; Friedman, "Civilizing the Masses."

60. E.g., Barme, "To Screw Foreigners Is Patriotic"; Cohen, "Remembering and Forgetting"; Garver, "More from the 'Say No Club'"; Gries, "Tears of Rage"; idem, *China's New Nationalism*; Zhao, "Chinese Intellectuals' Quest."

61. See Gries, *China's New Nationalism*, 7–8; Zhao, 735–38.

Insofar as this resurgent nationalism is focused on national wealth and state power on the world stage, it aligns closely with state goals. Indeed, some expressions of this antiforeign, culturally based nationalism, such as the popular book *China Can Say No* (*Zhongguo keyi shuo bu*), have received tacit state approval.[62] But mass enthusiasm for nationalist causes can sometimes surge to degrees and in ways not anticipated or controlled by the government. Witness the chaotic and sometimes violent protests in reaction to the U.S. bombing of China's Belgrade Embassy in 1999 and in the 2005 protests over treatment of the Sino-Japanese War in Japanese history textbooks. Popular expectations for the state to represent China's interests internationally and to realize long-held goals of wealth and power also generate popular opinion pressure that can influence state policy, as occurred during the Sino-American standoff over a downed U.S. surveillance plane on Hainan Island in 2001. Such rising popular nationalism threatens to create a situation where "struggling just to keep up with popular nationalist demands, the Party is slowly losing its hegemony over Chinese nationalism."[63] Through this new populist nationalism, the people, apart from the party, could develop a sense of collective identity and political subjectivity in ways similar to the urban mass nationalism of the Republican period.[64]

Moments of diplomatic crisis tend to focus nationalist enthusiasm on a single target and create an impression of conceptual unity. But regional and local expressions of cultural identity have complicated attempts to construct a conception of Han culture and Chinese national identity that is monolithic and historically continuous. Since the late 1980s, regions, provinces, and localities have competed to present their distinctive cultures as expressive of the essence of national culture.[65] Consequently, even bracketing the thorny issue of

---

62. Gries, *China's New Nationalism*, chap. 7.

63. Ibid., 121.

64. In drawing such parallels, though, we must be attentive to the dynamics of political power. Jeffrey Wasserstrom argues that the most apt comparison might be between the recent protests and those of the Nationalist period. In both instances, he suggests, student protests emerged in the context of a Leninist state that was ambivalent about antiforeign nationalism and unsettled by mass activism ("Student Protests in Fin-de-Siècle China").

65. See Friedman, "Reconstructing China's National Identity"; Oakes, "China's Provincial Identities."

China's 56 "national minorities," instead of a unified and stable national essence that the party-state can claim to represent, we see a widening field for dialogue about Chinese cultural identity that is rooted in plural local cultures. Moreover, the growing number of candidates for regional cultures that can be considered both civilized and Chinese will inevitably complicate efforts to define a unified mode of cultural citizenship that is connected to global patterns of civility.

Many of the elements of current citizenship discourse echo those of the Republican period. The competing versions of nationalism, a liberal democratic approach to political participation, elite civilizing projects, and social membership that is negotiated in the capitalist marketplace all parallel approaches to citizenship that were debated and sometimes enacted in China during the first half of the twentieth century. Moreover, the current environment of relatively open debate about the nature and practice of citizenship recalls the indeterminate nature of citizenship during the late Qing and Republican periods. During that time, Chinese elites drew together a conception of the nation as threatened and incomplete, a civic republican approach to political participation, and an organic vision of society to create an activist form of citizenship that after 1927 came to be guided by a single party-state. By exploring how complex engagements among competing groups with diverse goals and interests contributed to articulating citizenship in the Republican period, this book has demonstrated a conceptual model for how we can analyze the current reformulation of Chinese citizenship. That is, conceiving of citizenship as a complex formation composed of national identity, political rights and participation, socioeconomic membership, and cultural expression provides an analytic framework for understanding the meaning and practice of citizenship in twenty-first-century China and other world contexts.

As a historian, I venture no predictions regarding the course of Chinese citizenship in the twenty-first century. But this book's exploration of the discourse and practice of citizenship during the Republican period provides a basis for evaluating long-term trends over a century of change. Two key factors supported the emergence of a state-led civic republican approach to citizenship during the Republican period. One was the shared aspiration of the Nationalist and Communist parties to take a guiding role in almost every aspect of political, social, cultural, and economic life. A second was the almost constant outside political, economic, and military threats that underwrote those parties' efforts to impose a "single substantive idea of the common good" and the attempts by party leaders, intellectuals, and educators to mobilize people for public service.

Neither factor is currently operative. Presently, China occupies a much more secure and prominent place in the world than it did early in the twentieth century. Further, the Chinese Communist Party, though still able to intervene periodically with great power in social life, seems to be scaling back its aspirations as a guiding and mobilizing force. These changes make it almost impossible to imagine a return to the state-led civic republican approach to citizenship with a mobilized citizenry that characterized much of the Republican period and the Maoist era. Instead, current trends hold out the promise of a mode of citizenship with limits on the state's and national community's claims to the lives and action of individuals. With the passing of twentieth-century China's distinctive mode of civic republican citizenship, individual citizens' dedicated participation in civic life is rapidly coming to an end. Chinese society is gradually losing the ethos of spirited collective action and public service that was one of the most powerful products of the first effort by intellectuals, political leaders, students, and a range of other social groups to articulate a Chinese form of citizenship.

Appendixes

APPENDIX A

A Statistical Portrait of Lower Yangzi Region Secondary Schools and Students During the Republican Period

Table A1
Middle and Normal Schools in Jiangsu, Zhejiang,
Shanghai, and Nanjing, 1915, 1920, and 1935

Localities	1915	1920*	1935
Jiangsu	40	61	158
Zhejiang	35	43	95**
Shanghai†			124
Nanjing†			27
Total for Jiang-Zhe	75	104	404
Nationwide	683	822	2,756
Jiang-Zhe as percentage of total	10.98%	12.65%	14.66%

*These figures were based on a survey conducted in 1922 and 1923 by the Chinese Educational Improvement Society (Zhonghua jiaoyu gaijin she) and were supplemented by 1919 reports by the Ministry of Education.
**The figures for Zhejiang are for the 1934 school year.
†The figures for 1915 and 1920 are included in the totals for Jiangsu.
SOURCES: 1915: [Jiaoyubu], *Zhonghua minguo di san ci jiaoyu tongji tubiao*, 590, 595–96, 601–2. 1920: *Zhongguo jiaoyu tongji gailan*, 19–20, 28. 1935: Jiaoyubu tongjishi, *Quanguo jiaoyu tongji jianbian*, 36–37.

Table A2
Middle and Normal School Students in Jiangsu,
Zhejiang, Shanghai, and Nanjing

Localities	1915	1920*	1935
Jiangsu	6,206	13,737	36,432
Zhejiang	6,208	8,170	21,304**
Shanghai†			33,491
Nanjing†			8,976
Total for Jiang-Zhe	12,414	21,907	100,203
Nationwide	93,933	141,662	522,625
Jiang-Zhe percentage of total	13.22%	15.46%	19.17%

*These figures were based on a survey conducted in 1922 and 1923 by the Chinese Educational Improvement Society (Zhonghua jiaoyu gaijin she) and were supplemented by 1919 reports by the Ministry of Education. The figures for students in this column do not include those from missionary schools.
**The figures for Zhejiang are for the 1934 school year.
†The figures for 1915 and 1920 are included in the totals for Jiangsu.
SOURCES: See Table A1.

Capsule Biographies of
Selected Textbook Authors

Du Weitao (1901–?) was a native of Taiping, Anhui, but he was born in Huaiyin, Jiangsu. In 1916, he graduated from Anhui's Wuhu Lower-Level Commercial School, and in 1919 he graduated from Anhui Provincial Fourth Normal School in Xuancheng. After graduating from the physics and chemistry (*lihua*) division of Beiping Higher Normal School in 1923, Du taught at various primary, middle, and normal schools in Beijing, Anhui, Henan, and Jiangsu, and he participated in various forms of mass education. He later served as professor in a number of academies (*xueyuan*). (Xu Youchun et al., eds., *Minguo renwu da cidian*, 241.)

Fu Weiping (Yunsen) (dates unknown) served as an editor at the Commercial Press at least from the time of Wang Yunwu's reorganization of the editing department (*bianyisuo*) in 1922, but probably from much earlier. Fu was one of the lead editors for the Commercial Press's encyclopedia; he was editor or proofreader for primary-level history readers during the 1910s; and he wrote the only secondary-level integrated history textbook published for the New School System during the mid-1920s. (For Fu's role in Wang Yunwu's reorganized editing department, see Yang Yang, *Shangwu yinshuguan*, 102, 122–24. For evidence of his earlier editing activities, see Beijing tushuguan and Renmin jiaoyu chubanshe tushuguan, *Minguo shiqi zongshumu [1911–1949]: Zhong xiao xue jiaocai*, 85–88.)

Ge Suicheng (also called Kanglin; courtesy name Yifu) (1897–1978) was a native of Dongyang County, Zhejiang. In 1914, he graduated from Zhejiang Provincial Seventh Normal School, and in 1916 he tested into the Zhonghua Book Company as an apprentice, becoming an editor in the history and geography department in 1918. By 1932 he was director of the history and geography department at Zhonghua Book Company and had served as professor at several Shanghai-area universities. Ge was also one of the founding members of the Chinese Geography Study Society (Zhonghua dili xuehui) and the main editor of *Geography Quarterly* (*Dili jikan*). He continued to combine work in higher education and publishing throughout his career. (Xu Youchun et al., 1269.)

Gu Shusen (courtesy name Yinting) (1885–?) was a native of Jiading, Jiangsu. In his youth he studied in England at the University of London. After returning home, he served as the principal of Shanghai's China Vocational School (Zhonghua zhiye xuexiao), as secretary in the administrative office of the International Association (Guoji lianhehui) representative in China, and in a series of administrative posts. In December 1928, Gu became the head of the Nanjing Municipal Bureau of Education, and in June 1930 he became the head of the Department of Regular Education in the Ministry of Education. He later served in several other leadership positions within the ministry. (Xu Youchun et al., 1680.)

Lü Simian (courtesy name Chengzhi) (1884–1957) was a native of Wujin, Jiangsu, who was educated at home and at a private village school. During the late Qing and early Republic he taught history, geography, and language at a number of schools in the lower Yangzi region, including Dongwu University in Suzhou, the Changzhou Prefectural School, Nantong's National Language Junior College (Guowen zhuanke xuexiao), Jiangsu Provincial First Normal School, and Guanghua University. He remained at Guanghua University from 1926 through the start of the Sino-Japanese War in 1937. During the Republican period he also served as an editor at the Commercial Press and Zhonghua Book Company. Lü penned a number of works on history, including the well-received *Vernacular Chinese History* (*Baihua benguoshi*) at the start of the 1920s. (Xu Youchun et al., 331–32.)

Shu Xincheng (originally Weizhou) (1893–1960) was a native of Xupu, Hunan. He was educated in various local schools in Hunan, ultimately graduating from Hunan Provincial Higher Normal School in 1917. His first career was as an educator, teaching primarily in secondary schools in Changsha, Wusong, Nanjing, and Chengdu from 1917 to 1925. He was an active promoter of new educational methods, notably the Dalton Plan, which was popular among Western-influenced educators during this period. After publishing a number of books on educational history, theory, and practice during the mid-1920s, Shu joined the staff of the Zhonghua Book Company. While at Zhonghua Shu served as the executive editor for the encyclopedic dictionary *Cihai* and as the head of the editorial board, as well as managing the library. Under Shu's tutelage *Cihai* came to encompass not only classical scholarly language, but terms from drama, fiction, and the vernacular spoken language as well. (Bian Chunguang et al., eds., *Chuban cidian*, 856; Boorman and Howard, eds., *Biographical Dictionary of Republican China*, 135–37; Xu Youchun et al., 1166.)

Sun Benwen (originally Binfu) (1892–1979) was a native of Wujiang, Jiangsu. He started studying in an informal private school in 1896, and he attended a series of local primary schools beginning in 1905. In 1909, he entered the lower normal course of the Jiangsu Provincial Two-Level Normal School (Jiangsu liangji shifan xuetang), graduating in 1915. He started teaching at a primary school in Wujiang but then tested into Beijing University's philosophy program. Sun graduated in 1918 and taught language and philosophy at Nanjing Higher Normal Affiliated Middle School. In 1920, he won a fellowship to study at the University of Illinois, where he earned a master's degree in sociology. He continued his studies in sociology at Columbia University for two years. In 1925, he received a Ph.D. in sociology from New York University, and the same year he went to the University of Chicago to continue studies in social theory. In 1926, he returned to Shanghai and taught at Daxia and Fudan Universities. In 1929, he moved to Nanjing's Central University, where he served as a sociology professor and head of the department. He played a leading role in creating the Chinese Sociology Society (Zhongguo shehui xue she), in 1929, and the *Sociology Journal* (*Shehuixuekan*), which he edited. In May 1930, he became head of the Department of Higher Education in the Ministry of Education while continuing to teach at Central University. Sun joined the Nationalist Party in 1931. In 1932, he resigned his post at the Ministry of Education to concentrate on chairing the sociology department at Central University, and he continued to serve in high-ranking posts at the university throughout the Nanjing decade. (Xu Youchun et al., 779.)

Sun Lianggong (originally Guangce) (1894–1962) was a native of Longhui County, Hunan. Having studied in a private elementary school when young, he entered middle school in Changsha at seventeen, and in 1912 he entered the Middle School for Hunan People Sojourning in Hubei, graduating in 1916. He subsequently tested into the national language department (*guowenbu*) at Beijing Higher Normal School (Beijing gaodeng shifan). While in Beijing he joined the Comrades Society (Tongren she) and helped to organize the anarchist Work-Study Association (Gongxue hui), editing *May Seventh Monthly* (*Wuqi yuekan*) and cooperating with Xu Minghong and Zhou Yutong to publish *Common People's Education* (*Pingmin jiaoyu*) and *Work-Study Magazine* (*Gongxue zazhi*). He graduated in 1920, taught at Changsha First Normal School, and went to Shanghai in 1922 to teach at China Academy (Zhongguo gongxue). In 1924, Sun went to Japan to study, returning in 1928 to teach at Fudan University and the Jiangwan Labor University (Jiangwan laodong daxue) in Shanghai. In 1930, he became chair of the literature department at Fudan. For a time during the mid-1930s, he also served as a compiler at the National Institute for Compilation and Translation (Guoli bianyiguan). (Xu Youchun et al., 784.)

Tao Menghe (originally Lügong) (1889–1960) was born and raised in Tianjin, though his ancestral home was Shaoxing, Zhejiang. Tao was one of the first students of the Nankai School, graduating in its first class of normal-course students. He subsequently went to the London School of Economics to study sociology. After returning to China, he worked as an editor for the Commercial Press as well as serving as professor and dean of the College of Arts and Sciences at Beijing University. In 1926, with the support of a foundation, he established the Social Survey Department (Shehui diaocha bu) and served as its secretary. In 1934, this organization merged with the Social Science Institute (Shehui kexue yanjiusuo) of the Academia Sinica, and Tao served as head of the institute. In the subsequent decade, he conducted survey research throughout China, producing a monograph on county-level finance. As a nonaligned intellectual, he served in a number of administrative positions under both the Nationalist Party and the CCP. (Xu Youchun et al., 1081–82.)

Wang Zhongqi (courtesy name Boxiang) (1890–1975) was a native of Suzhou. He tested into the Suzhou Sino-Western School (Suzhou zhongxi xuetang) in 1906 and the following year entered the Suzhou Public Middle School (Suzhou gongli zhongxue). Starting in 1908 he helped to form and participated in several poetry, literature, and art societies with Gu Jiegang and Ye Shengtao, and he contributed to the production of several publications. After 1911, while teaching in the Suzhou area, he also was a corresponding member of a group for national studies (*guoxue*) at Beijing University. He taught at several secondary and college preparatory schools before joining the Commercial Press's editing department in 1922, where he served as a history and geography editor. In 1932, he joined Kaiming Bookstore as an editor. (Xu Youchun et al., 58.)

Xie Guan (courtesy name Liheng) (1880–1950) was a native of Wujin, Jiangsu. In his youth, he was educated at home, learning medicine and classics, and he studied history and geography at Dongwu University as an adult. Early in his career he undertook editing work in both education and medicine. (Xu Youchun et al., 1564.)

Zhang Qiyun (courtesy name Xiaofeng) (1901–85) was a native of Yin County, Zhejiang. He attended Zhejiang Provincial Fourth Middle School in Ningbo in 1915 and entered Nanjing Higher Normal School in 1919. After graduating in 1923 he took a position as editor at the Commercial Press. In 1927, he became a lecturer and then professor in the geography department at Central University. In 1935, Zhang became a member of the advisory board of Academia Sinica, and in 1936 he became professor and chair of the history and geography department at Zhejiang University. During and after the war, he served in a number of cultural and administrative posts under the Nationalist government. He went to Taiwan in 1949. (Xu Youchun et al., 927–28.)

Zheng Chang (courtesy name Wuchang) (1894–1952) was a native of Sheng County, Zhejiang. He joined Zhonghua Book Company as a literature and history editor in 1921 and became director of the art department (*meishubu*) in 1924. He served concurrently as professor at a number of China's top art academies. (Xu Youchun et al., 1474.)

Reference Matter

Bibliographic Essay

This book focuses on citizenship, but it explores that theme through analysis of secondary-level education in the lower Yangzi region. In part, the richness of available sources on curricula, pedagogy, course content, and local school practice recommended this school-centered approach to analyzing theories and practices of citizenship. Sources ranging from textbooks and government policy statements to school publications and student journals facilitate grounded analysis of what ideas circulated in schools and how teachers and students acted. These materials open a window onto the educational process during the Republican period, but they also provide resources for exploring many other dimensions of society, politics, and culture. I briefly review here the various categories of resources that have proven valuable to this project.

Archives and Documentary Collections on Education Policy and Implementation

Collections of reprinted government documents and school records published in the past two decades now make available to scholars a diverse archive of sources with which to track government education policy and educators' opinions during the Republican period. The finest example of these collections is *Zhongguo jindai xuezhi shiliao* (Historical materials for the modern Chinese school system), which was edited by Zhu Youhuan with help from a group of scholars and published in seven combined volumes by East China Normal University during the 1980s and early 1990s. The collection is divided by level and kind of schooling and spans the period from missionary educational projects and the first late Qing educational reforms during the nineteenth century to the establishment of the New School System in 1922. It makes available educators' essays on curricular organization and pedagogy, government policy statements, laws and regulations, and materials from selected local schools. Of similar organization and value is the collection of statistics, government policy directives, communications, and commentary from educators included in *Zhongguo jindai jiaoyushi ziliao huibian* (Compilation of China's modern education history materials) published by Shanghai Education Press. The volumes on education in the Second Historical Archive's series *Zhonghua minguoshi dang'an ziliao huibian* (Compilation of archival materials for Chinese Republican history) include major policy statements and also valuable interagency correspondence, reports, and statistical data. The two volumes on education in that collection's series 5 illuminate many different aspects of Nationalist government education policy.

The past two decades have seen publication of collections of reprinted archival and published materials from particular local schools that provide a local correlate for these major nationwide archival publishing projects. Most noteworthy is the fourteen-volume collection featuring famous Republican-period primary and secondary schools, the *Zhongguo mingxiao congshu* (Series on China's famous schools), first published in 1997 by People's Education Press in Beijing. These volumes contain a wealth of material culled from archives, rare publications, and reminiscences. Collected materials related to particular incidents and local schools are also incredibly valuable, such as the volume *Zhejiang yishi fengchao* (The disturbance at Zhejiang First Normal) published by Zhejiang University. Finally, although not related solely to education, Liu Zhemin's *Jin-xiandai chuban xinwen fagui huibian* (A collection of laws for modern and contemporary publishing and the press) is essential for anyone dealing with issues of publishing and censorship during the Republican period. Besides gathering in one place the legal corpus related to publishing, it also includes many valuable government communications and archival documents.

Just as valuable as these collections of government documents and publications was access to national and local archives. The Second Historical Archive makes available holdings for the Ministry of Education for the Nationalist and Beiyang government periods. Archives at Shanghai and Suzhou offer extensive collections of materials that inform analysis of central government policy, local (provincial-, municipal-, and county-level) implementation, and school-level practices. Both municipal archives contain extensive correspondence among government and party organs and between government agencies and local schools that illuminate the process of policy formation and implementation. They also relate the concrete conditions under which local officials and school administrators operated. Both archives hold extensive collections of local school publications as well, especially commemorative volumes and school reports of various kinds. These supplement the school-based publications available at regional libraries (see below). Also helpful to this project was the Nationalist Party History Archive in Taiwan. Archival materials from the Nationalist Party's Youth Bureau (Qingnianbu) provide an inside perspective on issues of youth mobilization during the period of the first United Front and the Northern Expedition.[1] The archive also holds extensive materials on topics such as Scouting and military training under Nationalist Party rule.

Textbook Collections

Textbooks provided one of the most valuable sources for this project, revealing in detail how abstract ideas were related to students. Besides their value as a source for education history, textbooks can shed light on various dimensions of political, cultural, and social change. *Minguo shiqi zong shumu (1911–1949): Zhong xiao xue jiaocai* (Republican-period general catalog [1911–49]: Secondary and primary school teaching materials) is based on the collections at the Beijing Municipal Library, People's Educational Press Library, and Beijing Normal University Library. Its listings provide a rough inventory of extant secondary and primary textbook titles published during the Republican period. In preparation from Shanghai's Lexicographical Publishing House (Shanghai Cishu chubanshe) is a comprehensive catalog of pre-1949 Chinese-language primary and secondary textbook titles held by institutions in the People's Republic and Taiwan. This publication will offer listings of extant primary and secondary

---

1. The archive also contains extensive materials for the other four of the "Five Bureaus," the bureaus for peasants, merchants, workers, and women, providing valuable resources for a wide range of social, cultural, and political history projects.

textbooks more complete and comprehensive than the *Minguo shiqi zong shumu*, as well as a detailed finding aid.

Beyond the libraries in Beijing that were the source institutions for the *Republican-Period General Catalog* volume, several libraries in China and the United States are noteworthy for their collections of Chinese-language textbooks from the late Qing and Republican periods. Shanghai Municipal Library contains considerable pre-1949 textbook holdings. The special collections department in Teachers College Library at Columbia University also holds an impressive array of several hundred late Qing and early Republican primary and secondary textbooks. The remarkable collection, which was gathered by Teachers College–affiliated researchers on trips to China during the Republican period, can provide basic resources for U.S.-based scholars seeking to use textbooks as part of their research. For perhaps the most extensive holdings of Republican-period textbooks gathered in one place, scholars can turn to Shanghai's Lexicographical Publishing House Library, which inherited the collection of the Zhonghua Book Company Library once managed by educational reformer Shu Xincheng. Because of Zhonghua's investment in the textbook market during the Republican period, this library contains copies of most series of primary and secondary textbooks published by both major and minor publishing companies from the 1900s through the 1940s.[2] The Lexicographical Publishing House Library's holdings of pre-1949 textbooks number more than twenty thousand volumes.

Local Periodicals and School Publications

All scholars working on Chinese education history are familiar with the major mass-circulation education journals of the early twentieth century, such as *Jiaoyu zazhi* (Education review), *Zhonghua jiaoyu jie* (Zhonghua's education circles), and *Xin jiaoyu* (New education). These are widely available in libraries throughout the United States, Europe, and East Asia. However, many other journals published locally, at the provincial, municipal, or county levels, reveal local officials' and educators' attitudes and practices and sometimes also con-

---

2. For a more extensive discussion of the holdings of this valuable library, see my "Research Note: Shanghai's Lexicographical Publishing House Library as a Resource for Research on Republican Period Popular Culture and Education."

tain valuable material from students. *Zhejiang jiaoyu jikan* (Zhejiang education quarterly), *Zhejiang jiaoyu yuekan* (Zhejiang education monthly), *Zhejiang qingnian* (Zhejiang youth), *Jiangsu xuesheng* (Jiangsu student), *Jiangsu jiaoyu* (Jiangsu education), and *Jiaoyu chao* (Tides in education) were all published by provincial bureaus of education or education associations. As such, they provide rich material on provincial education policies and the views of leading provincial educators. They also contain extensive material on local schools in the form of reports, accounts of specific incidents, and writings from teachers and students.

Even more valuable for revealing what was happening at the local level are periodicals, pamphlets, and books published by specific schools. School publications can be divided into three main categories: yearbooks (*niankan*), overviews (*yilan*), and commemorative volumes (*jiniance*) compiled and published periodically by the school's authorities; school journals published by the school administration; and student journals that were produced either by the student self-government association or by independent groups of students. Each kind of publication has its value and reveals different dimensions of life inside the school.

Yearbooks, surveys, and commemorative volumes, such as *Pudong zhongxuexiao nian zhou jiniankan* (Memorial volume for the twentieth anniversary of Pudong Middle School), *Jiangsu shengli Suzhou zhongxuexiao gaikuang* (The general situation of Jiangsu Provincial Suzhou Middle School), and *Zhenhua nü xuexiao sanshinian jiniankan* (The commemorative volume for 30 years of Zhenhua Girls' School), reveal the official view of the school. They often provide a basic historical overview of the school along with a chronicle of events. Sections on character-development education, instruction, and physical education allow in-depth analysis of a given school's curriculum, pedagogy, and training techniques. Other sections often detail school finances and management practices. Lower Yangzi region schools began publishing such volumes starting in the 1910s, but they became ubiquitous during the 1930s as schools founded in the late Qing began celebrating significant anniversaries. In some instances, secondary schools that were known for educational reforms also published pamphlets and books to publicize their policies and methods. For instance, teachers and administrators at Southeast University Affiliated Middle School compiled *Shixing xinxuezhi hou zhi dongda fuzhong* (Southeast University Affiliated Middle School after implementing the New School System), which detailed the educational reforms carried out under the leadership of Liao Shicheng at this prominent regional school. During the Nanjing decade, successful schools such as Jiangsu's Provincial Suzhou Middle School published numerous pamphlets on specific aspects of schooling, such as *Jiangsu shengli Suzhou zhongxue geke jiaoxue jindu biao* (Syllabi for instruction in each course at Jiangsu Provincial Suzhou Middle School) and *Xunyu guicheng* (Rules for character-development education).

School authorities also often published school journals that variously served as a news outlet and forum for the school community and as a way to raise the school's profile in local society. As with commemorative volumes and yearbooks, school journals became more common and published more frequently as the Republican period continued. By the early 1930s few large regional schools lacked a school journal. *Zhejiang yizhong zhoukan* (Zhejiang First Middle weekly), *Jiangsu shengli Shanghai zhongxue banyuekan* (Jiangsu Provincial Shanghai Middle School biweekly), *Songjiang nüzhong xiaokan* (Journal of Songjiang Girls' Middle), *Suzhong xiaokan* (Suzhou Middle's school journal), *Suzhou Zhenhua nüxue xiaokan* (Journal of Suzhou Zhenhua Girls' School), and *Yangzhong xiaokan* (Yangzhou Middle's school journal) are all examples of school journals that include voluminous material on the curriculum, pedagogy, and training techniques in local schools. These journals open a window onto the views and methods of teachers and administrators.

The most direct access to the student's voice comes through student journals. Some of these journals were produced by independent student groups or political organizations, for example, *Zhejiang xinchao* (Zhejiang new tide), the combined publication of students studying at Zhejiang Provincial First Middle School and Zhejiang Provincial First Normal School in Hangzhou, and *Zeren* (Responsibility), which was published with substantial contributions from Zhejiang First Normal graduates such as Xuan Zhonghua. Some of these journals are extant in local libraries, but significant excerpts can also be found in compilations including *Wusi yundong zai Zhejiang* (The May Fourth Movement in Zhejiang) and *Wusi shiqi de shetuan* (Social organizations of the May Fourth period). Many other student journals were published by student self-government organizations or academic classes. Individual issues and sometimes long runs of these journals are available at regional libraries. Notable examples include *Ningbo zhongxuesheng* (Ningbo Middle School students), *Sizhong xue-sheng* (Fourth Middle students), *Shaozhong* (Shaoxing Middle), *Jing-zhong xuesheng* (Jingye Middle students), *Minli xuesheng* (Minli students' voice), and *Jiangsu shengli di'er nüzi shifan xuexiao xiaoyouhui huikan* (Publication of the Alumni and Students' Association of Jiangsu Second Girls' Normal School).

These journals, because they were published by student organizations that were directly overseen by the school's teachers and administrators, were subject to some degree of censorship and restriction, especially during the 1930s. As a result, they need to be read critically. I have, for instance, tried to weed out student essays that clearly were written as class assignments or in response to prompts from teachers. But student writings in these journals did not necessarily merely echo the messages teachers delivered in the classroom. As we have seen, student journals often carried essays sharply critical of government authorities and presented ideas about culture, society, and politics that ran counter to social norms, the formal curriculum, or government policy. These journals, which overflow with criticism and independent thinking, have great value for helping us to understand the mind-set and motivations of students. They are especially valuable when used to discern the fundamental ideas and implicit assumptions that shaped students' more explicit arguments.

In this book, journals published by schools and students help us to understand the workings of local schools and citizenship education. Because teachers and sometimes students were important intellectuals in many towns and small cities, these journals also provide valuable resources for understanding the attitudes and opinions of local elites in a time of transition. Scholars interested in issues of local politics or social and cultural reform at the local level can find much of value in these sources.

Shanghai Municipal Library contains the most extensive holdings of local periodicals of this kind for the lower Yangzi region, including full or substantial partial runs of journals from the region's most prominent schools, such as Shanghai, Suzhou, and Yangzhou Middle Schools. Other regional libraries also hold valuable periodicals from local schools. In particular, Zhejiang Provincial Library in Hangzhou has an extensive collection of Republican-era periodicals from Zhejiang. Municipal libraries in regional cities including Suzhou, Yangzhou, and Nanjing similarly yield surprising finds in terms of student journals and local school publications. Although my primary research experience is limited to the lower Yangzi region, I find it likely that provincial and municipal libraries in other regions also hold significant collections of local school periodicals.

Memoirs and Interviews

Various forms of memoir literature and interviews are helpful in learning about schools' hidden curricula and informal processes, as well as students' extracurricular activities. Scholars of modern China are now thoroughly familiar with the strengths and limitations of *Wenshi ziliao* (Cultural and historical materials) and *Dangshi ziliao* (Party history materials) collections. Both the Shanghai and Zhejiang *Wenshi ziliao* collections have volumes dedicated to education in those areas: *Shanghai wenshi ziliao xuanji*, no. 59, and *Zhejiang jindai zhuming xuexiao he jiaoyujia* (Zhejiang's famous modern schools and educators), vol. 45, *Zhejiang wenshi ziliao*. Further, during the 1990s different areas of Zhejiang published collections of "revolutionary culture historical materials" (*geming wenhua shiliao*) focused on cultural work during the revolutionary process. These are valuable resources for studying social and cultural reform, as well as party formation, in local areas in part because, in addition to memoirs and interview-based articles, they also include reprints of original publications, government documents, and party communications and pronouncements.

This book also benefited enormously from interviews with alumni from many of the schools discussed here. Their reminiscences brought these schools to life, and they were able to evoke in full color the experience of Republican schooling from the student's perspective. More than three dozen alumni in both the People's Republic and Taiwan gave generously of their time to recall their school days. Unfortunately, time for these kinds of interviews is quickly running out. When I interviewed many of these alumni in the mid-1990s, nearly all of them were already in their seventies or eighties.

However, many prominent regional schools are also realizing that mortality is robbing them of a valuable historical resource, and some are conducting oral history projects to collect the stories and experiences of their alumni. As schools pass major anniversaries of 90 or

100 years, they are publishing some of these memoirs, reminiscences, and interviews in commemorative volumes, some of which are finding their way into local libraries. Although the reminiscences in commemorative volumes often lack the color and detail that comes across in face-to-face interviews, these volumes have the benefit of making these memories part of the permanent historical record.

Chinese-Language Scholarship on Education and Student Politics

Chinese scholars have produced numerous studies of Chinese students' political activism during the late Qing and Republican periods. The bibliographic essay in Jeffrey Wasserstrom's *Student Protests* details many of the documentary collections and studies of Chinese student protest movements published through the early 1990s. Publishing in this area is ongoing in the lower Yangzi region. For instance, in 1998 Hangzhou University Press published the *Hangzhou qingnian yundong zhi* (Annals of the youth movement in Hangzhou), and in 2002 the Shanghai Academy of Social Science's History Institute produced the massive *Shanghai qingnian zhi* (Shanghai youth gazetteer), which details youth activism and the formation and development of the Communist Youth League in Shanghai. Also continuing is publication of general histories of education reform and modernization in early twentieth-century China. Rather than try to survey this vast literature, in this section I highlight four studies that depart from the mainstream historiography, which tends to focus either on student activism or institutional history approaches to educational reform, and offer fresh interpretive perspectives.

Sang Bing's *Wanqing xuetang xuesheng yu shehui bianqian* (Late Qing modern school students and social change) locates the origins of modern Chinese student activism in the protests and political movements of the last decades of the Qing rather than in the May Fourth period. He thereby challenges the conventional periodization that marks 1919 as the beginning of China's student movement. Just as importantly, Sang situates student activism in relation to both institutional changes in the newly introduced modern schools and broader social and political trends. By analyzing this first generation of student activism, Sang reconstructs the initial emergence of modern students as a distinct social group and important political agents. Sang's impressive book, in many ways, sets the stage for this study and many others.

Taiwan scholar Huang Jinlin in *Lishi, shenti, guojia: jindai zhongguo de shenti xingcheng, 1895–1937* (History, body, nation-state: the formation of the modern Chinese body, 1895–1937) adopts a Foucauldian approach to develop a genealogy of the nationalized body in modern China. Central to this analysis is how the physical and moral training of education reshaped Chinese bodies to accord with the goals of nation-state formation in an economically and politically competitive world. For instance, Huang explores projects such as military-citizenship education (*junguomin jiaoyu*) and the 1920s movement for civic education that had national and geopolitical implications but concentrated on transforming the bodies of individual citizens.

Shanghai Academy of Social Science historian Li Huaxing was the lead author and project leader of the *Minguo jiaoyushi* (Republican education history), a comprehensive account of education during the Republican period. Li, as an intellectual and cultural historian, relates changes in the education system and curriculum to broader currents of thought and various trends of social and political change, illustrating how education was shaped by and contributed to these trends. Li and his collaborators avoid being mechanistic or reductionist in drawing these connections. For instance, even as they relate the emergence of the New School System to intellectual trends in the New Culture Movement, they also reconstruct a long gestation period of dialogue about educational reform that started in 1915, in step with the project of social reform. As a result, they suggest that educational reform efforts fed as well as built on trends in New Culture thought.

Similarly, Academia Sinica's Lü Fangshang in *Cong xuesheng yundong dao yundong xuesheng (Minguo banian zhi shibanian)* (From student movements to mobilizing students, 1919–29) describes a complex relationship between early Republican student activism and Nationalist and Communist Party efforts to mobilize students in the context of the Northern Expedition. Moreover, he uses the rich materials of the Nationalist Party History Archive to detail the competing approaches to youth mobilization that strained the Nationalist Party during the National Revolution and after. The result is a seminal work for understanding how the parties accommodated student activism and incorporated students as political agents.

All these works deal in some way with education, but they seek to relate educational reform to changes in other arenas of society and politics. In different ways, they all view the sphere of education as a context for intellectual, social, and political struggles, making schools and students central to modern Chinese history by using approaches similar to those I have attempted here. They offer valuable resources and insights for scholars interested in the interaction among schools, students, and citizenship.

Works Cited

Abe, Hiroshi. "Borrowing from Japan: China's First Modern Educational System." In *China's Education and the Industrialized World*, ed. Ruth Hayhoe and Marianne Bastid. Armonk, NY: M. E. Sharpe, 1987.

Alejandro, Roberto. *Hermeneutics, Citizenship, and the Public Sphere*. Albany: State University of New York Press, 1993.

Althusser, Louis. "Ideology and Ideological State Apparatuses (Notes Toward an Investigation)." In *Lenin and Philosophy and Other Essays*. New York: Monthly Review Press, 1971.

Anagnost, Ann. "The Politicized Body." In *Body, Subject, and Power in China*, ed. Angela Zito and Tani E. Barlow. Chicago: University of Chicago Press, 1994.

———. *National Past-Times: Narrative, Representation, and Power in Modern China*. Durham: Duke University Press, 1997.

———. "The Corporeal Politics of Quality (Suzhi)." *Public Culture* 16, no. 2 (2004): 189–208.

Anderson, Benedict. *Imagined Communities: Reflections on the Origin and Spread of Nationalism*. London: Verso Press, [1983] 1991.

Averill, Stephen C. "The Cultural Politics of Local Education in Early Twentieth-Century China." *Twentieth-Century China* 32, no. 2 (April 2007): 4–32.

Ayers, William. *Chang Chih-tung and Educational Reform in China*. Cambridge: Harvard University Press, 1971.

Bailey, Paul J. *Reform the People: Changing Attitudes Towards Popular Education in Early Twentieth Century China*. Vancouver: University of British Columbia Press, 1990.

———. "Active Citizen or Efficient Housewife? The Debate over Women's Education in Early Twentieth-Century China." In *Education, Culture, and Identity in Twentieth-Century China*, ed. Glen Peterson, Ruth Hayhoe, and Yongling Lu. Ann Arbor: University of Michigan Press, 2001.

Balakrishnan, Gopal, ed. *Mapping the Nation*. London and New York: Verso, 1996.

Balibar, Etienne, and Immanuel Wallerstein. *Race, Nation, Class: Ambiguous Identities*. London: Verso, 1991.

Barkan, Lenore. "Patterns of Power: Forty Years of Elite Politics in a Chinese County." In *Chinese Local Elites and Patterns of Dominance*, ed. Joseph W. Esherick and Mary Backus Rankin. Berkeley: University of California Press, 1990.

Barme, Geremie R. "To Screw Foreigners Is Patriotic." In *In the Red: On Contemporary Chinese Culture*. New York: Columbia University Press, 1999.

Bastid, Marianne. *Educational Reform in Early Twentieth-Century China*. Translated by Paul Bailey. Ann Arbor: Center for Chinese Studies, University of Michigan, 1988.

Bastid, Marianne, and Ruth Hayhoe, eds. *China's Education and the Industrialized World: Studies in Cultural Transfer*. Armonk, NY: M. E. Sharpe, 1987.

Beijing tushuguan, and Renmin jiaoyu chubanshe tushuguan, eds. *Minguo shiqi zongshumu (1911–1949): Zhong xiao xue jiaocai* (Republican-period general catalog [1911–1949]: Secondary and primary school teaching materials). Beijing: Shumu wenxian chubanshe, 1995.

Beiner, Ronald, ed. *Theorizing Citizenship*. Albany: State University of New York Press, 1995.

Bell, Catherine M. *Ritual Theory, Ritual Practice*. New York: Oxford University Press, 1992.

Benji quanti. "Wuben yiwu xuexiao de gaikuang" (The general situation of Wuben's common school). In *Shanghai xianli Wuben nüzi zhongxuexiao di'er jie biye jinianlu* (A record commemorating the second graduating class at Shanghai County Wuben Girls' Middle School) (July 1920).

"Benxiao duan xun" (Brief news items of our school). *Minli xuesheng* (January 1935): 21–22.

"Benxiao kangri jiuguo yundong zhi shishi banfa" (Methods of implementation for this school's Anti-Japanese National Salvation Movement). *Songjiang nüzhong xiaokan* 23 (November 5, 1931): 2–3.

Bian Chunguang et al., eds. *Chuban cidian* (A dictionary of publishing). Shanghai: Cishu chubanshe, 1992.

Bianji tongren. "Fakanci" (Inaugural preface). *Jiangsu shengli di'er shifan xuexiao xiaokan (Benxiao ershi jie jinian tekan)* 36 (May 5, 1925): 1.

Binns, Christopher A. P. "The Changing Face of Power: Revolution and Accommodation in the Development of the Soviet Ceremonial System: Part I." *Man*, n.s. 14, no. 4 (December 1979): 586–606.

Boorman, Howard, and Richard Howard, eds. *Biographical Dictionary of Republican China*. New York: Columbia University Press, 1970.

Borthwick, Sally. *Education and Social Change in China: The Beginnings of the Modern Era*. Stanford: Hoover Institution Press, 1983.

Bourdieu, Pierre. *Outline of a Theory of Practice*. Translated by Richard Nice. Cambridge: Cambridge University Press, 1977.

———. *Distinction: A Social Critique of the Judgement of Taste*. Translated by Richard Nice. Cambridge: Harvard University Press, 1984.

———. *The Logic of Practice*. Translated by Richard Nice. Stanford: Stanford University Press, 1990.

Bowles, Samuel, and Herbert Gintis. *Schooling in Capitalist America: Educational Reform and the Contradictions of Economic Life.* New York: Basic Books, 1976.

Boy Scouts of America. *Handbook for Scout Masters.* New York: Boy Scouts of America, 1913–14.

———. *Revised Handbook for Boys.* N.p.: Boy Scouts of America [1927]. 1st rev. ed., 1935.

Brokaw, Cynthia J. *The Ledgers of Merit and Demerit: Social Change and Moral Order in Late Imperial China.* Princeton: Princeton University Press, 1991.

———. "On the History of the Book in China." In *Printing and Book Culture in Late Imperial China,* ed. Cynthia J. Brokaw and Kai-wing Chow. Berkeley: University of California Press, 2005.

Brook, Timothy. "Family Continuity and Cultural Hegemony: The Gentry of Ningbo, 1368–1911." In *Chinese Local Elites and Patterns of Dominance,* ed. Joseph W. Esherick and Mary B. Rankin. Berkeley: University of California Press, 1976.

Brubaker, Rogers. *Citizenship and Nationhood in France and Germany.* Cambridge: Harvard University Press, 1992.

Bullard, Monte R. *The Soldier and the Citizen: The Role of the Military in Taiwan's Development.* Armonk, NY: M. E. Sharpe, 1997.

Cai Yuanpei. *Dingzheng zhongxue xiushen jiaokeshu* (Revised middle school moral cultivation textbook). Shanghai: Shangwu yinshuguan, [1912] 1913.

———. "Quanguo linshi jiaoyu huiyi kaihui ci" (Opening address of the Provisional National Educational Conference). In *Cai Yuanpei jiaoyu lunzhu xuan,* ed. Gao Pingshu. N.p.: Renmin jiaoyu chubanshe, 1991.

Cao Juren. *Wo yu wode shijie* (Me and my world). 2 vols. Taipei: Longwen chubanshe, 1990.

Cao Wenlin and Liu Heting. "Zhongxue zhi junshi jiaoyu" (Middle school military education). *Jiangsu shengli Shanghai zhongxuexiao banyuekan* 56 (October 1931): 1–3.

Cao Wenlin and Shen Yizhen. "Gaozhong xuesheng shenghuo junduihua zhi chubu jihua" (A preliminary plan for militarizing high school student life). *Jiangsu shengli Shanghai zhongxuexiao banyuekan* 76 (October 22, 1933): 26–28.

Carter, James H. *Creating a Chinese Harbin: Nationalism in an International City, 1916–1932.* Ithaca: Cornell University Press, 2002.

Cavendish, Patrick. "The 'New China' of the Kuomintang." In *Modern China's Search for a Political Form,* ed. Jack Gray. London: Oxford University Press, 1969.

Chan, Anita. *Children of Mao: Personality Development and Political Activism in the Red Guard Generation.* Seattle: University of Washington Press, 1985.

Chan, Anita; Richard Madsen; and Jonathan Unger. *Chen Village Under Mao and Deng.* Berkeley: University of California Press, 1992.

Chan, Ming K., and Arif Dirlik. *Schools into Fields and Factories: Anarchists, the Guomindang, and the National Labor University in Shanghai, 1927–1932.* Durham: Duke University Press, 1991.

Chan, Wing-tsit, trans. and comp. *A Source Book in Chinese Philosophy.* Princeton: Princeton University Press, 1963.

Chang Daozhi and Yu Jiaju. *Xuexiao fengchao de yanjiu* (Research on school disturbances). Shanghai: Jiaoyu zazhi she and Shangwu yinshuguan, 1925.

Chang, Hao. *Liang Ch'i-ch'ao and Intellectual Transition in China, 1890–1907*. Cambridge: Harvard University Press, 1971.

———. *Chinese Intellectuals in Crisis: Search for Order and Meaning, 1890–1911*. Berkeley: University of California Press, 1987.

Chang, Maria Hsia. *The Chinese Blue Shirt Society: Fascism and Developmental Nationalism*. Berkeley: Institute of East Asian Studies, University of California, 1985.

Chatterjee, Partha. *Nationalist Thought and the Colonial World: A Derivative Discourse?* Minneapolis: University of Minnesota Press, 1986.

———. *The Nation and Its Fragments: Colonial and Postcolonial Histories*. Princeton: Princeton University Press, 1993.

Chauncey, Helen R. *Schoolhouse Politicians: Locality and State During the Chinese Republic*. Honolulu: University of Hawaii Press, 1992.

Che Mingshen. "Wusa can'an de yiyi" (The significance of the May Thirtieth massacre). *Jiangsu shengli Shanghai zhongxue banyuekan* 71 (May 1, 1933): 29–32.

Chen Chaozong. "Zhongguo tongzijun de xianzai yu guoqu (xu)" (The past and present of the Chinese Scouts [continued]). *Tongzijun silingbu yuekan* 2 (March 1, 1929): 47–52.

Chen Chengze. *[Gongheguo jiaokeshu] Fazhi dayi* ([Republican textbook series] Outline of the legal system). Shanghai: Shangwu yinshuguan, 1913.

Chen Duxiu. "Shixing minzhi de jichu" (The basis for implementing popular rule). In *Chen Duxiu zhuzuo xuan* (A selection of Chen Duxiu's writings), 2: 28–39. Shanghai: Shanghai renmin chubanshe, 1993.

Chen Erchun. "Xin shenghuo yundong zhi wojian" (My views of the New Life Movement). In *Erliu ji jikan* (Publication of the class of 1937). June 1934.

Chen Gongmao. "Xuan Zhonghua 'Sha Xuantong' yi wen" (Xuan Zhonghua's article "Kill Xuantong"). *Zhejiang wenshi ziliao xuanji* 19 (1981): 82–83.

Chen Heting. "Linhai geming qingnian de yaolan—Yichou dushu she" (The cradle of Linhai's revolutionary youth—The 1925 Reading Society). *Zhejiang wenshi ziliao xuanji* 19 (1981): 147–59.

Chen Hengzhe. *[Xin xuezhi gaoji zhongxue jiaokeshu] Xiyangshi* ([New School System high school textbook] Western history). 2 vols. Shanghai: Shangwu yinshuguan, 1924, 1926.

Chen Jiashan. "Xuesheng zizhihui gongzuo baogao" (Work report of the student self-government association). *Shaozhong*, June 1933: 173–74.

Chen Jiang and Chen Yingnian, eds. *Shangwu yinshuguan jiushiwunian—wo he Shangwu yinshuguan, 1897–1992* (Ninety-five years of the Commercial Press—the Commercial Press and me, 1897–1992). Beijing: Shangwu yinshuguan, 1992.

Chen Jinjin. "Kangzhan qian Guomindang de jiaoyu zhengce (Minguo shisan nian zhi ershiliu nian)" (Nationalist Party education policy before the War of Resistance [1924–37]). *Guoshiguan guankan*, n.s. 13 (December 1992): 175–204.

Chen Lifu. "Dushu yundong zhi zhenyi" (The true meaning of the study movement). Recorded by Wen Lin. *Zhejiang qingnian* 1, no. 7 (May 1935): 3–9.

Ch'en, Li-fu. *The Storm Clouds Clear over China: The Memoir of Ch'en Li-fu.* Edited by Sidney H. Chang and Ramon H. Myers. Stanford: Hoover Institution Press, 1994.

Chen Lingmei and Ji Shujuan. "Benxiao ben xuenian dashiji" (A record of events at this school during this academic year). *Suzhou Zhenhua nüxue xiaokan* 1 (May 1927): 32–34.

Chen Shiying. "Cong Daji minjiaoguan dao Lianji minzu xiaoxue" (From the Daji Mass Education Center to the Lianji Nationality Primary School). *Lishui wenshi ziliao* 7 (1990): 123–29.

Chen Yuan et al., eds. *Shangwu yinshuguan jiushinian—wo he Shangwu yinshuguan, 1897–1987* (Ninety years of the Commercial Press—the Commercial Press and me, 1897–1987). Beijing: Shangwu yinshuguan, 1987.

Chen, Yung-fa. *Making Revolution: The Communist Movement in Eastern and Central China, 1837–1945.* Berkeley: University of California Press, 1986.

Chen Zhaokui. "Minzhong xuexiao baogao" (Report on the school for the masses). *Zhejiang shengli Hangzhou gaoji zhongxue xiaokan* 159 (December 25, 1936): 1584–86.

Chen Zhenbai. "Zhonghua minzu de shengming xian" (The lifeline of the Chinese nation). *Jiangsu shengli Shanghai zhongxue banyuekan* 83–84 (May 12, 1934): 3–4.

Cheng Jimei. *Tongzijun zuzhifa* (Organizational method for the Boy Scouts). Shanghai: Zhonghua shuju, 1922.

Cheng Liying. "Guanyu Shangzhongshi shizhengfu" (Regarding Shanghai Middle Municipality's government). *Jiangsu shengli Shanghai zhongxue banyukan* 83–84 (May 12, 1934): 14–15.

Chengrong. "Qingnian he huanjing" (Youth and environment). *Minli xunkan (xuesheng zhuanhao)*, June 1936: 5–6.

Chongbian riyong baike quanshu. 3 vols. Shanghai: Shangwu yinshuguan, 1934.

"Choukai shimin dahui xuzhi" (Essential knowledge for planning to hold a citizens' assembly). In *Minguo ershisi nian quanguo xin shenghuo yundong* (The nationwide New Life Movement in 1935), ed. Xin shenghuo yundong cujin zonghui, 528: 353–60; 529: 361–73. From *Jindai zhongguo shiliao congkan*, 3rd compilation, collection no. 53. Taipei: Wenhai chubanshe, n.d.

Chow, Tse-tsung. *The May 4th Movement: Intellectual Revolution in Modern China.* Cambridge: Harvard University Press, 1960.

———. *Research Guide to the May Fourth Movement: Intellectual Revolution in Modern China, 1915–1924.* Cambridge: Harvard University Press, 1963.

Chu Jinke. "Yu youren lun zhong ri jiaoshe shu" (A letter to friends discussing the Sino-Japanese negotiations). *Jiangsu shengli diwu zhongxuexiao zazhi* 2 (September 1915): 3–5.

———. "Shu benxiao zizhihui zhi dazhi" (Explaining the purpose of this school's self-government association). *Jiangsu shengli diwu zhongxuexiao zazhi* 2 (September 1915): 5–7.

Ch'ü, T'ung-tzu. *Local Government in China Under the Ch'ing.* Cambridge: Harvard University Press, 1962.

Claval, Paul. "From Michelet to Braudel: Personality, Identity and Organization of France." In *Geography and National Identity*, ed. David Hooson. Oxford: Blackwell, 1994.

Coble, Parks M., Jr. *The Shanghai Capitalists and the Nationalist Government, 1927–1937*. Cambridge: Harvard University Press, 1980.

——. *Facing Japan: Chinese Politics and Japanese Imperialism, 1931–1937*. Cambridge: Council on East Asian Studies, Harvard University, 1991.

Cochran, Sherman. "Commercial Penetration and Economic Imperialism in China: An American Cigarette Company's Entrance into the Market." In *America's China Trade in Historical Perspective*, ed. John K. Fairbank and Ernest May. Cambridge: Harvard University Press, 1986.

Cohen, Paul A. "Remembering and Forgetting National Humiliation in Twentieth-Century China." *Twentieth-Century China* 27, no. 2 (April 2002): 1–39.

Cong, Xiaoping. "Localizing the Global, Nationalizing the Local: The Role of Teachers' Schools in Making China Modern, 1897–1937." Ph.D. diss., University of California, Los Angeles, 2001.

Corrigan, Philip, and Derek Sayer. *The Great Arch: English State Formation as Cultural Revolution*. New York: Basil Blackwell, 1985.

Culp, Robert. "Elite Association and Local Politics in Republican China: Educational Institutions in Jiashan and Lanqi Counties, Zhejiang, 1911–1937." *Modern China* 20, no. 4 (October 1994): 446–77.

——. "Research Note: Shanghai's Lexicographical Publishing House Library as a Resource for Research on Republican Period Popular Culture and Education," *Republican China* 22, no. 2 (April 1997): 103–9.

——. "Self-determination or Self-discipline? The Shifting Meanings of Student Self-government in 1920s Jiangnan Middle Schools." *Twentieth-Century China* 23, no. 2 (April 1998): 1–39.

——. "'China—The Land and Its People': Fashioning Identity in Secondary School History Textbooks, 1911–1937." *Twentieth-Century China* 26, no. 2 (April 2001): 17–62.

——. "Setting the Sheet of Loose Sand: Conceptions of Society and Citizenship in Nanjing Decade Party Doctrine and Civics Textbooks." In *Defining Modernity: Guomindang Rhetorics of a New China, 1920–1980*, ed. Terry Bodenhorn. Ann Arbor: Center for Chinese Studies Publications Office, University of Michigan, 2002.

——. "Mediating Modernity: Textbook Publishing and the Spread of Ideas in Republican China." Paper presented at the annual meeting of the Association for Asian Studies, New York, March 2003.

——. "Rethinking Governmentality: Training, Cultivation, and Cultural Citizenship in Nationalist China." *Journal of Asian Studies* 65, no. 3 (August 2006): 529–54.

Dai Jitao. *Dai Jitao xiansheng wencun* (Dai Jitao's collected works). 4 vols. Taipei: Zhongguo Guomindang zhongyang weiyuanhui, 1959.

——. "Sun Wen zhuyi zhi zhexue de jichu" (The philosophical foundations of Sun Yat-senism). In *Sun Wen zhuyi lunji* (Essays in Sun Yat-senism). Taipei: Wenxing shudian, [May 1925] 1965.

Dai Ren (Jean-Pierre Drège). *Shanghai Shangwu yinshuguan* (Shanghai's Commercial Press). Translated by Li Tongshi of *La Commercial Press de Shanghai, 1897–1949* (Paris: Collège de France, Institute des Hautes Etudes Chinoises, 1978). Beijing: Shangwu yinshuguan, 1996.

Dajun. "Women de chulu" (Our outlet). *Minli xuesheng,* January 1935: 33–36.

Davis, Deborah S., ed. *The Consumer Revolution in Urban China.* Berkeley: University of California Press, 2000.

de Bary, Wm. Theodore. Introduction to *Self and Society in Ming Thought,* ed. Wm. Theodore de Bary and the Conference on Ming Thought. New York: Columbia University Press, 1970.

———. *Learning for One's Self: Essays on the Individual in Neo-Confucian Thought.* New York: Columbia University Press, 1991.

Deng Yuanzhong. *Guomindang hexin zuzhi zhenxiang: Lixingshe, Fuxingshe, ji suowei "Lanyishe" de yanbian yu chengzhang* (The real situation of the Nationalist Party's core organizations: the evolution and growth of the Society for Vigorous Practice, the Renaissance Society, and the so-called Blue Shirts). Taipei: Lianjing, 2000.

Dewey, John. *Lectures in China, 1919–1920: On Logic, Ethics, Education, and Democracy.* Translated and edited by Robert W. Clopton and Tsuin-chen Ou. Yangmingshan: Chinese Culture University Press, China Academy, 1985.

Dineng. "Jinri de zhongxuesheng" (Today's middle school students). *Shaozhong,* June 1933: 47–49.

"Disanjie xuesheng zizhihui ge yanjiu hui she renshu tongji biao" (Statistical chart of the numbers of students in each research society during the student self-government association's third session). In *Xuesheng zizhihui gaikuang* (The general situation of the student self-government association), ed. Jiangsu shengli Suzhou zhongxue. Suzhou: Jiangsu shengli Suzhou zhongxue, 1932.

"Dishijie shizhengfu jinxing fangzhen" (Current policies of the tenth session of the municipal government). *Jiangsu shengli Shanghai zhongxue banyuekan* 83–84 (May 12, 1934): 12–17.

Dickson, Bruce. "Do Good Businessmen Make Good Citizens? An Emerging Collective Identity Among China's Entrepreneurs." In *Changing Meanings of Citizenship in Modern China,* ed. Merle Goldman and Elizabeth J. Perry. Cambridge: Harvard University Press, 2002.

Dikötter, Frank. *The Discourse of Race in Modern China.* Stanford: Stanford University Press, 1992.

Dirlik, Arif. "The Ideological Foundations of the New Life Movement: A Study in Counterrevolution." *Journal of Asian Studies* 34, no. 4 (August 1975): 945–80.

———. *Revolution and History: Origins of Marxist Historiography in China, 1919–1937.* Berkeley: University of California Press, 1978.

———. *The Origins of the Chinese Communist Party.* New York: Oxford University Press, 1989.

———. *Anarchism in the Chinese Revolution.* Berkeley: University of California Press, 1991.

Dong Qicai. "Canjia quansheng tongzijun da jianyue ji" (A record of participating in the provincial Boy Scouts' general review). *Ningbo zhongxuesheng* 6 (June 1935): 14–16.

Dong Qijun. "Jing Hengyi yu Zhejiang shengli disi zhongxue" (Jing Hengyi and Zhejiang Provincial Fourth Middle School). *Ningbo wenshi ziliao* 1 (1983): 74–96.

Dong Shulin. "'Zhe yishi xuechao' de yingxiang" (The influence of "Zhejiang First Normal's study storm"). *Hangzhou wenshi ziliao* 2 (1983): 1–10.

Du Weitao and Zhang Liuquan. *Gaozhong gongmin* (High school civics). 3 vols. Shanghai: Zhonghua shuju, 1935.

———. *Chuzhong gongmin* (Lower middle school civics). 3 vols. Shanghai: Zhonghua shuju, 1935.

Duara, Prasenjit. *Rescuing History from the Nation: Questioning Narratives of Modern China*. Chicago: University of Chicago Press, 1995.

———. *Sovereignty and Authenticity: Manchukuo and the East Asian Modern*. Lanham, MD: Rowman and Littlefield, 2003.

Dunch, Ryan. *Fuzhou Protestants and the Making of Modern China, 1857–1927*. New Haven: Yale University Press, 2001.

———. "Mission Schools and Modernity: The Anglo-Chinese College, Fuzhou." In *Education, Culture, and Identity in Twentieth-Century China*, ed. Glen Peterson, Ruth Hayhoe, and Yongling Lu. Ann Arbor: University of Michigan Press, 2001.

———. "Science, Religion, and the Classics in the Texts and Curricula of Christian Higher Education in China to 1920." Paper prepared for the conference "The American Context of China's Christian Colleges," Wesleyan University, Middletown, CT, September 5–7, 2003. Cited with permission of the author.

Eastman, Lloyd E. *The Abortive Revolution: China Under Nationalist Rule, 1927–1937*. Cambridge: Harvard University Press, 1974.

———. "The Kuomintang in the 1930s." In *The Limits of Change: Essays on Conservative Alternatives in Republican China*, ed. Charlotte Furth. Cambridge: Harvard University Press, 1976.

———. "Nationalist China During the Nanking Decade, 1927–1937." In *The Cambridge History of China*, ed. John K. Fairbank and Albert Feuerwerker, vol. 13, part 2. New York: Cambridge University Press, 1986.

———. "The Rise and Fall of the Blueshirts: A Review Article." *Republican China* 12 (November 1987): 30–40.

Elias, Norbert. *The Civilizing Process: The History of Manners and State Formation and Civilization*. Cambridge: Blackwell, 1994.

Elman, Benjamin A. *Classicism, Politics, and Kinship: The Ch'ang-chou School of New Text Confucianism in Late Imperial China*. Berkeley: University of California Press, 1990.

———. *A Cultural History of Civil Examinations in Late Imperial China*. Berkeley: University of California Press, 2000.

Elman, Benjamin A., and Alexander Woodside, eds. *Education and Society in Late Imperial China, 1600–1900*. Berkeley: University of California Press, 1994.

Enloe, Cynthia. *Does Khaki Become You? The Militarisation of Women's Lives*. London: South End Press, 1983.

Er er shu bao she tong ren. "Er er shu bao she yan'ge" (The development of the Twenty-Two Reading Society). In *Zhejiang shengli di wu zhongxuexiao xianxing guicheng yilan*. [Shaoxing:] n.p., 1925.

Er er wu tongzijun shubao bianyi she. *Tongzijun chuji kecheng* (Boy Scouts elementary curriculum). Shanghai: Er er wu tongzijun shubao yongpin she, 1936.

"Erwanyu ren zhi guochi jinian (Zhenjiang)" (National humiliation commemoration of more than twenty thousand people in Zhenjiang). MGRB, May 11, 1923.

Esherick, Joseph W. *Reform and Revolution in China: The 1911 Revolution in Hunan and Hubei*. Berkeley: University of California Press, 1976.

Esherick, Joseph W., and Jeffrey N. Wasserstrom. "Acting out Democracy: Political Theater in Modern China." *Journal of Asian Studies* 49, no. 4 (November 1990): 835–65.

Evans, Harriet. "'Comrade Sisters': Gendered Bodies and Spaces." In *Picturing Power in the People's Republic of China: Posters of the Cultural Revolution*, ed. Harriet Evans and Stephanie Donald. Lanham, MD: Rowman and Littlefield, 1999.

———. "Marketing Femininity: Images of the Modern Chinese Woman." In *China Beyond the Headlines*, ed. Timothy B. Weston and Lionel Jensen. Lanham, MD: Rowman and Littlefield, 2000.

Fairbank, John K., ed. *The Chinese World-Order: Traditional China's Foreign Relations*. Cambridge: Harvard University Press, 1968.

Falasca-Zamponi, Simonetta. *Fascist Spectacle: The Aesthetics of Power in Mussolini's Italy*. Berkeley: University of California Press, 1997.

Fan Chongshan. "Dang lingdao de Yangzhou zaoqi geming douzheng" (Yangzhou's early party-led revolutionary struggles). In *Yangzhou shizhi ziliao*, 1 (1981): 52–58.

———. "Yangzhou zaoqi dang zuzhi de lingdaoren—Cao Qiqian" (A leader of Yangzhou's early party organization—Cao Qiqian). In *Yangzhou shizhi ziliao*, 1 (1981): 47–51.

———. "Yangzhou 'Wusa' fandi fengchao" (Yangzhou's "May Thirtieth" antiimperialist storm). *Yangzhou wenshi ziliao* 2 (1982): 1–8.

Fan Chongshan and Su Hehu. "'Wusi' yundong zai Yangzhou" (The "May Fourth" Movement in Yangzhou). *Yangzhou shizhi ziliao*, 1 (1981): 38–46.

Fan Xiaoliu. *Xinbian nü tongzijun chuji kecheng* (New edition Girl Scout elementary curriculum). Shanghai: Er er wu tongzijun shubao yongpin she, 1935.

———. *Xinbian nü tongzijun zhongji kecheng* (New edition Girl Scout intermediate curriculum). Shanghai: Er er wu tongzijun shubao yongpin she, 1936.

———. *Xinbian tongzijun chuji kecheng* (New edition elementary curriculum for Scouts). Shanghai: Er er wu tongzijun shubao yongpin she, 1937.

Feng Heyi. "Pochu mixin" (Eliminating superstition). *Sizhong xuesheng* 1 (January 1933): 37–41.

Feng Yuanhuai. "Ji xun de yiyi" (The significance of our class motto). In *Renshenji biyekan* (Yearbook of the 1932 graduating class). [Suzhou: Zhenhua nüzi zhongxue, 1932.]

Fewsmith, Joseph. "From Guild to Interest Group: The Transforming of Public and Private in Late Qing China." *Comparative Studies in Society and History* 25, no. 4 (1983): 617–40.

———. *Party, State, and Local Elites in Republican China: Merchant Organizations and Politics in Shanghai, 1890–1930*. Honolulu: University of Hawaii Press, 1985.

Fitzgerald, John. *Awakening China: Politics, Culture, and Class in the Nationalist Revolution*. Stanford: Stanford University Press, 1996.

Fogel, Joshua A., and Peter G. Zarrow. *Imagining the People: Chinese Intellectuals and the Concept of Citizenship, 1890–1920*. Armonk, NY: M. E. Sharpe, 1997.

Foucault, Michel. *Discipline and Punish: The Birth of the Prison*. Translated by A. Sheridan. New York: Vintage Books, 1979.

———. Afterword to *Michel Foucault: Beyond Structuralism and Hermeneutics*, ed. Hubert L. Dreyfus and Paul Rabinow. Chicago: University of Chicago Press, 1982.

Friedman, Edward. "Reconstructing China's National Identity: A Southern Alternative to Mao-Era Anti-imperialist Nationalism." *Journal of Asian Studies* 53, no. 1 (February 1994): 67–91.

Friedman, Edward; Paul G. Pickowicz; and Mark Selden. *Chinese Village, Socialist State*. New Haven: Yale University Press, 1991.

Friedman, Sara L. "Civilizing the Masses: The Productive Power of Cultural Reform Efforts in Late Republican-Era Fujian." In *Defining Modernity: Guomindang Rhetorics of a New China, 1920–1970*, ed. Terry Bodenhorn. Ann Arbor: Center for Chinese Studies, University of Michigan Press, 2002.

Fu Donghua. *[Fuxing chuji zhongxue jiaokeshu] Guowen* ([Revival lower middle school textbook] National language). 6 vols. Shanghai: Shangwu yinshuguan, 1933–35.

Fu Weiping. *[Fuxing chuji zhongxue jiaokeshu] Benguoshi* ([Revival lower middle school textbook] Chinese history). 4 vols. Shanghai: Shangwu yinshuguan, 1933.

Fu Yunsen. *[Gongheguo jiaokeshu] Xiyangshi* ([Republican textbook series] Western history). 2 vols. Shanghai: Shangwu yinshuguan, 1913.

———. *[Xiandai chuzhong jiaokeshu] Shijieshi* ([Modern lower middle school textbook] World history). 2 vols. Shanghai: Shangwu yinshuguan, 1924–25.

———. *[Xin xuezhi] Lishi jiaokeshu* ([New School System] History textbook). 2 vols. Shanghai: Shangwu yinshuguan, [1923] 1924–25.

Fujitani, Takashi. *Splendid Monarchy: Power and Pagentry in Modern Japan*. Berkeley: University of California Press, 1996.

Fung, Edmund S. K. *In Search of Chinese Democracy: Civil Opposition in Nationalist China, 1929–1949*. Cambridge: Cambridge University Press, 2000.

Gamble, Sidney D. *How Chinese Families Live in Peiping: A Study of the Income and Expenditure of 283 Chinese Families Receiving from $8 to $550 Silver per Month*. New York: Funk and Wagnalls, 1933.

Gao Bingsheng and Lou Xueli. "Kangzhan shiqi de shengli liangao—jian huai Zhang Yintong he quanti laoshi" (Provincial Associated High School during the War of Resistance—and remembering Principal Zhang Yintong and all the teachers). *Lishui wenshi ziliao* 7 (1990): 99–107.

Gao Maisheng and Zhang Shengyu. "Shishi ting ban benguo lishi jiaoxue jindu hou zhi yijian" (Ideas after implementing the teaching schedule for Chinese history promulgated by the [Jiangsu] Department [of Education]). *Suzhong xiaokan* 99 (April 1934): 1–3.

Gao Qi. *Zhongguo jiaoyushi yanjiu: xiandai fenjuan* (Studies in the history of Chinese education: the modern volume). Series edited by Chen Xuexun. Shanghai: Huadong shifan daxue chubanshe, 1994.

"Gaozhong tongxuehui weiyuanhui xize" (Detailed rules for the high school student assembly committee). *Zhejiang yizhong zhoukan* 27 (May 26, 1924): 1.

Garver, John W. "More from the 'Say No Club.'" *China Journal* 45 (January 2001): 151–58.

Ge Suicheng. *[Xin kecheng biaozhun shiyong] Chuzhong benguo dili* ([For use with the new curriculum standards] Lower middle school geography). 4 vols. Shanghai: Zhonghua shuju, 1933–34.

Geisert, Bradley Kent. "Toward a Pluralist Model of KMT Rule." *Chinese Republican Studies Newsletter* 11, no. 2 (February 1982): 1–10.

———. *Radicalism and Its Demise: The Chinese Nationalist Party, Factionalism, and Local Elites in Jiangsu Province, 1924–1931.* Ann Arbor: Center for Chinese Studies, University of Michigan, 2001.

Gellner, Ernest. *Nations and Nationalism.* Ithaca: Cornell University Press, 1983.

"Geming jinianri jianming biao" (Concise chart of revolutionary memorial days) [July 10, 1930]. *Chongbian riyong baike quanshu*, 3: 5787–92. Shanghai: Shangwu yinshuguan, 1934.

Gerth, Karl. *China Made: Consumer Culture and the Creation of the Nation.* Cambridge: Harvard University Asia Center, 2003.

Gilmartin, Christina Kelley. *Engendering the Chinese Revolution: Radical Women, Communist Politics, and Mass Movements in the 1920s.* Berkeley: University of California Press, 1995.

Giroux, Henry A. "Theories of Reproduction and Resistance in the New Sociology of Education: A Critical Analysis." *Harvard Educational Review* 53, no. 3 (1983): 257–93.

Gladney, Dru C. *Muslim Chinese: Ethnic Nationalism in the People's Republic.* Cambridge: Council on East Asian Studies, Harvard University, 1991.

Glosser, Susan L. *Chinese Visions of Family and State, 1915–1953.* Berkeley: University of California Press, 2003.

Goldman, Merle. "The Reassertion of Political Citizenship in the Post-Mao Era: The Democracy Wall Movement." In *Changing Meanings of Citizenship in Modern China*, ed. Merle Goldman and Elizabeth J. Perry. Cambridge: Harvard University Press, 2002.

———. *From Comrade to Citizen: The Struggle for Political Rights in China.* Cambridge: Harvard University Press, 2005.

Goldman, Merle, and Elizabeth J. Perry. "Introduction: Political Citizenship in Modern China." In *Changing Meanings of Citizenship in Modern China*, ed. Merle Goldman and Elizabeth J. Perry. Cambridge: Harvard University Press, 2002.

———, eds. *Changing Meanings of Citizenship in Modern China*. Cambridge: Harvard University Press, 2002.

Goldstein, Joshua S. *War and Gender: How Gender Shapes the War System and Vice Versa*. Cambridge: Cambridge University Press, 2001.

Goodman, Bryna. *Native Place, City, and Nation: Regional Networks and Identities in Shanghai, 1853–1937*. Berkeley: University of California Press, 1995.

———. "Improvisations on a Semicolonial Theme, or, How to Read a Celebration of Transnational Urban Community." *Journal of Asian Studies* 59, no. 4 (November 2000): 889–926.

———. "Democratic Calisthenics." In *Changing Meanings of Citizenship in Modern China*, ed. Merle Goldman and Elizabeth J. Perry. Cambridge: Harvard University Press, 2002.

Graham, Gael. *Gender, Culture, and Christianity: American Protestant Mission Schools in China, 1880–1930*. New York: Peter Lang, 1995.

Grieder, Jerome B. *Hu Shih and the Chinese Renaissance: Liberalism in the Chinese Revolution, 1917–37*. Cambridge: Harvard University Press, 1970.

Gries, Peter Hays. "Tears of Rage: Chinese Nationalist Reactions to the Belgrade Embassy Bombing." *China Journal* 46 (July 2001): 25–43.

———. *China's New Nationalism: Pride, Politics, and Diplomacy*. Berkeley: University of California Press, 2004.

Gu Jiegang, Fan Xiangshan, and Ye Shaojun. *[Xin xuezhi] Guoyu jiaokeshu* ([New School System] National language textbook). 6 vols. Shanghai: Shangwu yinshuguan, 1923–24.

Gu Jiegang and Wang Zhongqi. *[Xiandai chuzhong jiaokeshu] Benguoshi* ([Modern lower middle school textbook] Chinese history). 3 vols. Shanghai: Shangwu yinshuguan, [1923–24] 1926–27.

Gu Ruwen. "Xuesheng fuwu zizhi tuanti zhi wo jian" (My views on student service in self-government groups). *Shangzhongshi huikan* (1931): 1–2.

Gu Shusen and Pan Wenan. *[Xinzhu] Gongmin xuzhi* ([The newly written] Essential knowledge for citizens). 3 vols. Shanghai: Shangwu yinshuguan, [1923] 1924.

Guha, Ranajit. *Dominance Without Hegemony: History and Power in Colonial India*. Cambridge: Harvard University Press, 1997.

Guo Botang and Wei Bingxin. *[Xin zhuyi jiaokeshu] Gaozhong dangyi* ([A new principles textbook] High school party doctrine). 3 vols. Shanghai: Shijie shuju, [1929–30] 1931–32.

"Guochi jinian banfa" (Method for national humiliation commemorations). In *Chongbian riyong baike quanshu*, 5792–93. Shanghai: Shangwu yinshuguan, 1934.

"Guochi jinianri zhi guomin dahui (Jinhua)" (National Humiliation Memorial Day citizens' assembly in Jinhua). MGRB, May 14, 1923.

"Guochi jinianri zhi Shanghai" (Shanghai on National Humiliation Memorial Day). MGRB, May 10, 1925.

"Guochi jinian zhi youxing (Suzhou)" (Parade to commemorate national humiliation in Suzhou). MGRB, May 10, 1923.

Guoli bianyiguan. *Guomin zhengfu chengli yilai shending ji shixiao zhong xiao xue shifan zhiye xuexiao jiaoke tushu yilan* (An overview of secondary, primary, normal, and vocational school textbooks that have been authorized and lost authorization since the founding of the Nationalist government). [Nanjing:] Guoli bianyiguan, 1935.

Habermas, Jürgen. *Structural Transformation of the Public Sphere.* Translated by Thomas Burger. Cambridge: MIT Press, 1989.

Hall, Stuart. "Signification, Representation, Ideology: Althusser and the Post-Structuralist Debates." *Critical Studies in Mass Communication* 2, no. 2 (1985): 91–114.

———. "On Postmodernism and Articulation: An Interview with Stuart Hall." In *Stuart Hall: Critical Dialogues in Cultural Studies*, ed. David Morley and Kuan-Hsing Chen. London and New York: Routledge, 1996.

"Hang gejie jinian wujiu" (Each social sector in Hangzhou commemorates May Ninth). MGRB, May 10, 1926.

Hang Wei. "Shanghai zhongdeng xuexiao de fazhan guocheng" (The developmental process of Shanghai's secondary schools). *Shanghai wenshi ziliao xuanji* 59 (1988): 4–21.

"Hang xuesheng jinian wusi" (Hangzhou students commemorate May Fourth). MGRB, May 6, 1926.

"Hangyuan zhi guochi jinian hui" (Hangzhou's national humiliation memorial assembly). *Shenbao*, May 10, 1924.

Hangzhou diyi zhongxue xiaoqing qishiwu zhounian jiniance (Commemorative volume for Hangzhou First Middle School's 75th anniversary). [Hangzhou:] Hangzhou yizhong qishiwu zhounian xiaoqing choubei bangongshi, 1983.

Hangzhou shizhong yilan (An overview of Hangzhou Municipal Middle School). N.p.: n.p., 1932.

"Hangzhou wuyi jinianhui zhi shengkuang" (The great event of Hangzhou's May Day memorial meeting). MGRB, May 3, 1924.

"Hangzhou yizhong shubao fanmai qishe shi yi" (One matter regarding the start of the Hangzhou First Middle Book-Selling Society). *Zhejiang yizhong zhoukan* 28 (June 2, 1924): 1.

"Hangzhou zhi 'Wusi'" (Hangzhou's "May Fourth"). MGRB, May 7, 1925.

Harrell, Paula. *Sowing the Seeds of Change: Chinese Students, Japanese Teachers, 1895–1905.* Stanford: Stanford University Press, 1992.

Harrison, Henrietta. *The Making of the Republican Citizen: Political Ceremonies and Symbols in China, 1911–1929.* Oxford: Oxford University Press, 2000.

Hayford, Charles W. *To the People: James Yen and Village China.* New York: Columbia University Press, 1990.

Hayhoe, Ruth. "Cultural Tradition and Educational Modernization: Lessons from the Republican Era." In *Education and Modernization: The Chinese Experience*, ed. Ruth Hayhoe. Oxford: Pergamon Press, 1992.

He Bingsong et al., eds. *Zhejiang shengli diyi shifan xuexiao du'an jishi* (A true record of the poisoning case at Zhejiang Provincial First Normal School). Hangzhou: n.p., 1923.

He Qinghu. "Zizhu he huzhu" (Self-help and mutual aid). *Jizhong xuesheng* 4 (June 1935): 2–4.

He Shaozhang. *[Gongheguo jiaokeshu] Jingji dayi* ([Republican textbook series] Outline of economics). Shanghai: Shangwu yinshuguan, 1913.

Henriot, Christian. *Shanghai, 1927–1937: Municipal Power, Locality, and Modernization*. Berkeley: University of California Press, 1993.

Hevia, James L. *Cherishing Men from Afar: Qing Guest Ritual and the Macartney Embassy of 1793*. Durham: Duke University Press, 1995.

Hon, Tze-ki. "Ethnic and Cultural Pluralism: Gu Jiegang's Vision of a New China in His Studies of Ancient History." *Modern China* 22, no. 3 (July 1996): 315–39.

Honig, Emily. *Sisters and Strangers: Women in the Shanghai Cotton Mills, 1919–1949*. Stanford: Stanford University Press, 1986.

———. *Creating Chinese Ethnicity: Subei People in Shanghai, 1850–1980*. New Haven: Yale University Press, 1992.

———. "Maoist Mappings of Gender: Reassessing the Red Guards." In *Chinese Femninities/Chinese Masculinities*, ed. Susan Brownell and Jeffrey N. Wasserstrom. Berkeley: University of California Press, 2002.

Honig, Emily, and Gail Hershatter. *Personal Voices: Chinese Women in the 1980s*. Stanford: Stanford University Press, 1988.

Hooson, David, ed. *Geography and National Identity*. Oxford: Blackwell, 1994.

Hostetler, Laura. *Qing Colonial Enterprise: Ethnography and Cartography in Early Modern China*. Chicago: University of Chicago Press, 2001.

Howland, Douglas. *Borders of Chinese Civilization: Geography and History at Empire's End*. Durham: Duke University Press, 1996.

Hsiao, Kung-chuan. *Rural China: Imperial Control in the Nineteenth Century*. Seattle: University of Washington Press, 1960.

Hu Hongzhou. "Jin sui guochi yue zhi ganxiang" (Thoughts on this year's month of national humiliation). *Shangzhong shikan*, Shimin zuopin teji, June 5, 1934: 20–22.

Hu Huanyong. "Suzhou zhongxue jiaoxun heyi tan" (A discussion of uniting instruction and character-development education at Suzhou Middle School). *Jiangsu jiaoyu* 1, no. 10 (November 1932): 49–55.

Hu Liren. *Zhongji tongzijun* (Intermediate Scouting). Shanghai: Zhonghua shuju, [1936] 1941.

Hu Pu'an. "Guoxue zhi yanjiu" (Research on national learning). Lecture. Recorded by Huang Xijie. *Jiangsu shengli di'er shifan xuexiao xiaokan (Sanzhou jinian hao)* 30 (January 15, 1924): 16–20.

Hu Yuzhi. *[Xin shidai chuzhong] Sanmin zhuyi jiaokeshu, di yi ce, Minzu zhuyi* ([New era lower middle school] Three Principles of the People, vol. 1, nationalism). Shanghai: Xin shidai jiaoyushe and Shangwu yinshuguan, 1927.

Hu Zhide. "Du'an jilüe" (An account of the poisoning incident). In *Zhejiang shengli diyi shifan xuexiao du'an jishi* (A true account of the poisoning case at

Zhejiang Provincial First Normal School), ed. He Bingsong et al. Hangzhou: n.p., 1923.

Hu Zunong. "Gao tongxue shu" (A letter to fellow students). *Minli xunkan* 6 (May 10, 1936): 22–23.

———. "Gao tongxue shu, er" (A second letter to fellow students). *Minli xunkan* 8 (May 30, 1936): 21–23.

Huang, Jianli. *The Politics of Depoliticization in Republican China: Guomindang Policy Towards Student Political Activism, 1927–1949*. Bern: Peter Lang, 1996.

Huang Jinlin. *Lishi, shenti, guojia: Jindai zhongguo de shenti xingcheng, 1895–1937* (History, body, nation-state: The formation of the modern Chinese body, 1895–1937). Taipei: Lianjing, 2000.

Huang Xianhe. "Yi Min-Zhe bian kangri jiuwang ganbu xuexiao" (Remembering the Resist Japan National Salvation Cadre School on the Fujian-Zhejiang border). *Pingyang wenshi ziliao* 7 (1989): 1–5.

Huang Yong. "Huangpu xuesheng de zhengzhi zuzhi ji qi yanbian" (Whampoa students' political organizations and their evolution). In *Wenshi ziliao xuanji*, collection 11, ed. Zhongguo renmin zhengzhi xieshang weiyuanhui, Wenshi ziliao yanjiu weiyuanhui. [Beijing:] Zhongguo wenshi chubanshe, n.d.

Hung, Chang-tai. *War and Popular Culture: Resistance in Modern China, 1937–1945*. Berkeley: University of California Press, 1994.

Ichiko, Chuzo. "The Role of the Gentry." In *China in Revolution: The First Phase, 1900–1913*, ed. Mary C. Wright. New Haven: Yale University Press, 1968.

Israel, John. *Student Nationalism in China, 1927–1937*. Stanford: Stanford University Press, 1966.

———. *Lianda: A Chinese University in War and Revolution*. Stanford: Stanford University Press, 1998.

Israel, John, and Donald W. Klein. *Rebels and Bureaucrats: China's December 9ers*. Berkeley: University of California Press, 1976.

Ji Le. "Zhanshi Lishui wenyi xuanchuan huodong diandi" (Snippets of the artistic propaganda activities in wartime Lishui). *Lishui wenshi ziliao* 8 (1991): 86–87.

Ji Shaofu. *Zhongguo chuban jianshi* (An elementary history of Chinese publishing). Shanghai: Xuelin chubanshe, 1991.

Ji Tongyao, comp. *Guowen* (National language). Vol. 1. [Suzhou: Suzhou zhongxue, 1934.] Mimeographed.

Jia Fengzhen. "Nian nian lai benxiao changchu duanchu" (The strengths and weaknesses of the school these twenty years). *Jiangsu shengli di'er shifan xuexiao xiaokan (Benxiao ershi jie jinian tekan)* 36 (May 5, 1925): 2–3.

Jia [Shaoyi]. "Ben xiao gaozhong geji xuesheng zhunbei shixing junshi guanli" (Students in each class of this school's high school prepare for implementing military management). *Minli xunkan* 3 (April 10, 1936): 6–8; 4 (April 20, 1936): 4–6; 5 (April 30, 1936): 8–9.

Jiang Jiaxiang. "Fuzhong xuesheng shenghuo" (Student life at Southeast University's Affiliated Middle School). In *Shixing xinxuezhi hou zhi dongda fuzhong* (Southeast University Affiliated Middle School after implementing the New School System), ed. Liao Shicheng. Shanghai: Zhonghua shuju, [1924] 1926.

Jiang Lansun. "Bannian lai de xuesheng zizhihui" (A half year of student self-government). *Chuhang* 4 (1935): 109–18.

Jiang Liangfu and Zhao Jingshen, eds. *[Chuji zhongxue] Beixin wenxuan* ([Lower middle school] Beixin's literary selections). Shanghai: Beixin shuju, 1930–34.

Jiang Menglin. "Xuesheng zizhi" (Student self-government). *Xin jiaoyu* 2, no. 2 (October 1919): 118–21.

Jiang Weiqiao. *Jiangsu jiaoyu xingzheng gaikuang* (The administrative situation of Jiangsu education). Shanghai: Shangwu yinshuguan, 1924.

Jiang Weixian. "Ben xiao xuesheng zizhi hui de yan'ge" (The evolution of this school's student self-government association). In *Zhejiang shengli diyi zhong-xuexiao ershiwu nian jinian ce* (The 25th year commemorative volume of Zhejiang Provincial First Middle School). Hangzhou: n.p., 1923.

Jiang Xianduan. "Shimin dahui" (Citizens' assembly). In *Renshenji biyekan* (Yearbook of the 1932 graduating class). [Suzhou: Zhenhua nüzhong, 1932.]

Jiang Zhongzheng [Chiang Kai-shek]. "Xuesheng jizhong xunlian kaixue xunci" (Speech at the opening of the student collective military training). *Zhejiang qingnian* 1, no. 7 (May 1935): 1–2.

———. "Xuesheng ying liyong shujia fuwu shehui" (Students should use summer vacation to serve society). *Zhejiang qingnian* 2, no. 8 (June 1936): 1–2.

"Jiangsu ge zhongdeng xuexiao biyesheng chulu zhuangkuang tongji biao" (Statistical table of the career situations of graduates from each of Jiangsu's secondary schools). *Jiangsu sheng jiaoyuhui nianjian* 8 (1923): 1–16.

"Jiangsu ge zhongdeng xuexiao qinian jian biyesheng chulu zhuangkuang bijiao biao" (Table comparing the career paths of graduates from each of Jiangsu's secondary schools during [the past] seven years). *Jiangsu sheng jiaoyuhui nianjian* 8 (1923): 16–17.

"Jiangsu gexian tongzijun tuan tongji biao" (Statistical chart of the Boy Scout troops in each of Jiangsu's counties) (June 1927). Nationalist Party Historical Archives, Taipei, (Xunlian)bu 1821.

Jiangsu jiaoyuting, comp. *Jiangsu sheng xianxing jiaoyu faling huibian* (Compendium of current educational laws for Jiangsu Province). N.p.: Jiangsu sheng jiaoyuting, 1932.

[Jiangsu jiaoyuting]. "Yinianlai Jiangsu zhongdeng jiaoyu zhi huigu yu zhanwang" (A look back at Jiangsu's secondary education during the last year and a look forward). *Jiangsu jiaoyu* 2, nos. 1–2 (February 1933): 70–96.

"Jiangsu sheng ge zhongdeng xuexiao biyesheng chulu tongji biao" (Statistical table of the careers of graduates from each of Jiangsu Province's secondary schools). *Jiangsu sheng jiaoyuhui nianjian* 6 (1921): 1–14.

"Jiangsu sheng ge zhongdeng xuexiao sannian jian biyesheng chulu zhuangkuang tongji biao" (Statistical table of the career paths of graduates from each of Jiangsu Province's secondary schools during [the past] three years). *Jiangsu sheng jiaoyuhui nianjian* 6 (1921): 14–15.

Jiangsu sheng jiaoyuting. *Jiangsu jiaoyu gailan*. Vol. 1. [Zhenjiang:] Jiangsu sheng jiaoyuting bianshen shi, 1932. Reprint, *Minguo shiliao congkan*, no. 7, ed. Wu Xiangxiang and Liu Shaotang. Taipei: Zhuanji wenxue chubanshe, 1971.

Jiangsu sheng xuesheng jizhong xunlian gongzuo baogao (Work report for the Jiangsu Province student collective [military] training). N.p.: n.p., 1936.

Jiangsu sheng Yangzhou zhongxue jianxiao jiushi zhounian jiniance (Commemorative volume for the 90th anniversary of the founding of Jiangsu Province's Yangzhou Middle School). N.p.: n.p., 1992.

"Jiangsu shengli diba zhongxue renxu ji dashiji" (A chronicle of Jiangsu Provincial Eighth Middle School's class of 1922). In *Jiangsu shengli diba zhongxuexiao renxuji jiniance*. Yangzhou: n.p., 1922.

Jiangsu shengli di'er shifan xuexiao yilan (An overview of Jiangsu Provincial Second Normal School). N.p.: n.p., 1925.

Jiangsu shengli disan zhongxue zazhi (Magazine of Jiangsu Provincial Third Middle School) 3 (1920).

"Jiangsu shengli diwu zhongxuexiao xuesheng zizhi guiyue shixing xize" (Detailed rules for implementation of the Jiangsu Provincial Fifth Middle School student self-government regulations). *Jiangsu shengli diwu zhongxue zazhi*, 2 (September 1915): 6–7.

"Jiangsu shengli diwu zhongxuexiao xueye jiaoshou chengxu biao" (Chart of the academic syllabus for Jiangsu Provincial Fifth Middle School). *Jiangsu shengli diwu zhongxuexiao zazhi* 2 (September 1915).

"Jiangsu shengli diwu zhongxuexiao zizhi guiyue" (Jiangsu Provincial Fifth Middle School student self-government regulations). *Jiangsu shengli diwu zhongxue zazhi* 2 (September 1915): 4–6.

"Jiangsu shengli diwu zhongxuexiao zizhihui yishibu yizhi xize" (Detailed rules governing the consultative assembly and representatives of the Jiangsu Provincial Fifth Middle School student self-government association). *Jiangsu shengli diwu zhongxue zazhi* 2 (September 1915): 7–14.

Jiangsu shengli diyi shifan xuexiao, ed. *Ziyang xiang zizhi zhi* (The self-government system of Ziyang Township). Suzhou: Jiangsu shengli diyi shifan xuexiao, 1921.

———. *Niankan* (Yearbook). Suzhou: Jiangsu shengli diyi shifan xuexiao, 1923.

Jiangsu shengli diyi shifan xuexiao xuesheng ruxue xuzhi (Handbook for incoming students at Jiangsu Provincial First Normal School). [Suzhou:] n.p., 1926.

Jiangsu shengli diyi shifan xuexiao yaolan (A look at the essentials of Jiangsu Provincial First Normal School). [Suzhou:] n.p., 1919.

Jiangsu shengli Shanghai zhongxue chuban weiyuanhui. *Jiangsu shengli Shanghai zhongxue yilan* (An overview of Jiangsu Provincial Shanghai Middle School). Shanghai: Jiangsu shengli Shanghai zhongxue, 1930.

———. *Jiangsu shengli Shanghai zhongxue yilan* (An overview of Jiangsu Provincial Shanghai Middle School). Shanghai: Jiangsu shengli Shanghai zhongxue, 1933.

———. *Jiangsu shengli Shanghai zhongxue yilan* (An overview of Jiangsu Provincial Shanghai Middle School). Shanghai: Jiangsu shengli Shanghai zhongxue, 1936.

Jiangsu shengli Shanghai zhongxue chuzhongbu Shangzhongshi zhengfu, ed. *Shangzhongshi gaikuang* (The general situation of Shanghai Middle Municipality). Shanghai: n.p., 1936.

Jiangsu shengli Songjiang nüzi zhongxue yilan (An overview of Jiangsu Provincial Songjiang Girls' Middle School). [Songjiang:] n.p., 1929.

"Jiangsu shengli Songjiang nüzi zhongxuexiao xuesheng zizhihui zhangcheng" (Jiangsu Provincial Songjiang Girls' Middle School Student Self-government Association regulations). *Songjiang nüzhong xiaokan* 17 (May 6, 1931): 14–16.

Jiangsu shengli Suzhou zhongxue, ed. *Xuesheng zizhihui gaikuang* (The general situation of the student self-government association). Suzhou: Jiangsu shengli Suzhou zhongxue, 1932.

———. *Xunyu guicheng* (Rules for character-development education). N.p.: [Jiangsu shengli Suzhou zhongxue], 1932.

Jiangsu shengli Suzhou zhongxue jiaowu chu. *Jiangsu shengli Suzhou zhongxue geke jiaoxue jindu biao* (Syllabi for instruction in each course at Jiangsu Provincial Suzhou Middle School). Suzhou: Jiangsu shengli Suzhou zhongxue, 1933.

Jiangsu shengli Suzhou zhongxuexiao gaikuang (The general situation of Jiangsu Provincial Suzhou Middle School). N.p.: n.p., 1936.

Jiangsu shengli Wuxi zhongxue gailan (A general overview of Jiangsu Provincial Wuxi Middle School). N.p.: Jiangsu shengli Wuxi zhongxue, 1932.

Jiangsu shengli Wuxi zhongxue shifanke gaikuang (The general situation of the teacher training course at Jiangsu Provincial Wuxi Middle School). N.p.: n.p. [1930].

Jiangsu shengli Yangzhou zhongxue liushi zhounian jinian tekan bianji weiyuanhui, comp. *Jiangsu shengli Yangzhou zhongxue liushi zhounian xiaoqing jinian* (Commemoration of the 60th anniversary of Jiangsu Provincial Yangzhou Middle School). Taipei: Qingzhu muxiao Jiangsu shengli Yangzhou zhongxue chengli liushi zhounian jinian choubei weiyuanhui, 1987.

"Jiangsu shengli Yangzhou zhongxue niaokan" (A bird's-eye view of Jiangsu Provincial Yangzhou Middle School). *Jiangsu jiaoyu* 1, no. 1 (December 1931).

Jiaoyubu. "Zhong xiao xue xuesheng biye huikao zhanxing guicheng" (Provisional rules for secondary and primary students' graduation examinations) [May 26, 1932]. In *Diyici Zhongguo jiaoyu nianjian* (The first yearbook of Chinese education), ed. Jiaoyubu. Shanghai: Shangwu yinshuguan, 1934.

Jiaoyubu, ed. *Di yi ci Zhongguo jiaoyu nianjian* (The first yearbook of Chinese education). Shanghai: Shangwu yinshuguan, 1934. Reprint, 5 vols. Taipei: Shangwu yinshuguan, 1961.

[Jiaoyubu]. *Zhonghua minguo di san ci jiaoyu tongji tubiao* (The third set of educational statistics of the Chinese Republic). [Beijing:] n.p., 1915. In *Kindai Chūgoku kyōiku shi shiryō* (Materials on the history of education in modern China), 5 vols., ed. Taga Akigorō. Tokyo: Nihon gakujutsu shinkōkai, 1972.

Jiaoyubu bianshenchu. *Bianshen huiyi jilu* (Editing and review meeting records). 2 vols. January 11, 1929–March 19, 1932. ZDLD, Jiaoyubu 5, 1211.

Jiaoyubu canshichu. *Jiaoyu faling huibian*, diyiji (Compilation of educational laws, collection no. 1). Changsha: Shangwu yinshuguan, [1936] 1938.

Jiaoyubu tongjishi. *Quanguo zhongdeng xuexiao yilan biao, Zhonghua minguo ershisi niandu* (Tables providing an overview of the nation's secondary schools for 1935). Shanghai: Commercial Press, 1936.

Jiaoyubu tongjishi. *Quanguo jiaoyu tongji jianbian, Zhonghua minguo ershisi niandu* (Volume of national education statistics for 1935). Changsha: Shangwu yinshuguan, 1938.

Jin Wanxiang. "Guonan zhong nüzi yingyou de zeren" (The responsibilities women should have in a time of national crisis). *Songjiang nüzhong xiaokan* 26 (December 30, 1931): 8–9.

Jin Yiqun. "Ningbo qingyun de xianfeng—Qimeng he Peiying nüzhong" (The vanguard of the Ningbo youth movement—Qimeng and Peiying Girls' Middle Schools). *Zhejiang wenshi ziliao xuanji* 19 (1981): 130–37.

Jin Zhaozi. *[Xin zhongxue jiaokeshu] Chuji shijie shi* ([New middle school textbook] Lower-level world history). Shanghai: Zhonghua shuju, [1924] 1925.

———. *[Xin zhongxue jiaokeshu] Chuji benguoshi* ([New middle school textbook] Lower-level Chinese history). 2 vols. Shanghai: Zhonghua shuju, [1923] 1931.

———. *Xin zhonghua waiguo shi* (New Zhonghua foreign history). Shanghai: Zhonghua shuju, [1930] 1932.

———. "Wo zai Zhonghua shuju sanshi nian" (My 30 years at the Zhonghua Book Company). In *Xuelin manlu, siji* (Casual accounts of the forest of learning, collection no. 4), ed. Zhonghua shuju bianjibu. Beijing: Zhonghua shuju, 1981.

Jing, Jun. "Environmental Protests in Rural China." In *Chinese Society: Change, Conflict, and Resistance*, ed. Elizabeth J. Perry and Mark Selden. London and New York: Routledge, 2000.

Johnson, Chalmers A. *Peasant Nationalism and Communist Power: The Emergence of Revolutionary China, 1937–1945*. Stanford: Stanford University Press, 1962.

Judge, Joan. *Print and Politics: 'Shibao' and the Culture of Reform in Late Qing China*. Stanford: Stanford University Press, 1996.

———. "Citizens or Mothers of Citizens? Gender and the Meaning of Modern Chinese Citizenship." In *Changing Meanings of Citizenship in Modern China*, ed. Merle Goldman and Elizabeth J. Perry. Cambridge: Harvard University Press, 2002.

"Junshi guanli banfa dagang" (Outline of a method for military management). *Jiangsu shengli Shanghai zhongxuexiao banyuekan* 98 (November 30, 1935): 9–10.

"Junshi xunlian xiaoxi" (Military training news). *Jiangsu shengli Shanghai zhongxuexiao banyuekan* 35 (April 30, 1930): 20.

Kaiming huoye wenxuan zongmu (General index for Kaiming's loose-leaf literary selection series). Shanghai: Kaiming shudian, 1934.

Karl, Rebecca E. *Staging the World: Chinese Nationalism at the Turn of the Twentieth Century*. Durham: Duke University Press, 2002.

Kasson, John F. *Rudeness and Civility: Manners in Nineteenth-Century Urban America*. New York: Hill and Wang, 1990.

Keenan, Barry. *The Dewey Experiment in China: Educational Reform and Political Power in the Early Republic*. Cambridge: Harvard University Press, 1977.

———. *Imperial China's Last Classical Academies: Social Change in the Lower Yangzi, 1864–1911*. Berkeley: Institute of East Asian Studies, University of California, 1994.

———. "Lung-men Academy in Shanghai and the Expansion of Kiangsu's Educated Elite, 1865–1911." In *Education and Society in Late Imperial China, 1600–1900*, ed. Benjamin A. Elman and Alexander Woodside. Berkeley: University of California Press, 1994.

Kemp, G. S. F. "Boy Scouts Association of China" and "Tongzijun hui baogao" (Report of the Boy Scouts Association). *Xin Qingnian* 2, no. 5 (1917): 1–2.

King, Ambrose Y. C. "Individual and Group: A Relational Perspective." In *Individualism and Holism: Studies in Confucian and Taoist Values*, ed. Donald Munro. Ann Arbor: Center for Chinese Studies, University of Michigan, 1985.

Kirby, William C. *Germany and Republican China*. Stanford: Stanford University Press, 1984.

———. "Engineering China: Birth of the Developmental State, 1928–1937." In *Becoming Chinese: Passages to Modernity and Beyond*, ed. Wen-hsin Yeh. Berkeley: University of California Press, 2000.

Knopp, Guido. *Hitler's Children*. Translated by Angus McGeoch. Thrupp: Sutton Publishing, 2002.

Koch, H. W. *The Hitler Youth: Origins and Development, 1922–45*. New York: Dorset Press, [1975] 1988.

Koon, Tracy H. *Believe, Obey, Fight: Political Socialization of Youth in Fascist Italy, 1922–1943*. Chapel Hill: University of North Carolina Press, 1985.

Kraus, Richard Curt. *Class Conflict in Chinese Socialism*. New York: Columbia University Press, 1981.

Kuhn, Philip. "Local Self-government Under the Republic: Problems of Control, Autonomy, and Mobilization." In *Conflict and Control in Late Imperial China*, ed. Frederic Wakeman and Carolyn Grant. Berkeley: University of California Press, 1975.

Lang, Olga. *Chinese Family and Society*. N.p.: Archon Books, 1968.

Lee, Ching Kwan. "Pathways of Labor Insurgency." In *Chinese Society: Change, Conflict and Resistance*, ed. Elizabeth J. Perry and Mark Selden. London: Routledge, 2000.

Lee, Leo Ou-fan. *Shanghai Modern: The Flowering of a New Urban Culture in China, 1930–1945*. Cambridge: Harvard University Press, 1999.

Lefort, Claude. *The Political Forms of Modern Society: Bureaucracy, Democracy, Totalitarianism*. Edited by John B. Thompson. Cambridge: MIT Press, 1986.

Leung, Angela Ki Che. "Elementary Education in the Lower Yangtze Region in the Seventeenth and Eighteenth Centuries." In *Education and Society in Late Imperial China, 1600–1900*, ed. Benjamin A. Elman and Alexander Woodside. Berkeley: University of California Press, 1994.

Levenson, Joseph R. "The Suggestiveness of Vestiges: Confucianism and Monarchy at the Last." In *Confucianism and Chinese Civilization*, ed. Arthur F. Wright. Stanford: Stanford University Press, 1964.

———. *Confucian China and Its Modern Fate*. 3 vols. Berkeley: University of California Press, 1965.

Li Chongbi. "Fudan yiwu xiaoxue de guoqu ji jianglai" (The past and future of Fudan's common primary school). In *Fuzhong niankan* (Fudan Middle Yearbook). Shanghai: Fudan daxue fushu zhongxue xuesheng, 1934.

Li Huaxing et al., eds. *Minguo jiaoyushi* (Republican education history). Shanghai: Shanghai jiaoyu chubanshe, 1997.

Li Jigu. *Li shi chuzhong waiguoshi* (Mr. Li's lower middle school history of foreign countries). 2 vols. Shanghai: Shijie shuju, [1933–34] 1934–35.

Li Shaozhong. "Jizhong junxun qijian shenghuo de huiyi" (Memories of life during collective military training). *Zhejiang qingnian* 2, no. 1 (November 1935): 217–18.

Li Tinghan. *[Xin zhi] Benguo dili jiaoben* ([New system] Chinese geography textbook). 3 vols. Shanghai: Zhonghua shuju, 1914–15.

Li Yunfang. "Huzhu lun" (On mutual aid). *Jiangsu shengli Nanjing nüzi zhongxue xiaokan* 2 (April 1930): 23–24.

Li Zhendong. *[Gaozhong jiaokeshu] Gongmin: shehui wenti* ([High school textbook] Civics: social issues). Shanghai: Commercial Press, 1934.

Liang. "Hanjian lun" (On traitors). *Jingzhong qikan* 4 (July 1, 1937): 26–29.

"Liang xueshenghui zhi 'Wusi' jinianhui" (Two student unions' "May Fourth" memorial assemblies). MGRB, May 5, 1924.

Liao Shicheng et al., eds. *Shixing xinxuezhi hou zhi dongda fuzhong* (Southeast University Affiliated Middle School after implementing the New School System). Shanghai: Zhonghua shuju, [1924] 1926.

Lieberthal, Kenneth. *Governing China: From Revolution Through Reform*. New York: W. W. Norton, 1995.

Lin Mei. "1924–1927 nian Wenzhou xuesheng yundong" (The student movement in Wenzhou, 1924–27). *Wenzhou wenshi ziliao* 4 (1988): 161–66.

Linden, Allen B. "Politics and Education in China: The Case of the University Council, 1927–1928." *Journal of Asian Studies* 27, no. 4 (August 1968): 763–77.

———. "Politics and Higher Education in China: The Kuomintang and the University Community, 1927–1937." Ph.D. diss., Columbia University, 1969.

"Ling fa gao chuji zhongxue gongmin kecheng biaozhun" (Order to distribute high school and lower middle school civics curriculum standards). *Jiangsu jiaoyu* 3, no. 12 (December 1934): 171–74.

"Ling fa zhengshi kecheng biaozhun gongbu hou shending zhi zhong xiao xue jiaokeshu biao" (Order to distribute the list of approved primary and secondary school textbooks after the promulgation of the formal curriculum standards). *Jiangsu jiaoyu* 4, nos. 1–2 (February 1935): 266–76.

Ling Xuyou. "Tongjun luying zhi yiyi" (The significance of Scout camping). *Zhejiang qingnian* 2, no. 2 (December 1935): 203–4.

Liu Chengqing. *Chuji tongzijun shiyan jiaoben* (An experimental textbook for lower-level Boy Scouts). Shanghai: Shijie shuju, [1932] 1933.

Liu Dajun. "Zhongguo jindai tushu faxing tixi de jubian" (Dramatic change in China's modern book-distributing system). *Bianji xuekan* 5 (overall no. 49) (1996): 74–81.

Liu Dejue. "Xin shenghuo yu xin jingshen" (New life and new spirit). *Jiangsu xuesheng* 6, nos. 1–2 (April–May 1935): 149–51.

Liu Huru. *Xin shidai benguo dili jiaokeshu* (New era Chinese geography textbook). 2 vols. Shanghai: Xin shidai jiaoyushe, 1927–28.

Liu, Lydia H. *Translingual Practice: Literature, National Culture, and Translated Modernity, China, 1900–1937*. Stanford: Stanford University Press, 1995.

Liu Xunmu. "Xie zai wuyue jinian zhuanhao zhi qian" (Written before the special May commemoration issue). *Jiangsu shengli Shanghai zhongxue banyuekan* 71 (May 1, 1933): 10–14.

Liu Youxiang. "Di sibai tuan tongzijun guoqu he jianglai" (The past and future of Boy Scout troop number 400). In *Wunian gaikuang*. N.p.: Jiangsu Wuxi xianli chuji zhongxuexiao, 1934.

Liu Zhemin. *Jin-xiandai chuban xinwen fagui huibian* (Compilation of laws for modern and contemporary publishing and the press). Shanghai: Xuelin chubanshe, 1992.

"Liuzhou jinian jianyue xueshengjun ji tongzijun xunhua" (Lectures at the sixth anniversary review of the Student Army and the Boy Scouts). *Jiangsu shengli Shanghai zhongxuexiao banyuekan* 77 (November 15, 1933): 9–11.

Lou Tongsun. *[Xin shidai chuzhong] Sanmin zhuyi jiaokeshu, di er ce, minquan zhuyi* ([New era lower middle school] Three Principles of the People, vol. 2, people's sovereignty). Shanghai: Xin shidai jiaoyushe and Shangwu yinshuguan, 1928.

Louie, Kam, and Louise Edwards. "Chinese Masculinity: Theorizing *Wen* and *Wu*." *East Asian History* 8 (1994): 135–48.

Lowenthal, David. "European and English Landscapes as National Symbols." In *Geography and National Identity*, ed. David Hooson. Oxford: Blackwell, 1994.

Lü Fangshang. *Cong xuesheng yundong dao yundong xuesheng (Minguo banian zhi shibanian)* (From student movements to mobilizing students, 1919–29). Taipei: Zhongyang yanjiu yuan jindaishi yanjiusuo, 1994.

Lu Miaojin. "Dadao qingnian de tuifei bing" (Overcoming students' sickness of weakness and decadence). *Songjiang nüzhong xiaokan* 12 (June 20, 1930): 32–33.

Lü Simian. *[Fuxing gaoji zhongxue jiaokeshu] Benguoshi* ([Revival high school textbook] Chinese history). 2 vols. Shanghai: Shangwu yinshuguan, 1934–35.

Lü, Xiaobo, and Elizabeth J. Perry. "Introduction: The Changing Chinese Workplace in Historical and Comparative Perspective." In *Danwei: The Changing Chinese Workplace in Historical and Comparative Perspective*, ed. Xiaobo Lü and Elizabeth J. Perry. Armonk, NY: M. E. Sharpe, 1997.

Lu Ying. "Canjia quansheng tongzijun diliuci da jianyue da luying ji" (A record of participating in the Sixth Provincial Scouting review and jamboree). *Zhenhua jikan* 2, no. 2 (January 1936): 79–80.

Lukes, Steven. "Political Ritual and Social Integration." *Sociology* 9, no. 2 (1975): 289–308.

Luo Bin. "Tongzijun yewai huodong ji" (A record of Scouting outdoor activities). *Zhejiang qingnian* 3, no. 1 (November 1936): Qingnian yuandi, 3–5.

Luo Jiangyun. "Shizi yu guonan" (Literacy and the national crisis). *Nanjing nüzhong xuesheng zizhihui yuekan* 2 (December 15, 1933): 7–8.

Luo Yiwen. "Wu sa yu xuesheng" (May Thirtieth and students). *Minli xunkan* 8 (May 30, 1936): 12–14.

Lutz, Jessie Gregory. *China and the Christian Colleges, 1850–1950*. Ithaca: Cornell University Press, 1971.

Luykx, Aurolyn. *The Citizen Factory: Schooling and Cultural Production in Bolivia*. Albany: State University of New York Press, 1999.

Ma Zhengjun. "Shimin shenghuo gaikuang" (The general situation of citizens' lives). *Jiangsu shengli Shanghai zhongxue banyuekan* 83–84 (May 12, 1934): 18–19.

MacFarquhar, Roderick. *The Origins of the Cultural Revolution*. Vol. 2, *The Great Leap Forward, 1958–1960*. Oxford: Oxford University Press, 1983.

Mair, Victor H. "Language and Ideology in the Written Popularizations of the *Sacred Edict*." In *Popular Culture in Late Imperial China*, ed. David Johnson, Andrew J. Nathan, and Evelyn S. Rawski. Berkeley: University of California Press, 1985.

Mao Liyuan. "Wo suo zhidao de Jiangsu shengli Suzhou zhongxue" (The Jiangsu Provincial Suzhou Middle School that I knew). *Suzhou wenshi ziliao* 8 (1982): 1–4.

Marshall, T. H. *Citizenship and Social Class and Other Essays*. Cambridge: Cambridge University Press, 1950.

McDonald, Angus. *The Urban Origins of Rural Revolution: Elites and the Masses in Hunan Province, China, 1911–1927*. Berkeley: University of California Press, 1978.

McElroy, Sarah Coles. "Transforming China Through Education: Yan Xiu, Zhang Boling, and the Effort to Build a New School System, 1901–1927." Ph.D. diss., Yale University, 1996.

———. "Forging a New Role for Women: Zhili First Women's Normal School and the Growth of Women's Education in China, 1901–21." In *Education, Culture, and Identity in Twentieth-Century China*, ed. Glen Peterson, Ruth Hayhoe, and Yongling Lu. Ann Arbor: University of Michigan Press, 2001.

McGrath, Thomas E. "A Warlord Frontier: The Yunnan-Burma Border Dispute, 1910–1937." Unpublished paper. Cited with permission of the author.

Meisner, Maurice. *Marxism, Maoism, and Utopianism: Eight Essays*. Madison: University of Wisconsin Press, 1982.

Meskill, John. *Academies in Ming China: A Historical Essay*. Tucson: University of Arizona Press for the Association for Asian Studies, 1982.

Miller, David. *Citizenship and National Identity*. Cambridge: Polity Press, 2000.

Minguo sannian chun Zhonghua shuju gaikuang (Zhonghua Book Company's general situation in spring 1914). N.p.: [Zhonghua shuju], 1914.

Mitchell, Timothy. *Colonising Egypt*. Berkeley: University of California Press, 1991.

Morris, Andrew D. *Marrow of the Nation: A History of Sport and Physical Culture in Republican China*. Berkeley: University of California Press, 2004.

Mosse, George L. *The Nationalization of the Masses: Political Symbolism and Mass Movements in Germany from the Napoleonic Wars Through the Third Reich*. Ithaca: Cornell University Press, [1975] 1996.

Mouffe, Chantal. "Democratic Citizenship and the Political Community." In *Dimensions of Radical Democracy: Pluralism, Citizenship, Community*, ed. Chantal Mouffe. London and New York: Verso, 1992.

———, ed. *Dimensions of Radical Democracy: Pluralism, Citizenship, Community*. London and New York: Verso, 1992.

Munro, Donald J. *The Concept of Man in Contemporary China*. Ann Arbor: University of Michigan Press, 1977.

———. "The Family Network, the Stream of Water, and the Plant: Picturing Persons in Sung Confucianism." In *Individualism and Holism: Studies in Confucian and Taoist Values*, ed. Donald Munro. Ann Arbor: Center for Chinese Studies, University of Michigan, 1985.

———. *Images of Human Nature: A Sung Portrait*. Princeton: Princeton University Press, 1988.

Muxin. "Jiaoyu yu demokelaxi" (Education and democracy). *Jiaoyu zazhi* 11, no. 9 (September 20, 1919): 1–6.

Nanjing shi shiba niandu jiaoyu gaikuang tongji (Statistics for the educational situation in Nanjing in 1929). N.p.: Nanjing shi zhengfu jiaoyuju [1929].

Nanyang zhongxue (Catalogue of Nanyang Middle School, Shanghai). [Shanghai:] n.p., 1915–16.

Nathan, Andrew. *Chinese Democracy*. Berkeley: University of California Press, 1985.

Ni Qikun. "Huabei de weiji" (The crisis in North China). *Jingzhong qikan* 4 (July 1, 1937): 29–32.

Oakes, Tim. "China's Provincial Identities: Reviving Regionalism and Reinventing 'Chineseness.'" *Journal of Asian Studies* 59, no. 3 (August 2000): 667–92.

O'Brien, Kevin J. "Villagers, Elections, and Citizenship." In *Changing Meanings of Citizenship in Modern China*, ed. Merle Goldman and Elizabeth J. Perry. Cambridge: Harvard University Press, 2002.

Oldfield, Adrian. *Citizenship and Community: Civic Republicanism and the Modern World*. London and New York: Routledge, 1990.

Ong, Aihwa. "Cultural Citizenship as Subject-Making: Immigrants Negotiate Racial and Cultural Boundaries in the United States." *Current Anthropology* 37, no. 5 (December 1996): 737–62.

Ozouf, Mona. *Festivals and the French Revolution*. Translated by Alan Sheridan. Cambridge: Harvard University Press, 1988.

Pan Pi, Zhang Can, and Sha Ping, comps. *Subei kangri genjudi jishi* (Chronicle of the northern Jiangsu base area). Shanghai: Huadong ligong daxue chubanshe, 1998.

Pan Wenan. "Jinhou zhongdeng xuexiao zhi xunyu" (Future character-development education for secondary schools). *Zhongdeng jiaoyu* 3, no. 2 (August 1, 1924): 1–7.

Pang Guoliang. "Rensheng poujie" (Analyzing human life). In *Niankan* (Yearbook), ed. Jiangsu shengli diyi shifan xuexiao. Suzhou: Jiangsu shengli diyi shifan xuexiao, 1923.

Parish, William L., and Martin King Whyte. *Village and Family in Contemporary China*. Chicago: University of Chicago Press, 1978.

———. *Urban Life in Contemporary China*. Chicago: University of Chicago Press, 1984.

Peake, Cyrus H. *Nationalism and Education in Modern China*. New York: Columbia University Press, 1932.

Peng Kunyuan. "Xiandai nüzi yingyou de juewu" (The realization that modern women should have). *Songjiang nüzhong xiaokan* 59 (June 1, 1934): 15–16.

Pepper, Suzanne. *Civil War in China*. Reprint, Lanham, MD: Rowman and Littlefield, 1999.

Perdue, Peter C. "Boundaries, Maps, and Movement: Chinese, Russian, and Mongolian Empires in Early Modern Central Eurasia." *International History Review* 20, no. 2 (June 1998): 263–86.

Perry, Elizabeth J. *Shanghai on Strike: The Politics of Chinese Labor*. Stanford: Stanford University Press, 1993.

———. "From Paris to the Paris of the East—and Back: Workers as Citizens in Modern Shanghai." In *Changing Meanings of Citizenship in Modern China*, ed. Merle Goldman and Elizabeth J. Perry. Cambridge: Harvard University Press, 2002.

———. "Moving the Masses: Emotion Work in the Chinese Revolution." *Mobilization* 7, no. 2 (2002): 111–28.

———. *Patrolling the Revolution: Worker Militias, Citizenship, and the Modern Chinese State*. Lanham, MD: Rowman and Littlefield, 2006.

Perry, Elizabeth J., and Xun Li. *Proletarian Power: Shanghai in the Cultural Revolution*. Boulder: Westview Press, 1997.

Perry, Elizabeth J., and Mark Selden. *Chinese Society: Change, Conflict, and Resistance*. London and New York: Routledge, 2000.

Perry, Elizabeth J., and Jeffrey Wasserstrom, eds. *Popular Protest and Political Culture in Modern China: Learning from 1989*. Boulder, CO: Westview Press, 1992.

Peterson, Glen. *The Power of Words: Literacy and Revolution in South China, 1949–95*. Vancouver: University of British Columbia Press, 1997.

Petrone, Karen. *Life Has Become More Joyous, Comrades: Celebrations in the Time of Stalin*. Bloomington: Indiana University Press, 2000.

"Pingmin yexiao gaikuang" (General situation of the common people's night school). In *Yinianlai zhi Yangzhong* (Yangzhou Middle during the last year), ed. Zhongyang daxuequli Yangzhou zhongxuexiao. [Yangzhou:] Zhongyang daxuequli Yangzhou zhongxue chuban weiyuanhui, 1928.

Pocock, J. G. A. *The Machiavellian Moment: Florentine Political Thought and the Atlantic Republican Tradition*. Princeton: Princeton University Press, 1975.

———. "The Ideal of Citizenship Since Classical Times." In *Theorizing Citizenship*, ed. Ronald Beiner. Albany: State University of New York Press, 1995.

Pomeranz, Kenneth. "Ritual Imitation and Political Identity in North China: The Late Imperial Legacy and the Chinese National State Revisited." *Twentieth-Century China* 23, no. 1 (November 1997): 1–30.

Poon, Shuk Wah. "Refashioning Festivals in Republican Guangzhou." *Modern China* 30, no. 2 (April 2004): 199–227.

Price, Don C. "From Civil Society to Party Government: Models of the Citizen's Role in the Late Qing." In *Imagining the People: Chinese Intellectuals and the Concept of Citizenship, 1890–1920*, ed. Joshua A. Fogel and Peter G. Zarrow. Armonk, NY: M. E. Sharpe, 1997.

Pudong zhongxue xiaoshi bianxie zu. "Pudong zhongxue jianshi" (A concise history of Pudong Middle School). In *Shanghai wenshi ziliao xuanji* 59 (1988): 201–18.

Pudong zhongxuexiao nian zhou jiniankan (Memorial volume for the twentieth anniversary of Pudong Middle School). [Shanghai:] Pudong zhongxuexiao nian zhou jinian choubeihui, 1926.

Pusey, James Reeve. *China and Charles Darwin*. Cambridge: Harvard University Press, 1983.

Qi Shuzu. "Dui zhongxuesheng jiang xunlian yu zizhi" (Talking to middle school students about training and self-government). *Jiangsu xuesheng* 8, no. 1 (December 1936): 98–106.

"*Qianjiang pinglun* fakan zhiqu" (The purpose of launching the *Qian River Critic*). In *Wusi yundong zai Zhejiang*, ed. Zhonggong Zhejiang shengwei dangxiao dangshi jiaoyan shi. [Hangzhou:] Zhejiang renmin chubanshe, 1979.

Qiang Yuanben. "Zujie zai zhongguo" (Concession areas in China). *Jingzhong qikan* 4 (July 1, 1937): 35–45.

"Qingniantuan teji" (Special compilation on the Youth Group). *Zhejiang shengli Hangzhou gaoji zhongxue xiaokan* 154 (October 10, 1936): 1471–76.

Qiu Qiqin. "Jixun shenghuo yizhou ji" (A record of life during one week of collective military training). *Minli xunkan* 6 (May 10, 1936): 21–22.

Quanguo jiaoyuhui lianhehui xinxuezhi kecheng biaozhun qicao weiyuanhui. *Xinxuezhi kecheng biaozhun gangyao* (Outline of the curriculum standards for the New School System). 3rd ed. N.p.: n.p., 1924.

"Quanguo tongzijun di er ci da jianyue da luying" (The second national Scouts' review and jamboree). *Jiaoyu zazhi* 26, no. 12 (December 10, 1936): 126–29.

Rankin, Mary Backus. *Early Chinese Revolutionaries: Radical Intellectuals in Shanghai and Chekiang, 1902–1911*. Cambridge: Harvard University Press, 1971.

———. *Elite Activism and Political Transformation in China*. Stanford: Stanford University Press, 1986.

Rawski, Evelyn Sakakida. *Education and Popular Literacy in Ch'ing China*. Ann Arbor: University of Michigan Press, 1979.

Reed, Christopher A. *Gutenberg in Shanghai: Chinese Print Capitalism, 1876–1937*. Vancouver: University of British Columbia Press, 2004.

"Relie qingzhu lingxiu wushi shouchen" (Enthusiastically celebrate the leader's 50th birthday). *Zhejiang shengli Hangzhou gaoji zhongxue xiaokan* 156 (November 10, 1936): 1507–8.

Ridley, Charles Price; Paul H. B. Godwin; and Dennis J. Doolin. *The Making of a Model Citizen in Communist China*. Stanford: Hoover Institution Press, 1971.

Rigger, Shelley. "Nationalism Versus Citizenship in the Republic of China on Taiwan." In *Changing Meanings of Citizenship in Modern China*, ed. Merle Goldman and Elizabeth J. Perry. Cambridge: Harvard University Press, 2002.

Rogaski, Ruth. *Hygienic Modernity: Meanings of Health and Disease in Treaty-Port China*. Berkeley: University of California Press, 2004.

Rosaldo, Renato. "Cultural Citizenship and Educational Democracy." *Cultural Anthropology* 9, no. 3 (August 1994): 402–11.

Rosenthal, Michael. *The Character Factory: Baden-Powell and the Origins of the Boy Scout Movement*. New York: Pantheon Books, 1986.

Rowe, William T. *Hankow: Conflict and Community in a Chinese City, 1796–1895*. Stanford: Stanford University Press, 1989.

————. "The Public Sphere in Modern China." *Modern China* 16, no. 3 (July 1990): 309–29.

————. *Saving the World: Chen Hongmou and Elite Consciousness in Eighteenth-Century China.* Stanford: Stanford University Press, 2001.

Ru Gengbo. "Nanjing luying shenghuo yi duanpian" (A piece on life while camping in Nanjing). *Zhejiang qingnian* 3, no. 8 (June 1937): Qingnian yuandi, 5–8.

Rui Jiarui. *Xuesheng zizhi xuzhi* (Essential knowledge about student self-government). Shanghai: Shangwu yinshuguan, [1921] 1922.

Sahlins, Peter. *Boundaries: The Making of France and Spain in the Pyrenees.* Berkeley: University of California Press, 1989.

Sang Bing. *Wanqing xuetang xuesheng yu shehui bianqian* (Late Qing school students and social change). Taipei: Daohe chubanshe, 1991.

Schein, Louisa. *Minority Rules: The Miao and the Feminine in China's Cultural Politics.* Durham: Duke University Press, 2000.

Schneider, Laurence A. *Ku Chieh-kang and China's New History: Nationalism and the Quest for Alternative Traditions.* Berkeley: University of California Press, 1971.

Schoppa, R. Keith. *Chinese Elites and Political Change: Zhejiang Province in the Early Twentieth Century.* Cambridge: Harvard University Press, 1982.

————. "Contours of Revolutionary Change in a Chinese County, 1900–1950." *Journal of Asian Studies* 51, no. 4 (November 1992): 770–96.

————. *Blood Road: The Mystery of Shen Dingyi in Revolutionary China.* Berkeley: University of California Press, 1995.

Schwarcz, Vera. *The Chinese Enlightenment: Intellectuals and the Legacy of the May 4th Movement of 1919.* Berkeley: University of California Press, 1986.

Schwartz, Benjamin. *In Search of Wealth and Power: Yen Fu and the West.* Cambridge: Harvard University Press, 1964.

————. "The Reign of Virtue: Some Broad Perspectives on Leader and Party in the Cultural Revolution." *Party Leadership and Revolutionary Power in China,* ed. John Wilson Lewis. Cambridge: Cambridge University Press, 1970.

————, ed. *Reflections on the May Fourth Movement: A Symposium.* Cambridge: Harvard University Press, 1972.

Scott, James C. *Seeing Like a State: How Certain Schemes to Improve the Human Condition Have Failed.* New Haven: Yale University Press, 1998.

"Shanghai gexiao xueshengjun diaocha biao (jia)" (Questionnaire regarding the student armies at each school in Shanghai [A]). *Jiaoyu zazhi* 20, no. 12 (December 20, 1928).

Shanghai Minli zhongxue sanshizhou jiniankan (Memorial volume for the 30th anniversary of Shanghai's Minli Middle School). [Shanghai:] n.p. [1933].

Shanghai Minli zhongxuexiao yichou nian zhangcheng (The 1925 regulations for Shanghai's Minli Middle School). [Shanghai:] n.p., 1925.

Shanghai qiuxue zhinan (A guidebook for choosing among Shanghai's schools). 2 vols. Shanghai: Tianyi shuju, 1921.

"Shanghai shi gaozhong yishang xuexiao xiaozhang huiyi lu" (Record of the meeting of principals of Shanghai's high schools and above). February 28, 1936. SMA Q235-1-548.

"Shanghai shi geji xuexiao ge minzhong tuanti juxing jinianzhou banfa" (Method for each level of school and mass organization in Shanghai to hold the weekly memorial meeting). May 9, 1929. SMA Q235-1-141.

Shanghai shi jiaoyuju. "Shanghai shi zhong xiao xue shixing xin shenghuo banfa dagang" (Outline of methods for the implementation of the New Life Movement by Shanghai's secondary and primary schools). May 19, 1934. SMA Q235-1-323.

Shanghai shi Nanyang zhongxue jianxiao 100 zhounian (Centennial album of Shanghai Nanyang High School). [Shanghai:] n.p. [1996].

Shanghai shi Songjiang xian di'er zhongxue jianxiao jiushi zhounian jiniance (Commemorative volume for the 90th anniversary of the founding of Shanghai Municipality Songjiang County Second Middle School). [Songjiang:] n.p., 1994.

Shanghai shili Wuben nüzi zhongxuexiao gaikuang (The general situation of Shanghai Municipal Wuben Girls' Middle School). [Shanghai:] n.p., 1934.

"Shanghai xueshenghui zhi wusi jinian" (The Shanghai Student Union's May Fourth memorial). *Shenbao*, May 5, 1924.

"Shangzhongshi zuzhi" (The organization of Shanghai Middle School's municipal government), *Jiangsu shengli Shanghai zhongxue banyuekan*, 83–84 (May 12, 1934): 26–49.

Shao, Qin. *Culturing Modernity: The Nantong Model, 1890–1930*. Stanford: Stanford University Press, 2004.

Shao Weizheng. "Yaqian nongmin xiehui shimo" (The beginning and the end of the Yaqian Peasant Association). In *Dangshi yanjiu ziliao*, vol. 5. Chengdu: Sichuan renmin chubanshe, 1985.

"Shaoxing minzhong jinian wusi" (The Shaoxing masses commemorate May Fourth). MGRB, May 7, 1925.

Shaoxing shi wenhuaju and Zhonggong Shaoxing shiwei dangshi bangongshi. *Shaoxing geming wenhua shiliao huibian* (Compilation of revolutionary culture historical materials from Shaoxing). Beijing: Tuanjie chubanshe, 1992.

Shen Baoqi. *Tongzijun benji kecheng* (Scouting basic curriculum). Shanghai: Zhonghua shuju, [1923] 1927.

Shen Guanqun. "Woguo zhongdeng jiaoyu zhi shi de jiantao" (A review of the history of our country's secondary education). In *Geming wenxian*, vol. 57, *Kangzhan qian zhongdeng jiaoyu*. Taipei: Zhongguo Guomindang zhongyang weiyuanhui dangshi shiliao bianzuan weiyuanhui, 1971.

Shen Xingyi. *[Xin zhongxue jiaokeshu] Chuji guoyu duben* ([New middle school textbook] Lower-level national language reader). 3 vols. Shanghai: Zhonghua shuju, [1924–25] 1924–26.

Shen Yiwen and Xiao Xianning, eds. With Zhang Quan and Chen Anhong. *Jiangsu sheng Yangzhou zhongxue* (Jiangsu Province's Yangzhou Middle School). Zhongguo mingxiao congshu (Series on China's famous schools). Beijing: Renmin jiaoyu chubanshe, [1997] 2002.

Shen Yizhen. "Jiaoxun heyi" (Uniting instruction and character-development education). *Jiangsu shengli Shanghai zhongxuexiao banyuekan* 76 (October 22, 1933): 15–19.

Shen Zhilin. "Nüzi jiuguo de fangfa" (Women's method of saving the nation). *Zhenhua jikan*, Chuangkanhao (March 1934): 51–52.

Shen Ziqiang, Zhao Zijie, Xu Bonian, and Huang Meiying, eds. *Zhejiang yishi fengchao* (The disturbance at Zhejiang First Normal). Hangzhou: Zhejiang daxue chubanshe, 1990.

Sheng. "Song junxun tongxue" (Seeing off the [collective] military-training students). *Minli xunkan* 2 (March [30] 1936): 12–14.

Sheng Youxuan. "Wumen xuesheng de zeren" (The responsibility of we the students). *Jiangsu shengli di'er shifan xuexiao xiaokan (Benxiao ershi jie jinian tekan)* 36 (May 5, 1925): 50–53.

"Shengli dijiu zhongxuexiao shicha baogao shu" (Report on the inspection of [Zhejiang] Provincial Ninth Middle School). *Zhejiang jiaoyu jikan* 2 (June 30, 1926): 7–13.

Sheridan, James. *China in Disintegration: The Republican Era in Chinese History.* New York: Free Press, 1971.

Sheridan, Mary. "The Emulation of Heroes." *China Quarterly* 33 (January 1968): 47–72.

Shi Yide. "Dadao qingnian de tuifei bing" (Overcoming students' sickness of weakness and decadence). *Songjiang nüzhong xiaokan* 12 (June 20, 1930): 31–32.

"Shinianlai zhi zhongdeng jiaoyu gaishu" (An overview of ten years of secondary education) [Chongqing, 1940]. In *Geming wenxian*, vol. 57, *Kangzhan qian zhongdeng jiaoyu*. Taipei: Zhongguo Guomindang zhongyang weiyuanhui dangshi shiliao bianzuan weiyuanhui, 1971.

Shu Xincheng. *[Xin zhongxue jiaokeshu] Chuji gongmin keben* ([New middle school textbook] Lower-level civics textbook). 3 vols. Shanghai: Zhonghua shuju, 1923–24.

———. *Wo he jiaoyu* (Me and education). Reprint, Taipei: Longwen chubanshe, 1990.

Siu, Helen F. *Agents and Victims in South China: Accomplices in Rural Revolution.* New Haven: Yale University Press, 1989.

Slack, Jennifer Daryl. "The Theory and Method of Articulation in Cultural Studies." In *Stuart Hall: Critical Dialogues in Cultural Studies*, ed. David Morley and Kuan-hsing Chen. London and New York: Routledge, 1996.

Smethurst, Richard. *A Social Basis for Prewar Japanese Militarism: The Army and the Rural Community.* Berkeley: University of California Press, 1974.

Smith, Anthony D. *National Identity.* Reno: University of Nevada Press, 1991.

Smith, Nathaniel B. "The Idea of the French Hexagon." *French Historical Studies* 6, no. 2 (Autumn 1969): 139–55.

Smith, Richard J. "Mapping China's World: Cultural Cartography in Late Imperial Times." In *Landscape, Culture, and Power in Chinese Society*, ed. Wen-hsin Yeh. Berkeley: Institute of East Asian Studies, University of California, 1998.

Snow, Edgar. *Red Star over China.* Rev. ed. New York: Grove Press, 1973.

Solinger, Dorothy. *Contesting Citizenship in Urban China: Peasant Migrants, the State, and the Logic of the Market.* Berkeley: University of California Press, 1999.

Song Xuewen. "Qing lai taolun guoxue yanjiushe de jige wenti" (Please come and discuss several problems of the National Learning Research Society). *Jiangsu shengli di'er shifan xuexiao xiaokan* 30 (January 15, 1924): 20–28.

"Songjiang nüzhong Nuli tuan biye jiniankan" (Graduation memorial publication of the Songjiang Girls' Middle Endeavor Troop). *Songjiang nüzhong xiaokan* 12 (June 20, 1930).

Sontag, Susan. "Fascinating Fascism." In idem, *Under the Sign of Saturn*. New York: Anchor Books / Doubleday, [1980] 1991.

Spencer, Herbert. "The Social Organism." In *Essays Scientific, Political, and Speculative*, vol. 1. New York: D. Appleton, 1901.

Stachura, Peter D. *Nazi Youth in the Weimar Republic*. Santa Barbara, CA: Clio Books, 1975.

Stranahan, Patricia. *Underground: The Shanghai Communist Party and the Politics of Survival, 1927–1937*. Lanham, MD: Rowman and Littlefield, 1998.

Strand, David. *Rickshaw Beijing: City People and Politics in the 1920s*. Berkeley: University of California Press, 1989.

———. "Citizens in the Audience and at the Podium." In *Changing Meanings of Citizenship in Modern China*, ed. Merle Goldman and Elizabeth J. Perry. Cambridge: Harvard University Press, 2002.

Su Yiri. *[Xin shidai chuzhong] Sanmin zhuyi jiaokeshu, disan ce, Minsheng zhuyi* ([New era lower middle school] Three Principles of the People textbook, vol. 3, People's livelihood). [Shanghai:] Xin shidai jiaoyu she and Shangwu yinshuguan, [1927] 1928.

Sun Benwen. *Gaoji zhongxue gongmin: Shehui wenti* (High school civics: Social issues). Nanjing: Zhengzhong shuju, 1935.

Sun Bojian. *[Fuxing chuji zhongxue jiaokeshu] Gongmin* ([Revival lower middle school textbook] Civics). 3 vols. Shanghai: Shangwu yinshuguan, [1933] 1934–35.

Sun Jingkun. "Shenghuo junshihua" (Militarization of life). *Jizhong xuesheng* 4 (June 1935): 3.

Sun Lianggong and [Shen] Zhongjiu. *Chuji zhongxue guoyuwen duben* (Lower middle school national language reader). 6 vols. Shanghai: Minzhi shuju, [1922–24] 1926.

Sun, Yat-sen. *San Min Chu I: The Three Principles of the People*. Translated by Frank W. Price. Edited by L. T. Chen. Shanghai: China Committee, Institute of Pacific Relations, 1927.

———. *Fundamentals of National Reconstruction [Jianguo Dagang]*. Taipei: China Cultural Service and Sino-American Publishing Co., 1953.

Sun Yixin. "Zhongxue tongzijun wenti" (Middle school Scouting problems). *Jiangsu shengli Shanghai zhongxuexiao banyuekan* 56 (October 1931): 1–2.

Suzhou Zhenhua nü xuexiao. *Zhenhua shenghuo* (Zhenhua life). Suzhou: Zhenhua nü xuexiao, 1934.

———, ed. *Zhenhua nü xuexiao sanshinian jiniankan* (The commemorative volume for 30 years of Zhenhua Girls' School). *Zhenhua jikan* 2, nos. 3–4 (August 1936).

Suzhou zhongxue gaikuang (1928 niandu) (Chuzhongbu) (The general situation of Suzhou Middle School lower middle school division during the 1928 academic year). [Suzhou: Jiangsu shengli Suzhou zhongxue], 1929.

Tai Shuangqiu. "Xunyu shishi de yizhong jieguo" (One kind of result from implementing character-development education). *Zhongdeng jiaoyu* 4 (December 1922): 1–11.

Taizhou diqu wenhua ju geming wenhua shiliao zhengji bangongshi, and Taizhou diqu wenwu guanli weiyuanhui, comps. *Taizhou geming wenhua shiliao xuanbian* (Selected revolutionary culture historical materials from Taizhou). N.p.: n.p., 1992.

Tan Shaozeng. "Junxun qijian shenghuo huiyi" (Memories of life during military training). *Zhejiang qingnian* 1, no. 12 (October 1935): 145–47.

Tanaka, Stefan. *Japan's Orient: Rendering Pasts into History.* Berkeley: University of California Press, 1993.

Tang Xiaobing. *Global Space and the Nationalist Discourse of Modernity: The Historical Thinking of Liang Qichao.* Stanford: Stanford University Press, 1996.

Tao An. "Wo ren zai xiao ruhe xiuyang zhi shehui fuwu shi fang neng xing qi suo zhi?" (How should we be cultivated at school such that when it comes time for social service we can do what is intended?). *Songjiang nü zhong xiaokan* 27 (January 20, 1932): 9–10.

Tao Menghe. [*Xinxuezhi gaozhong jiaokeshu] Shehui wenti* ([New school system high school textbook] Social issues). Shanghai: Shangwu yinshuguan, 1924.

Tao Zhixing [Xingzhi]. "Xuesheng zizhi wenti zhi yanjiu" (Research into the issue of student self-government). *MGRB*, January 10–11, 1920, *Juewu.*

Thøgerson, Stig. *A County of Culture: Twentieth-Century China Seen from the Village Schools of Zouping, Shandong.* Ann Arbor: University of Michigan Press, 2002.

Thompson, E. P. "Time, Work-Discipline, and Industrial Capitalism." *Past and Present* 38 (December 1967): 56–97.

Thompson, Roger. *China's Local Councils in the Age of Constitutional Reform, 1898–1911.* Cambridge: Harvard University Press, 1995.

Thongchai, Winichakul. *Siam Mapped: A History of the Geo-body of a Nation.* Honolulu: University of Hawaii Press, 1994.

———. "The Quest for 'Siwilai': A Geographical Discourse of Civilizational Thinking in the Late Nineteenth and Early Twentieth-Century Siam." *Journal of Asian Studies* 59, no. 3 (August 2000): 528–49.

Tianmin, trans. "Demokelaxi yu xuexiao guanli" (Democracy and school administration). *Jiaoyu zazhi* 11, no. 9 (1919): 1–8.

Tien, Hung-Mao. *Government and Politics in Kuomintang China, 1927–1937.* Stanford: Stanford University Press, 1972.

Tong Bozhang. *Gongmin xue* (The study of civics). Changzhou: Xinqun shushe, 1923.

"Tongzijun jiaoyu cujin hui jinxing zhuangkuang" (The efforts of the Scouting Education Promotion Association). *Zhejiang sheng jiaoyuhui yuekan* 7 (February 1925): 35–36.

"Tongzijun zhi zuzhi" (The Boy Scouts' organization). *Zhejiang yizhong zhoukan* 2 (October 8, 1923): 1.

Townsend, James. "Chinese Nationalism." *Australian Journal of Chinese Affairs*, no. 27 (January 1992): 97–130.

Tsang, Chiu-sam. *Nationalism in School Education in China*. Hong Kong: Progressive Education Publishers, 1967.

Tsin, Michael. *Nation, Governance, and Modernity in China: Canton, 1900–1927*. Stanford: Stanford University Press, 1999.

Turner, Bryan S. "Contemporary Problems in the Theory of Citizenship." In *Citizenship and Social Theory*, ed. Bryan S. Turner. London: Sage Publications, 1993.

"Tushuguan goudao xinshu" (New books purchased by the library). *Zhejiang yizhong zhoukan* 18 (March 24, 1924).

"Tushuguan qishi" (Library announcement). *Zhejiang yizhong zhoukan* 2 (October 8, 1923): 1.

"Tushuguan qishi" (2) (Library announcement). *Zhejiang yizhong zhoukan* 9 (November 26, 1923): 1.

"Tushuguan tonggao" (A notice from the library). *Zhejiang yizhong zhoukan* 2 (October 8, 1923).

"Tushuguan tonggao" (2) (A notice from the library). *Zhejiang yizhong zhoukan* 6 (November 5, 1923).

Van de Ven, Hans J. *From Friend to Comrade: The Founding of the Chinese Communist Party, 1920–1927*. Berkeley: University of California Press, 1991.

Van Slyke, Lyman P. *Enemies and Friends: The United Front in Chinese Communist History*. Stanford: Stanford University Press, 1967.

Vishnyakova-Akimova, Vera Vladimirovna. *Two Years in Revolutionary China, 1925–1927*. Translated by Steven I. Levine. Cambridge: East Asia Research Center, Harvard University, 1971.

Vogel, Ezra F. "From Friendship to Comradeship: The Change in Personal Relations in Communist China." *China Quarterly* 21 (January–March 1965): 46–60.

Wagner, Rudolf G. "The Canonization of May Fourth." In *The Appropriation of Cultural Capital: China's May Fourth Project*, ed. Milena Dolezelova-Velingerova and Oldrich Kral, with Graham Sanders. Cambridge: Harvard University Asia Center, 2001.

Wakeman, Frederic, Jr. *Policing Shanghai, 1927–1937*. Berkeley: University of California Press, 1995.

———. "A Revisionist View of the Nanjing Decade: Confucian Fascism." In *Reappraising Republican China*, ed. Frederic Wakeman, Jr., and Richard Louis Edmonds. Oxford: Oxford University Press, 2000.

———. *Spymaster: Dai Li and the Chinese Secret Service*. Berkeley: University of California Press, 2003.

Walzer, Michael. "The Civil Society Argument." In *Theorizing Citizenship*, ed. Ronald Beiner. Albany: State University of New York Press, 1995.

[Wang] Binglong. "Rensheng de yiyi" (The significance of human life). *Minli xuesheng* (January 1935): 37–38.

Wang Chichang. "Shifan xuexiao xuesan, si xuenian guowen jiaoxue de guanjian" (My humble thoughts regarding national language instruction for third- and fourth-year normal school classes). In *Niankan* (Yearbook), ed. Jiangsu shengli diyi shifan xuexiao. Suzhou: Jiangsu shengli diyi shifan xuexiao, 1923.

Wang Duanteng. "Qingnian yinggai renshi de liangge wenti" (Two problems that youths should recognize). *Jizhong xuesheng* 4 (June 1935): 4–5.

Wang, Fan-shen. "Evolving Prescriptions for Social Life in the Late Qing and Early Republic: From Qunxue to Society." In *Imagining the People: Chinese Intellectuals and the Concept of Citizenship, 1890–1920*, ed. Joshua A. Fogel and Peter G. Zarrow. Armonk, NY: M. E. Sharpe, 1997.

Wang Gaiyai. *Benguoshi* (Chinese history). 2 vols. Suzhou: Suzhou Middle School, 1934. Mimeographed lecture notes.

Wang, Guanhua. *In Search of Justice: The 1905–1906 Chinese Anti-American Boycott.* Cambridge: Havard University Asia Center, 2001.

Wang Guohao and Wu Jue. "Minli zhongxue jianshi" (A historical sketch of Minli Middle School). *Shanghai wenshi ziliao xuanji* 59 (1988): 337–40.

Wang Jie. "Ruhe dapo xisu" (How to break through habits and customs). *Shangzhong shikan*, Shimin zuopin teji (June 5, 1934): 12–13.

Wang, Liping. "Creating a National Symbol: The Sun Yatsen Memorial in Nanjing." *Republican China* 21, no. 2 (April 1996): 23–63.

Wang, Q. Edward. *Inventing China Through History: The May Fourth Approach to Historiography.* Albany: State University of New York Press, 2001.

Wang Ruyu. "Ni xinzhi zhongxue shiyong guowen jiaoxue cao'an" (A draft syllabus for new system national language instruction for use in middle schools). *Zhejiang yizhong zhoukan* 25–26 (May 12–19, 1924).

Wang Shouhua. "Riji liangze" (Two diary entries). In *Hangzhou diyi zhongxue xiaoqing qishiwu zhounian jiniance* (Hangzhou First Middle School 75th anniversary celebration memorial volume). [Hangzhou:] Hangzhou yizhong qishiwu zhounian xiaoqing choubei bangongshi, 1983.

Wang Ti'an. "Zhong ri tixi sheng zhong de huabei wenti" (The North China problem amid the sound of [calls for] Sino-Japanese Aid). *Jizhong xuesheng* 5 (June 1936): 1–4.

Wang Weihua. "Yichan he xin qingnian ziji de geming" (Inheritance and the new youth's own revolution). In *Nanyang zhongxue xinyou ji jiniance* (Memorial volume for Nanyang Middle School's class of 1921). [Shanghai:] n.p., 1921.

Wang Xiangqin. "Xuesheng shuqi fuwutuan gaikuang" (The general situation of the student summer-service troops). *Sheng xian wenshi ziliao* 3 (1986): 134–35.

Wang Xuancheng, comp. "Kangri shiqi Jinhua de fuyun—ji Zhejiang baoyuhui de yixie huodong" (The Jinhua women's movement during the period of resistance to Japan—recording some activities of the Zhejiang Nursery Association). *Jinhua wenshi ziliao* 2 (1986): 142–50.

Wang Yankang. "Zhongxuexiao jiji xunyu zhi sizhong shishi" (Four kinds of implementation of active character-development education for middle schools). *Zhongdeng jiaoyu* 3, no. 2 (August 1, 1924): 1–16.

Wang Yiyai and Zhou Lisan. *[Chuji zhongxue] Benguo dili* ([Lower middle school] Chinese geography). 4 vols. Nanjing: Zhengzhong shuju, 1935.

Wang Yongquan. "Gaozhong biyesheng de zeren" (The responsibility of high school graduates). In *Pudong zhongxue nian'er jie biye jiniankan* (Memorial publication of the 22nd graduating class of Pudong Middle School), ed. Nian-er jie biye jiniankan bianji weiyuanhui. N.p.: n.p., 1935.

Wang, Zheng. *Women in the Chinese Enlightenment: Oral and Textual Histories.* Berkeley: University of California Press, 1999.

Wang Zhiyi. *Kaiming shudian jishi* (A record of Kaiming Publishing). In *Bianji huiyilu zhuanji congshu.* Taiyuan: Shanxi renmin chubanshe, 1991.

Wang Zhongqi. *[Xiandai chuzhong jiaokeshu] Benguo dili* ([Modern lower middle school textbook] Chinese geography). 2 vols. Shanghai: Shangwu yinshuguan, [1923–24] 1924–25.

Wang Zicheng. "Huiyi Minzhi shuju" (Remembering Minzhi Book Company). *Chuban shiliao* 2 (December 1983): 113–15.

Wasserstrom, Jeffrey N. *Student Protests in Twentieth-Century China: The View from Shanghai.* Stanford: Stanford University Press, 1991.

———. "Student Protests in Fin-de-Siècle China." *New Left Review* 237 (September/October 1999): 52–76.

Watson, James L. "Rites or Beliefs? The Construction of a Unified Culture in Late Imperial China." In *China's Quest for National Identity*, ed. Lowell Dittmer and Samuel S. Kim. Ithaca: Cornell University Press, 1993.

Wedeen, Lisa. *Ambiguities of Domination: Politics, Rhetoric, and Symbols in Contemporary Syria.* Chicago: University of Chicago Press, 1999.

Wei Bingxin and Xu Yingchuan. *[Xin zhuyi jiaokeshu] Chuzhong dangyi* ([A new principles textbook] Lower middle school party doctrine). 6 vols. Shanghai: Shijie shuju, 1929.

Wei Zhen. "Xin shenghuo yundong he qiangpo jiaoyu de xiaoguo" (The results of the New Life Movement and compulsory education). *Minli xunkan (xuesheng zhuanhao)*, June 1936: 11–13.

Weilutsu. "Foreword." In *Pudong zhongxue nian'er jie biye jiniankan* (Memorial volume of the 22nd graduating class of Pudong Middle School), ed. Nian'er jie biye jiniankan bianji weiyuanhui. N.p.: n.p., 1935.

Wen Liangzhi. "Ershi niandai Zhenjiang jiaoyujie de yichang fengbo" (A tempest in Zhenjiang's educational circles during the 1920s). *Zhenjiang wenshi ziliao* 13 (1987): 164–68.

"Wenhua shiliao huibian" bianweihui. *Jinhua shi geming wenhua shiliao huibian* (Compilation of revolutionary culture historical materials for Jinhua city). Hangzhou: Hangzhou daxue chubanshe, 1991.

Weston, Timothy B. *The Power of Position: Beijing University, Intellectuals, and Chinese Political Culture, 1898–1929.* Berkeley: University of California Press, 2004.

Westbrook, Robert B. *John Dewey and American Democracy.* Ithaca: Cornell University Press, 1991.

Westheimer, Joel, and Joseph Kahne. "Education for Action: Preparing Youth for Participatory Democracy." In *Teaching for Social Justice*, ed. William Ayers, Jean Ann Hunt, and Therese Quinn. New York: Teachers College Press, 1998.

White, Lynn T., III. *Policies of Chaos: The Organizational Causes of Violence in China's Cultural Revolution.* Princeton: Princeton University Press, 1989.

Whyte, Martin King. *Small Groups and Political Rituals in China.* Berkeley: University of California Press, 1974.

Williams, Brackette F. *Stains on My Name, War in My Veins: Guyana and the Politics of Cultural Struggle.* Durham: Duke University Press, 1991.

Willis, Paul. *Learning to Labour: How Working Class Kids Get Working Class Jobs.* Farnborough: Saxon House, 1977.

Wilson, Richard W. *Learning to Be Chinese: The Political Socialization of Children in Taiwan.* Cambridge: MIT Press, 1970.

"Wo lixiang zhong de geren shenghuo" (My ideal individual life). In *Erwuji niankan.* [Suzhou: Zhenhua nüzhong, 1933.]

Wright, Mary C. *The Last Stand of Chinese Conservatism: The T'ung-chih Restoration, 1862–1874.* Stanford: Stanford University Press, 1957.

———, ed. *China in Revolution: The First Phase, 1900–1913.* New Haven: Yale University Press, 1968.

Wu Lieh Tso [Wu Liezuo]. "The Fundamental Training for Youth." In *Fuzhong niankan* (Fudan Middle Yearbook). Shanghai: Fudan daxue fushu zhongxue xuesheng, 1934.

Wu, S. L. [Wu Shilie]. "Fight for the Children." In *Fuzhong niankan* (Fudan Middle Yearbook). Shanghai: Fudan daxue fushu zhongxue xuesheng, 1934.

Wu Sihong. "Dageming shiqi de Shaoxing xianli nüzi shifan" (Shaoxing County Girls' Normal during the period of the Great Revolution). In *Shaoxing geming wenhua shiliao huibian,* ed. Shaoxing shi wenhuaju and Zhonggong Shaoxing shiwei dangshi bangongshi. Beijing: Tuanjie chubanshe, 1992.

Wu Tao. "Xuesheng zhi tongbing" (Students' common failings). *Zhejiang yizhong zhoukan* 15 (January 6, 1924): 4.

"Wu wu jinianri Pan Juzhang dui jizhong junxun xuesheng xunhua ji" (A record of [Shanghai Education] Bureau Director Pan's lecture to collective military-training students on the May Fifth Memorial Day). *Minli xunkan* 7 (May 20, 1936): 10–12.

Wu xian jiaoyuju xunling, no. 349, "Xuesheng kewai yuedu kanwu diaocha biao" (Survey form for student extracurricular readings). April 1933. SMA2 Jia5.422.

Wu xian jiaoyuju xunling, no. 18, "Zuijin fandong kanwu diaochao biao" (Survey form for the most recent counterrevolutionary reading materials). February 27, 1936. SMA2 Jia5.431.

[Wu] Yanyin. "Zhuchi zhongdeng jiaoyu zhe jinhou zhi juewu" (The realization for the future of those who manage secondary education). *Jiaoyu zazhi* 11, no. 7 (July 20, 1919): 1–5.

Wu Yaolin. *Tongzijun quanshu* (The complete book of the Boy Scouts). Shanghai: Liming shuju, 1935.

"Wujiu jinian zhi qingxing (Jiaxing)" (May Ninth memorial situation in Jiaxing). MGRB, May 13, 1923.

"Wujiu jinianri zhi youxing (Changzhou)" (Parades for the May Ninth Memorial Day in Changzhou). MGRB, May 11, 1923.

Wuxi xian jiaoyuju. *Wuxi jiaoyu zuijin gaikuang* (The most recent situation of Wuxi education). Wuxi: Wuxi xian jiaoyuju, 1932.

Wuxing xianli nüzi chuji zhongxue chuban weiyuanhui, ed. *Wuxing xianli nüzi chuji zhongxue zuijin gaikuang* (The most recent circumstances of Wuxing County Girls' Lower Middle School). Hangzhou: Wuxing xianli chuji zhongxue, 1934.

Xia Mianzun and Ye Shengtao, eds. *Guowen baiba ke* (One hundred and eight lessons for national language). 3 vols. Shanghai: Kaiming shudian, 1935–38. Reprint, Beijing: Renmin jiaoyu chubanshe, 1987.

"Xianshi Kongzi danchen jinian banfa" (Method for commemorating the birthday of the first teacher, Confucius) [July 5, 1934]. In *Jiaoyu faling huibian*, Diyiji, ed. Jiaoyubu canshichu. Changsha: Shangwu yinshuguan, [1936] 1938.

Xiang Jutan. *Gongmin xue* (The study of civics). Nanjing: Zhonghua shuju, 1923.

Xiao Zhizhi. "Li Gengsheng zhuan" (Biography of Li Gengsheng). In *Zhonghua minguoshi ziliao conggao: Renwu zhuanji*, ed. Zhongguo shehui kexueyuan jindaishi yanjiusuo zhonghua minguoshi yanjiushi et al. Beijing: Zhonghua shuju, 1973–86.

Xiaohe. "Xianzai qingnian suo biju de tiaojian" (The conditions that today's youth must prepare). *Minli xunkan (Xuesheng zhuanhao)* (June 1936): 4–5.

Xiaokan bianji weiyuanhui. *Songjiang gaoji zhiye zhongxue* (Songjiang Technical High School). Songjiang: Jiangsu shengli Songjiang gaoji yingyong huaxue ke zhiye xuexiao, 1937.

Xiaoshan xianzhi bianzuan weiyuanhui. *Xiaoshan xianzhi* (Xiaoshan County gazetteer). Shanghai: Zhejiang renmin chubanshe, 1987.

Xie Guan. *[Gongheguo jiaokeshu] Benguo dili* ([Republican textbook series] Chinese geography). 2 vols. Shanghai: Shangwu yinshuguan, 1913.

———. *[Gongheguo jiaokeshu] Waiguo dili* ([Republican textbook series] Foreign geography). 2 vols. Shanghai: Shangwu yinshuguan, [1914] 1919.

Xin shenghuo yundong cujin zonghui, comp. *Minguo ershisi nian quanguo xin shenghuo yundong* (The nationwide New Life Movement in 1935), nos. 528–29. In *Jindai zhongguo shiliao congkan*, 3rd compilation, collection no. 53. Taipei: Wenhai chubanshe, n.d.

Xingzhi. "Zhongdeng xuexiao xunyu zhi shangque" (A discussion of secondary school character-development education). *Jiangsu xuesheng* 5, no. 4 (January 10, 1935): 13–20.

Xu Fuzeng. "Tuanti shenghuo de yaosu" (Essential elements of group life). *Songjiang nüzhong xiaokan* 23 (November 5, 1931): 11.

Xu Minzhong. "Jiaxing zhongdeng xuesheng shenghuo zhuangkuang ji qi zhiyuan de diaocha" (A survey of the living conditions and aspirations of Jiaxing's secondary students). *Zhongxuesheng* 44 (April 1, 1934): 89–110.

Xu Peihuang. "Zhongxuesheng kehou de xiuyang" (Middle school students' extracurricular cultivation). *Jiangsu xuesheng* 1, no. 1 (October 1932): 8–9.

Xu Shouyuan. "Guanyu Shangzhongshi zhengfu shijie jinian de hua" (Regarding the tenth-session commemoration of the Shanghai Middle municipal government). *Jiangsu shengli Shanghai zhongxue banyuekan* 83–84 (May 12, 1934): 17–18.

Xu Xingzhi. "Dang chengli shiqi Zhejiang de gong nong yundong" (Zhejiang's peasant and workers' movements during the period of party formation) [March 1957]. In *"Yida" qianhou: Zhongguo gongchandang diyici daibiao dahui qianhou ziliao xuanbian* (Before and after "The First Party Congress": Selected materials from before and after the Chinese Communist Party's First Congress of Representatives), ed. Zhongguo shehui kexueyuan xiandaishi yanjiushi and Zhongguo geming bowuguan dangshi yanjiushi. N.p.: Renmin chubanshe, 1980.

Xu Yangben. "Wo duiyu qingniantuan de xiwang" (My hopes for the Youth Group). *Hang chu* 5 (1936): 18–19.

Xu Yaoliang. "Guanyu Shangzhongshi zhengfu de hua" (Some words concerning Shanghai Middle municipal government). *Jiangsu shengli Shanghai zhongxue banyuekan* 83–84 (May 12, 1934): 23–24.

Xu Youchun et al., eds. *Minguo renwu da cidian* (Biographical dictionary of Republican China). Shijiazhuang: Hebei renmin chubanshe, 1991.

"Xuexiao xuenian xueqi ji xiuxue riqi guicheng" (Rules for the school year, semester, and holidays) [September 3, 1912; revised February 20, 1914]. *Jiaoyu fagui huibian* (1919).

"Xuexiao yishi guicheng" (Rules for school ceremonies) [November 6, 1912]. *Jiaoyu fagui huibian* (1919).

"Xunyu tanhuahui ji jiaozhiyuan xuesheng lianxi huiyi jishi" (A record of a discussion meeting about character-development education and a joint meeting of teachers and students). *Zhejiang yizhong zhoukan* 6 (November 5, 1923): 1.

"Yaqian nongcun xiao xuexiao xuanyan" (The manifesto of Yaqian's village school). *Dangshi yanjiu ziliao*. Vol. 5. Chengdu: Sichuan renmin chubanshe, 1985.

"Yaqian nongmin xiehui xuanyan" (The manifesto of Yaqian's peasant association). *Dangshi yanjiu ziliao*. Vol. 5. Chengdu: Sichuan renmin chubanshe, 1985.

Yang Changchun. "Guoli Bianyiguan shulüe" (An account of the National Institute for Compilation and Translation). In *Chubanshi yanjiu* (Research in the history of publishing), ed. Ye Zaisheng. Beijing: Zhongguo shuji chubanshe, 1995.

Yang Donglin. "Wode lishiguan" (My historical perspective). *Zhejiang yizhong zhoukan* 19 (March 31, 1924): 3; 20 (April 7, 1924): 2–3.

Yang Hongsheng. "Wode riji" (My diary). *Shangzhong shikan*, Shimin zuopin teji (June 5, 1934): 21–23.

Yang Lieyu. "Xie zai qiantou" (Introduction). *Zhejiang shengli Hangzhou gaoji zhongxue xiaokan* 165 (April 25, 1937): 1681–82.

Yang, Mayfair Mei-hui. *Gifts, Favors, and Banquets: The Art of Social Relationships in China*. Ithaca: Cornell University Press, 1994.

Yang Ronan. "Ji Zhenhua nüzi zhongxue" (Remembering Zhenhua Girls' Middle School). *[Suzhou]wenshi ziliao xuanji* 8 (1982): 35–51.

Yang, Wen Pin [Wenbin]. "The Students' Part in the National Crisis." *Minli xuesheng* (June 1933): 29–30.

Yang Xianjiang. "Xuesheng zizhi heyi biyao" (Why student self-govenment is necessary). MGRB, March 1, 1920, *Juewu*.

Yang Yang. *Shangwu yinshuguan: minjian chubanye de xingshuai* (Commercial Press: the rise and fall of the private press). Shanghai: Shanghai shiji chuban jituan and Shanghai jiaoyu chubanshe, 2000.

Yao Shaohua. *[Xin kecheng biaozhun shiyong] Chuzhong benguo lishi* ([For use with the new curriculum standards] Lower middle school Chinese history). 4 vols. Shanghai: Zhonghua shuju, 1933–34.

Yee, Cordell D. K. "Cartography of China." In *Cartography in Traditional East and Southeast Asian Societies. The History of Cartography*, vol. 2.2, ed. J. B. Harley and David Woodward. Chicago: University of Chicago Press, 1994.

Yeh Sheng-tao. *Schoolmaster Ni Huan-chih*. Translated by A. C. Barnes. Peking: Foreign Languages Press, 1978.

Yeh, Wen-hsin. *The Alienated Academy: Culture and Politics in Republican China, 1919–1937*. Cambridge: Council on East Asian Studies, Harvard University, 1990.

———. *Provincial Passages: Culture, Space, and the Origins of Chinese Communism*. Berkeley: University of California Press, 1996.

"Yiwu xiaoxue xiaoshi" (A history of the Common School). In *Fudan daxue fuzhong yijiu nianji biye jiniankan* (The Fuhtan Middle School annual). Shanghai: Students of Fuhtan Middle School, 1930.

Young, Iris Marion. "Polity and Group Difference: A Critique of the Ideal of Universal Citizenship." In *Theorizing Citizenship*, ed. Ronald Beiner. Albany: State University of New York Press, 1995.

Young, Marion. "Chicken Little in China: Some Reflections on Women." In *Marxism and the Chinese Experience: Issues in Contemporary Chinese Socialism*, ed. Arif Dirlik and Maurice Meisner. Armonk, NY: M. E. Sharpe, 1989.

Yu Fu. "Tan fuxing nongcun" (Reviving the villages). In *Pudong zhongxue nian'er jie biye jiniankan* (Memorial volume of the 22nd graduating class of Pudong Middle School), ed. Nian'er jie biye jiniankan bianji weiyuanhui. N.p.: n.p., 1935.

Yu Hetong. "Huabei teshuhua yu zhong ri gongtong fanggong liang da wenti" (The two great questions of North China's special status and China and Japan's common defense against communism). *Jingzhong xuesheng* 3 (January 1, 1937): 12–13.

Yu Xiaoyao and Liu Yanjie, eds. *Lufei Kui yu Zhonghua shuju* (Lufei Kui and the Zhonghua Book Company). Beijing: Zhonghua shuju, 2002.

Yu Yueyin. "Kangri zhanzheng shiqi Xinchang zhongxue de geming douzheng pianduan" (Passages about revolutionary struggle at Xinchang Middle School during the Anti-Japanese War period). In *Shaoxing geming wenhua shiliao huibian*, ed. Shaoxing shi wenhuaju and Zhonggong Shaoxing shiwei dangshi bangongshi. Beijing: Tuanjie chubanshe, 1992.

Yu Zhenjian. "Zhongxuesheng de zeren" (The responsibility of middle school students). *Jiuzhong xuesheng* (January 1931): 31–32.

Yu Zhongyin. "Fendou ba! Pengyou!" (Let's struggle! My friends!). *Jizhong xuesheng* 4 (June 1935): 7.

Yu Zixiang. "Zhongxuesheng yu xiangcun (luntan)" (Middle school students and the rural villages [a forum]). *Jiuzhong xuesheng* (January 1931): 89–92.

Yuan Yi. "Zhong ri dili shang zhi jiaoshe" (Sino-Japanese geographic negotiations). *Jiangsu shengli di'er nüzi shifan xuexiao xiaoyouhui huikan* 8 (May 1919): 5–6.

Yuan Yi. "Zizhi zhuyi de xunlian" (Training in self-governmentalism). In *Zizhi hui chengli jiniance* (The memorial volume of the founding of the student self-government association), ed. Zhejiang shengli diyi shifan xuexiao xuesheng zizhi hui. [Hangzhou:] n.p., 1919.

"*Yuesheng* fakanci" (Inaugural preface for the *Voice of Shaoxing*). In *Wusi yundong zai Zhejiang*, ed. Zhonggong Zhejiang shengwei dangxiao dangshi jiaoyanshi. [Hangzhou:] Zhejiang renmin chubanshe, 1979.

Zanasi, Margherita. "Chen Gongbo and the Construction of a Modern Nation in 1930s China." In *Nation Work: Asian Elites and National Identities*, ed. Timothy Brook and Andre Schmid. Ann Arbor: University of Michigan Press, 2000.

Zarrow, Peter. *Anarchism and Chinese Political Culture*. New York: Columbia University Press, 1990.

———. "Introduction: Citizenship in China and the West." In *Imagining the People*, ed. Joshua Fogel and Peter Zarrow. Armonk, NY: M. E. Sharpe, 1997.

———. "Political Ritual in the Early Republic of China." In *Constructing Nationhood in Modern East Asia*, ed. Kai-wing Chou, Kevin M. Doak, and Poshek Fu. Ann Arbor: University of Michigan Press, 2001.

"*Zeren* fakanci" (Inaugural preface to *Responsibility*). In *Wusi yundong zai Zhejiang*, ed. Zhonggong Zhejiang shengwei dangxiao dangshi jiaoyanshi. [Hangzhou:] Zhejiang renmin chubanshe, 1979.

"Zenyang liyong women de shujiaqi (Wenti taolunhui, diwuci taolun)" (How to use our summer vacation [Discussion forum, discussion no. 5]). *Zhongxuesheng* 16 (June 1931): 99–108.

Zeng Youde. "Shangzhong shizhengfu dishi jie jinianhui" (Memorial meeting for the tenth session of the Shanghai Middle municipal government). *Jiangsu shengli Shanghai zhongxue banyuekan* 83–84 (May 12, 1934): 28.

Zerubavel, Yael. *Recovered Roots: Collective Memory and the Making of Israeli National Tradition*. Chicago: University of Chicago Press, 1995.

Zhanqian [Lu Renji]. "Wusi yundong de huigu" (A look back at the May Fourth Movement). *Jiangsu shengli Shanghai zhongxue banyuekan* 71 (May 1, 1933): 20–25.

Zhang Bihua. "Nü qingniantuan huodong jilüe" (A record of girls' Youth Group activities). *Zhejiang qingnian* 3, no. 9 (July 1937): Qingnian yuandi: 8–9.

Zhang Daoren. "Jiangsu shengli diyi zhongxuexiao xuesheng zizhihui gaizu yilai zhi gaikuang" (The situation since the reorganization of Jiangsu Provincial First Middle School's student self-government association). *Zhongdeng jiaoyu* 3, no. 2 (August 1, 1924): 1–9 [page numbers discontinuous].

Zhang Jinglu. *Zhongguo xiandai chuban shiliao* (Historical materials for China's modern publishing). 5 vols. Shanghai: Zhonghua shuju, 1954.

Zhang Junchou. "Tongzijun jiujing shi shenma?" (What exactly is Scouting?). *Wuxian jiaoyu yuekan* 1, no. 6 (March 1, 1922): 1–6.

Zhang Lunqing. "Wode song dao" (My praise and entreaty). *Jiangsu shengli Shanghai zhongxue banyuekan* 83–84 (May 12, 1934): 6–8.

Zhang Minghe. "Canjia Shanghai quan shi tongzijun disanci da jianyue ji da luying suoji" (A brief account of participating in the Third Shanghai Municipal Boy Scout review and jamboree). *Shangzhongshi huikan* (1931): 54–55.

Zhang Nianzu. "Zhongdeng xuexiao xunyu zhi yanjiu" (A study on secondary school character-development education). *Zhongdeng jiaoyu* 2, no. 2 (June 1, 1923): 1–19.

Zhang Qiyun. *[Xin xuezhi gaoji zhongxue jiaokeshu] Benguo dili* ([New School System high school textbook] Chinese geography). 2 vols. Shanghai: Shangwu yinshuguan, [1926, 1928] 1928.

Zhang Quanping. "Guoxue shangdui hui gaikuang" (The general situation of the association to discuss national learning). *Suzhou Zhenhua nüxue xiaokan* 1 (May 1927): 35.

Zhang Shengyu. "Wunian lai duiyu guowen jiaoxue shang zhi yixie jingyan" (Some experience regarding national language instruction during the past five years). *Niankan* (Yearbook), ed. Jiangsu shengli diyi shifan xuexiao. Suzhou: Jiangsu shengli diyi shifan xuexiao, 1923.

Zhang Xiang. *[Xinzhi] Xiyangshi jiaoben* ([New system] Western history textbook). 2 vols. Shanghai: Zhonghua shuju, [1914] 1920.

Zhang Xiujun. "Feichang shiqi zhong qingnian xuesheng yingshou de xunlian" (The training young students should receive during the crisis period). *Jiangsu xuesheng* 8, no. 1 (December 1936): 59–61.

[Zhang] Yuanxin. "Nei Meng wenti" (The Inner Mongolia problem). *Minli xuesheng*, Tedahao (January 1, 1935): *luntan* 17–23.

Zhao Bangheng et al. *Tongzijun gaoji keben* (Advanced Boy Scout handbook). Shanghai: Dadong shuju, 1936.

———. *Tongzijun zhongji keben* (Intermediate Boy Scout handbook). Shanghai: Dadong shuju, 1936.

Zhao Qikun. "Xu" (Preface). In *Chuji tongzijun shiyan jiaoben* (An experimental textbook for lower-level Boy Scouts), ed. Liu Chengqing. Shanghai: Shijie shuju, [1932] 1933.

Zhao Qingqing and Qian Jiheng, comps. *Hangzhou sizhong 95 zhounian xiaoqing* (Celebration of the 95th anniversary of Hangzhou Fourth Middle). [Hangzhou:] n.p., 1994.

Zhao, Suisheng. "Chinese Intellectuals' Quest for National Greatness and Nationalistic Writings in the 1990s." *China Quarterly* (1997): 725–45.

Zhao Yusen. *[Gongheguo jiaokeshu] Benguoshi* ([Republican textbook series] Chinese history). 2 vols. Shanghai: Shangwu yinshuguan, [1913] 1920.

———. *[Xinzhu] Benguoshi* ([The newly written] Chinese history). 2 vols. Shanghai: Shangwu yinshuguan, 1922–24.

Zhao Zijie, Xu Shaoquan, and Li Weijia. "Xuan Zhonghua." In *Zhonggong dangshi renwu zhuan*, vol. 12, ed. Zhonggong dangshi renwu yanjiuhui. Xi'an: Shaanxi renmin chubanshe, 1983.

"Zhejiang diyi zhongxue xuesheng zizhihui chengli xuanyan" (Declaration of the establishment of Zhejiang First Middle School's student self-government association). MGRB, January 20, 1920.

"*Zhejiang diyi zhongxuexiao xuesheng zizhihui banyuekan* fakanci" (Inaugural preface of the *Biweekly Journal of the Zhejiang First Middle School Student Self-government Association*) [January 17, 1920]. In *Wusi yundong zai Zhejiang,* ed. Zhonggong Zhejiang shengwei dangxiao dangshi jiaoyanshi. [Hangzhou:] Zhejiang renmin chubanshe, 1979.

"Zhejiang gexian xuexiao xuesheng zizhihui zhangcheng, mingce" (Student self-government association rules and name lists for schools in each county of Zhejiang). March 1936. ZDLD 11.2837.

Zhejiang jiaoyu jianzhi (A brief chronicle of Zhejiang education). [Hangzhou:] Zhejiang renmin chubanshe, 1988.

Zhejiang renwu jianzhi (Brief accounts of Zhejiang's [historical] characters). Vol. 3. Hangzhou: Zhejiang renmin chubanshe, 1986.

Zhejiang sheng jiaoyuting. *Zhejiang sheng ershier niandu di'er xueqi zhong xiao xue biye huikao shiti da'an* (Questions and answers from the Zhejiang Province secondary and primary graduation comprehensive examination for second semester 1933). N.p.: n.p. [193?].

Zhejiang sheng Lishui diqu geming wenhua shiliao zhengbian bangongshi. *Lishui diqu geming (jinbu) wenhua shiliao huibian* (Compilation of revolutionary [and progressive] culture historical materials from the Lishui region). Lishui: Zhejiang sheng xinwen chubanju, 1992.

"Zhejiang sheng qingniantuan dagang" (Outline of the Zhejiang Province Youth Group). *Zhejiang qingnian* 2, no. 7 (May 1936): 202–11.

Zhejiang sheng zhengxie wenshi ziliao weiyuan hui. *Zhejiang jindai zhuming xuexiao he jiaoyujia* (Zhejiang's famous modern schools and educators). Vol. 45 of *Zhejiang wenshi ziliao*. Hangzhou: Zhejiang renmin chubanshe, 1991.

Zhejiang shengli di'er zhongxue yilan (An overview of Zhejiang Provincial Second Middle School). [Jiaxing:] n.p., 1929.

Zhejiang shengli disi zhongxue yilan (An overview of Zhejiang Provincial Fourth Middle School). [Ningbo:] n.p., 1930.

Zhejiang shengli diwu zhongxuexiao xianxing guicheng yilan (An overview of the current rules at Zhejiang Provincial Fifth Middle School). N.p.: n.p., 1925.

Zhejiang shengli diyi shifan xuexiao xuesheng zizhi hui. *Zizhi hui chengli jiniance* (The memorial volume of the founding of the student self-government association). [Hangzhou:] n.p., 1919.

"Zhejiang shengli diyi zhongxue gaojibu tongxuehui jianzhang" (Rules for the fellow students' association of Zhejiang Provincial First Middle School's High School Division). *Zhejiang yizhong zhoukan* 10 (December 3, 1923): 1.

Zhejiang shengli diyi zhongxuexiao ershiwu nian jiniance (The 25th year commemorative volume of Zhejiang Provincial First Middle School). Hangzhou: n.p., 1923.

"Zhejiang shengli diyi zhongxuexiao xuesheng zizhi hui huizhang" (Rules for the self-government association of Zhejiang Provincial First Middle School). In *Zhejiang shengli diyi zhongxuexiao ershiwu nian jiniance* (The 25th year commemorative volume of Zhejiang Provincial First Middle School). Hangzhou: n.p., 1923.

Zhejiang shengli gaoji zhongxue xiaokan (School publication of Zhejiang Provincial High School). [Hangzhou:] n.p., 1932.

Zhejiang shengli Hangzhou chuji zhongxue bianji weiyuanhui, ed. *Zhejiang shengli Hangzhou chuji zhongxue liuzhou jiniance* (Memorial volume of the sixth anniversary of Zhejiang Provincial Hangzhou Lower Middle School). Hangzhou: n.p., 1935.

Zhejiang shengli Hangzhou nüzi zhongxue. *Zhejiang shengli Hangzhou nüzi zhongxue wuzhou jiniankan* (Fifth anniversary memorial publication of Zhejiang Provincial Hangzhou Girls' Middle School). Hangzhou: Zhejiang shengli Hangzhou nüzi zhongxue, 1936.

Zhejiang shengli Jinhua zhongxue chuban weiyuanhui. *Zhejiang shengli Jinhua zhongxue yilan* (An overview of Zhejiang Provincial Jinhua Middle School). N.p.: Zhejiang shengli Jinhua zhongxue xiaofei hezuo she, 1936.

"*Zhejiang xinchao* fakanci" (Inaugural preface to *Zhejiang New Tide*). In *Wusi yundong zai Zhejiang*, ed. Zhonggong Zhejiang shengwei dangxiao dangshi jiaoyanshi. [Hangzhou:] Zhejiang renmin chubanshe, 1979.

"Zhexi nongcun xunli" (Tour of the villages in western Zhejiang). *Zhejiang shengli Hangzhou gaoji zhongxue xiaokan* 165 (April 25, 1937): 1681–88.

Zhenhua nüxuexiao sanshi nian jiniankan (Memorial volume for 30 years of Zhenhua Girls' School). Suzhou: Zhenhua nüxuexiao, 1936.

Zheng Chang. *[Chuji zhongxue yong] Xin zhonghua benguoshi* ([For use in lower middle schools] New Zhonghua Chinese history). 2 vols. Shanghai: Xin guomin tushushe and Zhonghua shuju, 1930–31.

———. *Chuzhong waiguoshi* (Lower middle school history of foreign countries). 2 vols. Shanghai: Zhonghua shuju, 1934.

Zheng Haozhang. *Shanghai tongzijun shi* (History of the Shanghai Boy Scouts). N.p.: Zhongguo tongzijun lianhuan she, 1932.

———. *Tongzijun xing shan ri lu* (A daily record of a Boy Scout's good deeds). Shanghai: Zhonghua shuju, 1936.

Zheng Ruochuan. "Woguo nongcun jingji pochan zhi yuanyin ji qi jiuji de fangfa" (The reasons for the bankruptcy of our country's village economy and the methods to relieve it). *Songjiang nüzhong xiaokan* 54 (March 15, 1934): 5–7.

Zheng, Yuan. "The 'Partification' of Education: A Pivotal Turn in Modern Chinese Education, 1924–1929." *Twentieth-Century China* 25, no. 2 (April 2000): 33–53.

Zhong Yulong. *[Xinzhi] Benguoshi* ([New system] Chinese history). 3 vols. Shanghai: Zhonghua shuju, [1914] 1914–15.

Zhonggong Jiangsu shengwei dangshi gongzuo weiyuanhui and Jiangsu sheng minzhengting. *Jiangsu geming lieshi zhuan xuan bian: kangri zhanzheng shiqi* (Selected biographies of Jiangsu revolutionary martyrs: the Anti-Japanese War period). Nanjing: Jiangsu renmin chubanshe, 1988.

Zhonggong Jing'an quwei dangshi ziliao zhengji weiyuanhui bangongshi and Zhonggong Jing'an quwei jiaoyu gongzuo weiyuanhui. *Shanghai shi Jing'an qu jiaoyu xitong geming douzheng shiliao* (Historical materials of the revolutionary struggle in Shanghai Jing'an District's educational system). Shanghai: Fudan daxue, 1989.

Zhonggong Shanghai shiwei dangshi ziliao zhengji weiyuanhui. *Huo hong de qing chun—Shanghai jiefang qian zhongxue xuesheng yundong shishi xuanbian* (Flaming youth—compilation of historical facts regarding Shanghai's pre-Liberation middle school students' movement). Shanghai: Shanghai waiyu jiaoyu chubanshe, 1994.

Zhonggong Zhejiang shengwei dangshi ziliao zhengji yanjiu weiyuanhui and Zhonggong Xiaoshan xianwei dangshi ziliao zhengji yanjiu weiyuanhui, eds. *Yaqian nongmin yundong* (The Yaqian peasant movement). Beijing: Zhonggong dangshi ziliao chubanshe, 1987.

Zhonggong Zhejiang shengwei dangxiao dangshi jiaoyanshi, ed. *Wusi yundong zai Zhejiang* (The May Fourth Movement in Zhejiang). [Hangzhou:] Zhejiang renmin chubanshe, 1979.

Zhongguo di'er lishi dang'anguan, ed. *Zhonghua minguoshi dang'an ziliao huibian* (Compilation of archival materials for Chinese Republican history). Series 5, part 1, Jiaoyu. 2 vols. Nanjing: Jiangsu guji chubanshe, 1994.

Zhongguo Guomindang disanci quanguo daibiao dahui huiyi jilu (Record of the Third Party Congress of the Chinese Nationalist Party). N.p.: Zhongyang zhixing weiyuanhui mishuchu, 1930.

Zhongguo Guomindang Zhejiang sheng zhixing weiyuanhui xunlianbu, ed. *Dangyi jiaoyu (Xunlian congshu zhi ba)* (Party doctrine education [No. 8 of the training series]). Hangzhou: Zhongguo Guomindang Zhejiang sheng zhixing weiyuanhui xunlianbu, 1929.

Zhongguo Guomindang zhongyang weiyuanhui dangshi shiliao bianzuan weiyuanhui, ed. *Geming wenxian, di 54 ji, kangzhan qian jiaoyu zhengce yu gaige* (Revolutionary documents, collection 54, educational policies and reforms before the War of Resistance). Taibei: Zhongguo Guomindang zhongyang weiyuanhui dangshi shiliao bianzuan weiyuanhui, 1971.

Zhongguo Guomindang zhongyang zhixing weiyuanhui. "Xuesheng zizhi hui zuzhi dagang shixing xize" (Specific rules for the implementation of the organizational outline for student self-government associations). *Jiaoyu zazhi* 23, no. 1 (January 12, 1931): 189.

———. "Xuesheng zizhihui zuzhi dagang" (Organizational outline for student self-government associations) [January 23, 1930]. In *Di yi ci Zhongguo jiaoyu nianjian* (The first yearbook of Chinese education), ed. Jiaoyubu. Shanghai: Shangwu yinshuguan, 1934. Reprint, 5 vols. Taipei: Shangwu yinshuguan, 1961.

Zhongguo Guomindang zhongyang zhixing weiyuanhui xuanchuanbu. *Xuesheng funü wenhua de tuanti zuzhi yuanzi xuanchuan dagang* (Propaganda outline of the organizing principles for student, women's, and cultural groups). N.p.: n.p., 1930.

Zhongguo Guomindang zhongyang zhixing weiyuanhui xunlianbu. *Jiaoyu zongzhi biaozhun ji shishi fang'an (cao'an)* (Plan for educational standards and their implementation [draft]). [Nanjing:] n.p., 1928.

———. *Fagui huikan* (Compilation of laws and rules). Collection 1, vol. 2. [Nanjing:] Zhongguo Guomindang zhongyang zhixing weiyuanhui xunlianbu, 1930.

Zhongguo jiaoyu tongji gailan, Zhonghua jiaoyu gaijin she congshu, di si zhong (An overview of Chinese statistics on education, No. 4 in the Chinese Educational Improvement Society series). Shanghai: Shangwu yinshuguan, 1923.

Zhongguo tongzijun chuji kecheng (Elementary curriculum for the Chinese Boy Scouts). Nanjing: Zhongguo Guomindang zhongyang zhixing weiyuanhui xunlianbu, 1930.

Zhongguo tongzijun xiehui. *Tongzijun chubu* (Basics of Scouting). Shanghai: Shangwu yinshuguan, [1918] 1931.

Zhongguo tongzijun zonghui choubeichu, ed. *Zhongguo tongzijun zonghui choubeichu gongzuo baogao* (Work report of the planning office of the General Association of the Chinese Boy Scouts). Nanjing: n.p., 1934.

Zhonghua minguo fagui daquan (Complete laws of the Chinese Republic). Shanghai: Shangwu yinshuguan, 1936.

Zhonghua quanguo tongzijun xiehui, ed. *Tongzijun guilü* (Scouting regulations). Shanghai: Shangwu yinshuguan, [1918] 1921.

Zhongyang daxuequli Shanghai zhongxuexiao mishuchu chuban weiyuanhui, comp. *Zhongyang daxuequli Shanghai zhongxuexiao yilan* (An overview of Central University District Shanghai Middle School). Shanghai: Zhongyang daxuequli Shanghai zhongxuexiao, 1928.

Zhongyang daxuequli Songjiang zhongxue yilan bianji weiyuanhui, ed. *Songjiang zhongxue yilan* (An overview of Songjiang Middle School). Songjiang: Songjiang zhongxue, 1929.

Zhongyang daxuequli Taicang zhongxuexiao, ed. *Yilan* (An overview). Suzhou: Zhongyang daxuequli Taicang zhongxue, 1929.

Zhongyang daxuequli Yangzhou zhongxue chuban weiyuanhui, ed. *Yinianlai zhi Yangzhong* (Yangzhou Middle during the last year). [Yangzhou:] Zhongyang daxuequli Yangzhou zhongxuexiao, 1928.

Zhongyang zhixing weiyuanhui qingnianbu. "Zhonghua minguo guomin zhengfu gonghan, di bashijiu hao" (Official letter no. 89 of the Nationalist Government of the Republic of China) (November 18, 1925). Nationalist Party History Archives, Taipei, Wubu dang'an no. 10349.

Zhou Fohai. "Ruhe zhengdun xuefeng? (Yi) Gao quan Su xuesheng" (How to correct school discipline? [1] Statement to all of Jiangsu's students). *Jiangsu jiaoyu* 1, no. 5 (June 3, 1932): 1–4.

———. "Women zenyang qu zuoren?" (How can we conduct ourselves?). *Jiangsu xuesheng* 1, no. 1 (October 1, 1932): 1–5.

Zhou Gengsheng. *[Xin xuezhi] Gongmin jiaokeshu* ([New School System] Civics textbook). Vols. 1–2. Shanghai: Shangwu yinshuguan, [1923] 1926.

Zhou Houshu. "Yangzhou zhongxue jiaoxun heyi hou zhi chubu shishi" (Yangzhou Middle School's initial implementation of the system of joining of instruction and character-development education). *Jiangsu jiaoyu* 1, no. 10 (November 1932): 56–66.

Zhou Jianwen. "Wode huigu" (My recollections). *Jiangsu shengli Shanghai zhongxue banyuekan* 83–84 (May 12, 1934): 33–60.

———. "Dijiujie de minzhong yexiao" (The night school for the masses during the ninth session [of the Shanghai Middle Municipality]). *Jiangsu shengli Shanghai zhongxue banyuekan* 83–84 (May 12, 1934): 61–65.

———. "Feichang shiqi zhong qingnian xuesheng yingshou de xunlian" (The training young students should receive during the crisis period). *Zhongxue-*

sheng 66 (June 1936): 197–200. Reprinted in *Jiangsu xuesheng* 8, no. 1 (December 1936): 49–55.

Zhou Zhijian and Zhou Xican. "Kangzhan chuqi Zhonggong Sheng xian dixiadang de jianli yu fazhan" (The establishment and development of Sheng County's underground party in the early period of the Resistance War). *Sheng xian wenshi ziliao* 3 (1986): 69–78.

Zhu Ji. "Junxun duiyu guojia minzu de qiantu" (Military training in relation to the future of the state and the nation). *Minli xunkan* [1] (chuangkanhao), March 20, 1936: 9–10.

Zhu Jianmang. *Chuzhong guowen* (Lower middle school national language). 6 vols. Shanghai: Shijie shuju, 1929–30.

Zhu Junda. "Bannian lai zhi tongzijun" (The Boy Scouts during the last half year). *Zhongdeng jiaoyu* 2 (April 1922): 3–8.

Zhu Lianbao. "Guanyu Shijie shuju de huiyi" (My recollections of World Book Company). *Chuban shiliao* 2 (overall no. 8) (1987): 52–68.

Zhu Wenshu and Song Wenhan. *[Xin kecheng biaozhun shiyong] Chuzhong guowen duben* ([For use with the new curriculum standards] Lower middle school national language reader). 6 vols. Shanghai: Zhonghua shuju, 1935–36.

Zhu Yixin. *Zhushi chuzhong benguoshi* (Mr. Zhu's lower middle school Chinese history). 4 vols. Shanghai: Shijie shuju, 1933.

Zhu Youhuan et al., eds. *Zhongguo jindai xuezhi shiliao* (Historical materials for the modern Chinese school system). 7 vols. Shanghai: Huadong shifan daxue chubanshe, 1983–92.

Zhu Yunyu. "Qingdao cunwang yu quanguo zhi guanxi" (Qingdao's survival and its relation to the nation as a whole). *Jiangsu shengli di'er nuzi shifan xuexiao xiaoyouhui huikan* 8 (May 1919): 15–16.

Zhuang Zeding. *Zhonghua zhongxue fazhi jiaokeshu* (Zhonghua's middle school legal system textbook). Shanghai: Zhonghua shuju, [1911] 1913.

Zou Zhuoli. *[Xin shidai zonghe bianzhi] Sanmin zhuyi jiaoben* ([New era synthetic organization] Three Principles of the People textbook). 3 vols. Shanghai: Shangwu yinshuguan, [1927] 1928–30.

Index

Harvard East Asian Monographs
(*out-of-print)

Harvard East Asian Monographs

Harvard East Asian Monographs

Harvard East Asian Monographs